D0622287

EXPLANATORY TRANSLATION

SYNTHESE LIBRARY

STUDIES IN EPISTEMOLOGY,

LOGIC, METHODOLOGY, AND PHILOSOPHY OF SCIENCE

Managing Editor:

JAAKKO HINTIKKA, *Boston University, U.S.A.*

Editors:

DIRK VAN DALEN, *University of Utrecht, The Netherlands*
DONALD DAVIDSON, *University of California, Berkeley, U.S.A.*
THEO A.F. KUIPERS, *University of Groningen, The Netherlands*
PATRICK SUPPES, *Stanford University, California, U.S.A.*
JAN WOLEŃSKI, *Jagiellonian University, Kraków, Poland*

VOLUME 312

VEIKKO RANTALA
University of Tampere,
Tampere, Finland

EXPLANATORY TRANSLATION

Beyond the Kuhnian Model of Conceptual Change

KLUWER ACADEMIC PUBLISHERS
DORDRECHT / BOSTON / LONDON

A C.I.P. Catalogue record for this book is available from the Library of Congress.

ISBN 1-4020-0827-9

Published by Kluwer Academic Publishers,
P.O. Box 17, 3300 AA Dordrecht, The Netherlands.

Sold and distributed in North, Central and South America
by Kluwer Academic Publishers,
101 Philip Drive, Norwell, MA 02061, U.S.A.

In all other countries, sold and distributed
by Kluwer Academic Publishers,
P.O. Box 322, 3300 AH Dordrecht, The Netherlands.

Printed on acid-free paper

All Rights Reserved
© 2002 Kluwer Academic Publishers
No part of this work may be reproduced, stored in a retrieval system, or transmitted
in any form or by any means, electronic, mechanical, photocopying, microfilming, recording
or otherwise, without written permission from the Publisher, with the exception
of any material supplied specifically for the purpose of being entered
and executed on a computer system, for exclusive use by the purchaser of the work.

Printed in the Netherlands.

FOR MIKKO AND JANNE

TABLE OF CONTENTS

PART ONE

THE PRAGMATICS AND HERMENEUTICS OF CONCEPTUAL CHANGE

PART TWO

THE LOGIC AND PRAGMATICS OF SCIENTIFIC CHANGE

PART THREE

THE FORMAL BASIS OF THE CORRESPONDENCE RELATION

PREFACE

One of the most controversial issues in recent philosophy, especially in the philosophy of science since the middle of the twentieth century, has evidently been the problem of whether the meaning of an expression tends to change according to its context of use. The present book is more or less connected with the same problem. It is an outgrowth of my long-term interest in subjects concerning conceptual change, particularly in problems related to the philosophy of science, definability, and translation. This interest derives its origin from my early studies of mathematics and physics, and it has ever since been nourished by my teachers, colleagues, and students in philosophy. Over the years, my philosophical and logical curiosity has been gradually shifting from foundational studies of science and mathematics to other fields, such as aesthetics and cognition. In spite of that, I have constantly returned to my old subjects, but now enriched with applications to those more recent areas of interest.

As the subtitle of the book implies, one starting point in my studies of translation is to critically evaluate Thomas S. Kuhn's well-known proposals concerning the role of translation in attempts to understand conceptual changes and conversions occurring in science, and in culture more generally. Another source is my extensive collaboration in the eighties with David Pearce, in our studies of scientific change – in our efforts to explicate such concepts as correspondence and reduction. These are concepts that historians and philosophers of science have often considered important, and occasionally scientists themselves. David found out, among other things, that the ideas created in that research can be used to approach the notion of commensurability in a new and fertile way, whereas one aim of the present volume is to show that some of the ideas are even relevant within a more general framework, so as to apply to many areas of nonscientific discourse as well.

I am greatly indebted to David for the ideas he brought to our joint enterprise. They made this book possible. In addition to David, there are four other persons whom I owe a special dept of gratitude. From Jaakko Hintikka, my teacher in philosophy, I have learned to appreciate the manner of keeping an eye on both logical and pragmatic aspects when investigating philosophical and methodological problems. Furthermore, many features of the notions of translation and correspondence studied in the present essay can be regarded as generalizing certain elements of his early work on definability. Later on, research done by former students and colleagues Cynthia Grund and Tere Vadén has been an impressive source of inspiration. Cynthia was the first to show me the intellectual allure of the philosophy of art. In addition, she kindly helped me to improve the language of the present volume. Tere, in turn, prompted my interest in the philosophy of cognition, one consequence of which can be seen in this book. I like to believe that the book would be much poorer without the enclosed studies of cognition, part of which derives from our research cooperation in the field. I am most grateful, however, to my wife, Pirkko, for her unselfish encouragement and understanding of this endeavor.

My research for this book was mainly supported by my own University and the Academy of Finland, but to a large extent also by the Academy of Sciences of the Czech Republic. I am very indepted to the colleagues working for the Department of Logic of the Czech Academy, with whom I have been collaborating in a joint research project on conceptual representation.

I am grateful to many colleagues, especially Ari Virtanen, Antti Vesikari, and Jarmo Niemelä, for helping me in computer problems, related to my attempts to get the manuscript ready for the press.

Some material of the book is based on my earlier work, part of which was written with David Pearce or with Tere Vadén. It has appeared in various journals, confcrcncc procccdings, and anthologies, but now revised for the current purpose. A great deal of the material concerning both foundational studies and applications is new or more recent.

The following articles or passages thereof are incorporated in this volume, with some revisions and corrections. I am indebted to the respective publishers and coauthors for permission to reprint the material:

Pearce, D. and Rantala, V., 'The Logical Study of Symmetries in Scientific Change', in P. Weingartner and H. Czermak (eds.), *Epistemology and Philos-*

ophy of Science, Hölder-Pichler-Tempsky, Vienna, 1983, pp. 330-332. Copyright © 1983 by Hölder-Pichler-Tempsky; reproduced here by kind permission of Hölder-Pichler-Tempsky and David Pearce.

Pearce, D. and Rantala, V., 'A Logical Study of the Correspondence Relation, *Journal of Philosophical Logic* 13 (1984), 47-84. Reproduced here by kind permission of David Pearce.

Rantala, V., 'The Old and the New Logic of Metascience', *Synthese* 39 (1978), 233-247.

Rantala, V., 'Scientific Change and Counterfactuals', in I. Hronszky, M. Fehér and B. Dajka (eds.), *Scientific Knowledge Socialized*, Hungarian Academy of Sciences, Budapest, 1988, pp. 399-408. Copyright © 1988 by Akadémiai Kiadó; reproduced here by kind permission of the Hungarian Academy of Sciences.

Rantala, V., 'Counterfactual Reduction', in K. Gavroglu, Y. Goudaroulis and P. Nicolacopoulos (eds.), *Imre Lakatos and Theories of Scientific Change*, Kluwer Academic Publishers, Dordrecht, 1989, pp. 347-360.

Rantala, V., 'Definitions and Definability', in J. H. Fetzer, D. Shatz and N. Schlesinger (eds.), *Definitions and Definability: Philosophical Perspectives*, Kluwer Academic Publishers, Dordrecht, Boston, London, 1991, pp. 135-159.

Rantala, V., 'Reduction and Explanation: Science vs. Mathematics', in J. Echeverria, A. Ibarra and T. Mormann (eds.), *The Space of Mathematics*, De Gruyter, Berlin, New York, 1992, pp. 47-59. Copyright © 1992 by Walter de Gruyter & Co.; reproduced here by kind permission of Walter de Gruyter & Co.

Rantala, V., 'Translation and Scientific Change', in W. E. Herfel, W. Krajewski, I. Niiniluoto and R. Wójcicki (eds.), *Theories and Models in Scientific Processes* (Poznan Studies in the Philosophy of the Sciences and the Humanities, vol. 44), Rodopi, Amsterdam, Atlanta, GA, 1995, pp. 249-268. Copyright © 1995 by Editions Rodopi B. V.; reproduced here by kind permission of Editions Rodopi B. V.

Rantala, V., 'Understanding Scientific Change', in P. I. Bystrov and V. N. Sadovsky (eds.), *Philosophical Logic and Logical Philosophy*, Kluwer Academic Publishers, Dordrecht, Boston, London, 1996, pp. 3-15.

Rantala, V., 'Metaphor and Conceptual Change', *Danish Yearbook of Philosophy*, Vol. 31 (1996), 181-190. Copyright © 1995 by Danish Yearbook of Philosophy; reproduced here by kind permission of the Danish Yearbook of Philosophy.

Rantala, V., 'Explanatory Translation. Beyond Functionalism and Reductionism', in M. L. Dalla Chiara, K. Doets, D. Mundici and J. van Benthem (eds.), *Logic and Scientific Methods. Volume One of the Tenth International Congress of Logic, Methodology and Philosophy of Science, Florence, August 1995*, Kluwer Academic Publishers, Dordrecht, Boston, London, 1997, pp. 399-41.

Rantala, V., 'Connectionism and the Emergence of Propositional Knowledge', in G. L. Farre and T. Oksala (eds.), *Emergence, Complexity, Hierarchy, Organization* (Acta Polytechnica Scandinavica, Ma 91), The Finnish Academy of Technology, Espoo, 1988, pp. 254-262. Copyright 1998 by The Finnish Academy

of Technology; reproduced here by kind permission of The Finnish Academy of
Technology.

Rantala, V., 'On the Emergence of Conceptual Knowledge', in O. Majer (ed.), *Topics in Conceptual Analysis and Modelling*, Filosofia, Prague, 2000, pp. 83-91.

Rantala, V., 'Translation and Conceptual Change in Art', *ibid.*, pp. 105-115.

Rantala, V., 'Reduction and Emergence', in T. Childers and J. Palomäki (eds.), *Between Words and Worlds. A Festschrift for Pavel Materna*, Filosofia, Prague, 2000, pp. 156-163.

Rantala, V., 'Carnap's General Syntax and Modern Logic', in T. Childers and O. Majer (eds.), *Logica* '01, Filosofia, Praha, 2002, forthcoming.

Rantala, V. and Vadén, T., 'Idealization in Cognitive Science. A Study in Counterfactual Correspondence', in M. Kuokkanen (ed.), *Idealization VII: Structuralism, Idealization and Approximation* (Poznan Studies in the Philosophy of the Sciences and the Humanities, Vol. 42), Rodopi, Amsterdam, Atlanta, GA, 1994, pp. 179-198. Copyright © 1994 by Editions Rodopi B. V.; reproduced here by kind permission of Editions Rodopi B. V. and Tere Vadén.

INTRODUCTION

The essay is a systematic attempt to understand cognitive characteristics of translation by embedding its logical and analytic features in an appropriate pragmatic and hermeneutic environment. To explain communication breakdowns that often result from conceptual disparities between two cultures, certain devices of the latter kinds must even be explicitly integrated with analytic considerations. Others can be left in the background. From the analytic side I employ such things as the notion of possible world and some ideas and methods related to possible worlds semantics and general model theory. A simple – though not entirely noncontroversial – pragmatic environment will be provided by my attempts to explore translation in the framework of speech acts, in a generalized sense of the word. What makes it simple is that it will only be used as a cognitive setting without any engagement in sophisticated aspects of speech act theory; and what may make it controversial is the generalization of the notion of ordinary speech act and its association with the idea of possible world. The nature of hermeneutic elements, on the other hand, and the way in which they cooperate with logical and pragmatic elements, will become apparent when we explore certain principles that seem to be involved in the understanding of speech acts.

The concept of translation that I investigate in this book will be called explanatory, for reasons to be explained below; but it is not translation in the standard sense of the word since it admits of meaning change. As we shall see, it can take various forms and degrees of precision, and therefore it can occur in contexts of different kinds: from everyday discourse to literary texts to scientific change. I try to argue that it has something in common with Thomas Kuhn's earlier conceptions of scientific change, or more generally, with his views of language learning.

The book is organized according to increasing logical complexity. Part One emphasizes pragmatic and hermeneutic elements of conceptual change and requires no formal sophistication whatsoever on the part of the reader. This part is self-contained, and it presents my basic ideas connected with explanatory translation and conceptual change, especially with respect to natural languages. Part Two presupposes some knowledge of model-theoretic notation, but it does not contain any formal argumentation. The reader who is not familiar with this notation is advised to consult Part Three, which provides nearly everything that is needed. It also formally works out some results exhibited in Part Two. Part Three is formal, but the reader who does not want to become involved in such details can omit this part without sacrificing much of Part Two. A great deal of the developments in Parts Two and Three is based on my earlier work and it has greatly benefited from my earlier collaboration with David Pearce and from his highly important ideas on intertheoretic relations. Most of Part Two is more recent or new, in particular the discussions concerning cognitive science and emergence and the applications to quantum mechanics and chemistry. Much of the research concerning cognitive science that is presented in this part and the next one is based on Tere Vadén's valuable ideas and on my cooperation with him.

The contents of the book are as follows. In Part One, the motivation for and the basic ideas of the notion of nonstandard translation are put forward and some applications and examples presented. Chapter 1 introduces the problems of conceptual change to be explored in the subsequent chapters. This is mainly done here in reference to the history and philosophy of science. The chapter sets the stage for my task by dealing with questions of scientific progress and with the idea of the Correspondence Principle and its possible generalizations for other cultural developments.

More specifically, I investigate here Kuhn's earlier objections to the Correspondence Principle and try to point out that his proposal concerning reinterpretations of scientific terms has an aspect that is analogous to what I have in mind when I in subsequent chapters reconstruct intertheoretic and intercultural relations in terms of a nonstandard notion of translation. He does not make clear, however, precisely what he means and in his later work favors a more standard notion of translation, so one cannot be certain about what he had in mind in this matter.

In Chapter 2, nonstandard translation is investigated in greater detail. The standard view holds that, ideally, a translation must preserve meaning, sometimes perhaps reference. Since this requirement is useless, for obvious reasons, I replace it by the condition that there be syntactic and semantic transformations which, together with certain principles of interpretation, take care of conceptual change. This generalized notion of translation has a formal predecessor in mathematical logic, and some recent forms of reduction occurring in the philosophy of science are also of that kind.

What I primarily try to show in this chapter is that the elements of the nonstandard notion of translation introduced here are implicit in everyday discourse. Therefore, I explore translation and interpretation in the context of ordinary speech acts. If, in particular, a given translation is of explanatory import, it will be called explanatory, and corrective if it corrects what is uttered by the speaker. I also make explicit certain principles of interpretation that seem to function in ordinary discourse, and present some examples from which we can see how these principles actually work. What is learned here is in later chapters enlarged so as to apply to more theoretical and global contexts.

Chapter 3 explores examples and applications of local translation, that is, assuming that a hearer translates single or just few utterances – rather than longer texts or even whole languages – having more comprehensive parts of a language only as a contextual element. This is typical for speech act contexts to which the name 'speech act' applies in the original, narrow sense of the word. I try to systematically sort out the different types of local translation and apply the result to understand better what is going on when people use metaphors or advocate different theories of perception or different views of artistic change. Though these fields of application seem to be conceptually far from each other in many ways, it turns out, I hope, that the present notion of translation makes their similarities and differences more transparent.

Chapter 4 turns the reader's attention to global translation. In the previous chapter, explanatory translation was considered in the local sense. However, when translating scientific or literary texts syntactic and semantic aspects must be explored more globally. In this chapter, I briefly investigate interpretation and nonstandard translation in connection with narratives, and, furthermore, consider special characteristics of the notions of truth and satisfaction relative to narratives, by keeping an eye on possible worlds semantics.

Part Two continues to study global translation, this time in reference to scientific theories and logics. When earlier chapters sort out the components of nonstandard translation in the local sense and then generalize the reconstruction for global cases, Chapter 5 shows how this generalization leads us to the notion of a correspondence relation of theories. The most intricate kind of correspondence is the so-called counterfactual correspondence, since it (intentionally) involves contrary-to-fact auxiliary assumptions. It will be seen in later chapters that this notion, especially its special case, limiting case correspondence, plays an important role in many actual cases of scientific change, which Kuhn calls revolutionary. The correspondence relation is considered in this chapter both in connection with nonformal theories, such as historical narratives, and formalized ones.

In the literature there has been some discussion about connections between correspondence and symmetries, i.e., invariance properties, in science. In Chapter 5, I also try to place the discussion in our model-theoretic framework. Furthermore, the concept of correspondence and that of translation in the more general sense are related to some problems of definability, on one hand, and to the intricate notion of emergence, on the other.

In Chapter 6 we investigate whether or in what sense correspondence, as defined in the preceding chapter, is of any explanatory import in science. In particular, I consider what I call counterfactual explanation, which involves contrary-to-fact assumptions and counterfactual conditionals. It is pointed out that on relevant pragmatic and logical conditions, such an explanation can be derived whenever a counterfactual correspondence is available. Since the notion of translation employed here is more liberal than the standard one, it can be used to go beyond Kuhn's criticism in the study of intertheoretic explanation. In this chapter, also some questions of explanation concerning mathematical theories and logics are investigated in the light of the correspondence relation.

Chapter 7 – which is the longest chapter – is devoted to case studies. We investigate some cases of scientific change, many of which are well known from, and much discussed in, recent history of science and recent history of logic, and point out that our notions are applicable to them. Whether an application of the correspondence relation, particularly of the counterfactual one, is of explanatory import in these cases is very much dependent on relevant pragmatic and hermeneutic conditions. I also sketch here a limiting case correspondence between symbolic and connectionist representations in cognitive science. A large part of this chapter investigates problems of cognition, of which the most interesting one (to me) is how propositional and nonpropositional knowledge might be related to each other. What is found out here is then applied to questions of epistemic logic and to the problem of emergence.

Most of the formal investigations and foundational studies of this book are gathered in Part Three. In its Chapter 8 I present the logical notation employed in some earlier sections and the formal and mathematical principles needed in the exact treatment of the case studies worked out in the next chapter. The idea is that this essay would be logically self-contained to the extent to which it is possible within reasonable limits. I attempt to choose exact tools that are able to cooperate with philosophical and pragmatic investigations. This conforms to the program I suggested above, that is, that logical investigations ought to be integrated with explanations which are pragmatic, paradigmatic, and hermeneutic. I begin the chapter by discussing the role of logic – especially the import of some concepts provided by abstract logic – in metascientific study, keeping an eye on this program. I also make a historical digression where we can see that Rudolf Carnap's logical aims were largely similar to mine.

Chapter 9 is formally more demanding. Some of the case studies of Chapter 7, mainly presented there in a nonformal manner, remained sketchy since the relevant syntactic and semantic entities cannot be exactly delineated. Some can be worked out in model-theoretic terms, as I point out in Chapter 9, by using many-sorted infinitary logics and nonstandard analysis. This is methodologically desirable in that it makes it possible to employ ordinary mathematical language and yet guarantees that we have a formal language and general model theory at our disposal, in the framework of which relevant notions can be represented whenever needed. Due to the restricted methodological and logical aims of Part Three, sophistications pertaining to more specific model-theoretic results cannot be found in this or the previous chapter, but this part offers a framework in which such results could be worked out.

The Appendix of the book contains a survey of some basic concepts and results of definability. Its origin is in my early cooperation

with Jaakko Hintikka and in his work on definitions and related topics. Questions concerning definability are often important in treatments of conceptual change, and its various kinds are akin to the notion of correspondence, as indicated in Chapter 5. Furthermore, in
model-theoretic applications it is crucial to know in which logics relevant classes of models can be defined. This part of the book consists
of both nonformal and model-theoretic considerations of definability
and, furthermore, some criticism of historically well-known issues in
the field.

—— PART ONE ——

THE PRAGMATICS AND HERMENEUTICS
OF
CONCEPTUAL CHANGE

1

PROLOGUE: THE CORRESPONDENCE PRINCIPLE

It used to be argued by many scholars that physics develops or it should develop so as to make valid what they call the Correspondence Principle. It was also said that in order for a scientific change in a more general sense to be continuous or cumulative (in some intuitive meanings of the words) the principle should apply to it. As we shall see later, we can even consider this principle from a more universal point of view and insist that something similar is needed in order to make conceptual and cultural changes understandable inspite of meaning variance.

In the present chapter, I shall mainly investigate Thomas S. Kuhn's earlier criticism of the view that scientific progress is cumulative and his proposal that if those intertheoretic relations which were claimed to exemplify the Correspondence Principle are to be more than mere formal relations, some terms occurring in the supplanted theories must be reinterpreted in the framework of their successors to take account of the fact that their meanings change. I point out that this proposal, rather than his later account of translation, can be construed as having something important in common with what I have in mind when reconstructing intertheoretic and intercultural relations by using a notion of translation that might be relevant if the principle is generalized.

1.1. A HEURISTIC PRINCIPLE

A heuristic rule, which Bohr (1920) calls *Korrespondenzprinzip*, was extensively used in the early developments of quantum mechanics. It says that a future theory is to be a generalization of a classical theory in that the former yields the latter as a special, or limiting, case in an appropriate sense.[1] In his description of how quantum mechanics was created, van der Waerden (1967) remarks that the principle had

3

guided physicists during the transition period from 1918 to 1925 and that a number of important results had been obtained by systematic guessing which was based on the Correspondence Principle.

Bohr was mainly interested in the role that the principle played in quantum mechanics, and it was originally mentioned in that context, but it has been considered crucial for developments in other branches of modern physics. Thus, for instance, Einstein used it as a guide in his search for relativity theory. Heisenberg (1958) and Born (1969) generalized it for the whole of physics. Heisenberg argues for the principle by saying that it plays an important role as a methodological principle, and Born maintains that the older theories have always been included in the new theories as limiting cases.

According to Heisenberg, modern physics can be grouped into four sets of concepts: (1) Newtonian mechanics, (2) phenomenological thermodynamics, (3) special relativity, electrodynamics, optics, and the like, and (4) quantum mechanics.[2] Now (1) is contained in (3) and (4) as the limiting cases where the velocity of light is considered infinite and Planck's constant infinitely small, respectively, and (2) is connected with the other sets. Heisenberg then goes on to argue that the independent existence of (3) and (4) suggests that there will be a fifth set in physics such that (3) and (4) are its limiting cases, and it is likely that it will be found in connection with the theory of elementary particles. Heisenberg in fact suggests that in the future correspondence relations could even obtain more generally, for in order to understand organic life it might be necessary to go beyond quantum theory and construct a new coherent set of concepts to which physics and chemistry would belong as limiting cases.

Similar views on the role of the Correspondence Principle in physics and in science more generally have been presented by several philosophers of science.[3] Post (1971), for instance, thinks of it as the most important heuristic restriction for developments in science, since to be acceptable a new theory should account for the success of its predecessor by being transformed into that theory under those conditions on which it has been well confirmed. The new theory should explain the well-confirmed part of its predecessor.[4] Post even argues that the principle is not only heuristic and normative but it is a historical fact that the successors have always carried out that task. There is one exception, however, according to Post. The principle fails in the case of quantum mechanics, for, contrary to what is said in some textbooks, quantum mechanics is not transformable into classical mechanics except locally, that is, with respect to certain subtheories. This is to be considered, says Post, a shortcoming of quantum mechanics rather than a breakdown of the principle itself, and therefore indicates that a better theory is needed.

1.2. REDUCTION AND THE CONTINUITY OF SCIENTIFIC CHANGE

The Correspondence Principle in the sense discussed by Bohr and other physicists was, as we saw, emphasized as a heuristic device. It has also been considered in the context of justification: as soon as a theory has a successor, the latter can be tested by checking whether the former is its special or limiting case. If it is, it is one indication that the respective scientific change is progressive.

According to some authors, the Correspondence Principle refers to a kind of intertheoretic relations, namely the correspondence relation, which can be regarded – perhaps subject to some reservations, however – as a kind of a general reduction relation. There are several reasons why reduction has been extensively discussed in the philosophy of science and in science itself. For example, it is often assumed that behind an observed, or otherwise given, phenomenon there exists a more fundamental reality to which the phenomenon can be reduced and which can be employed to explain and understand it. Furthermore, it is usually thought that scientific research is not feasible if it cannot be reduced to methods that in some sense are objective and reliable. Philosophy and science abound in historical examples and consequences of these ontological and methodological forms of reductionism. Such are radical empiricism and rationalism, the idea that the axiomatic method is reliable (these examples represent methodological reductionism deriving from the struggle for epistemic certainty), reductive materialism and idealism, the discussion concerning the reduction of biology to physics (which, in turn, represent ontological reductionism), and discoveries of elementary particles (which are a consequence of a kind of ontological reductionism). We may see that reduction has been discussed at various levels, at the levels of theories, knowledge, methods, and ontology. An important tacit idea has been, however, that if there is an ontological reduction, it should be mirrored at the theoretical level, and *vice versa*. In what follows, I shall mainly be concerned with the theoretical level.

The notion has also been important in debates concerning scientific change since it is often held – or, rather, it used to be a common view in the philosophy of science before Kuhn and other critics – that one indication of scientific progress is that theories are reducible, in some accurate, approximate, or limiting sense, to their successors, so that the latter theories are more comprehensive and more advanced than the former. A reduction, in turn, was thought to imply an explanation: If a theory is reducible to its successor, or to another more comprehensive theory, it follows that the latter explains the former in (something like) the sense of deductive-nomological explanation. Because the explanation is something that increases understanding, we should, after the reduction, be in a better position to see the nature of the reduced theory, and the nature of the change in question.

To place this discussion about reduction in its proper context, let us next make a quick survey of some earlier views concerning the question of what scientific progress might mean and whether it has been progressive. By progress I shall mean here progress within a given science, that is, I shall consider theories belonging to the same branch of science.

An important characteristic of progress mentioned in the literature is that scientific knowledge grows *cumulatively*, which means that old theories and the knowledge they represent survive, to some extent at least, when new and better ones appear, so that the latter extend the domain of scientific knowledge. Another characteristic is that a theory is *reducible* to its successor, which means that the old theory is not supplanted by the new one but can be thought of as being included in the new one as a special case. A third feature is described by suggesting that a new theory *explains* its predecessor, so that at least the most important principles, or laws, of the old theory can be deduced from those of its successor together with some auxiliary hypotheses. These characteristics go hand in hand, whereas the following two criteria are based on somewhat different ideas (but how far they are from the first three depends on how they are interpreted). One of the criteria is that a science is progressive if new theories *solve* the *problems* their predecessors do plus some further ones, or, rather, if they solve more or better problems,[5] and the other that progress means that scientific knowledge *approaches the truth*, that is, new theories are better than their predecessors if they are closer to the truth and hence describe and explain the world more accurately.[6]

These are, perhaps, the main characteristics of scientific progress that one can find in the literature, where it is maintained, moreover, that they apply to developments in 'mature' sciences, such as modern physics. Different authors defend different characteristics. For example, many empiricist philosophers, such as philosophers close to logical empiricism or its heir, the so-called Received View, held the view that the first three features are typical of modern science,[7] whereas some of their critics emphasize the fourth or fifth feature.[8] It seems that a crucial presupposition for the validity of the first three features would be that theories are well confirmed in some sense, so that they are relatively immune to rejection. That they are well confirmed implies, for instance, that they yield good predictions concerning phenomena to which they are applicable. That a new theory is adopted means then that the domain of application is extended.

In this book, I shall only consider the former view. One of its most outstanding representatives is Nagel (1961) whose classical concept of reduction has usually been assumed when the view has been advocated. In short, that a theory T is reducible to another theory T' means in a logical sense that the laws of T are deducible from the laws of T' together with appropriate auxiliary assumptions, some of which may link the languages of the two theories to each other. It follows that the reducing theory T' then explains the reduced theory T in the sense of deductive-nomological explanation, provided, of course, that the auxiliary assumptions satisfy appropriate theoretical and pragmatic conditions of adequacy.

According to Nagel, it is an undeniable feature of modern science that theories have been reduced to more inclusive theories, and he assumes that reduction will play an important role in the future. Among standard examples of reduction in the Nagelian sense that are usually mentioned are the reduction of rigid body mechanics to classical particle mechanics, of Kepler's laws of planetary motion to Newton's gravitational theory, and of classical particle mechanics to relativistic particle mechanics.

1.3. CONTINUITY CHALLENGED

However, Kuhn (1962), Feyerabend (1975), and other critics of the
Received View have attacked Nagel's ideas, and therefore also the validity of the Correspondence Principle, by arguing, for example, that
the meanings of scientific terms may change when theories change, or
a complete translation establishing a connection between the terms of
the two theories is not always possible, whence there in fact exist no
proper reductions in many actual cases where reductions were claimed to exist. Hence, no intertheoretic explanations are available in such
cases. The relation of Kepler's laws and Newton's gravitational theory and that of classical particle mechanics and relativistic particle mechanics exemplify the type of scientific change, radical change, which
Kuhn calls revolutionary, to which the Nagelian concepts of reduction and explanation and the Correspondence Principle are not applicable. Since this holds, it has been argued, we have to reject the view
that in such cases any scientific progress has taken place in the sense
of cumulation, reduction, or explanation; and the criticism also seems
to challenge the view that there has been progress in the sense that the
problem solving power or the truthlikeness of theories has increased.

Another kind of criticism of the Correspondence Principle is presented by Bunge (1973), who objects, for example, that the nature of
the limiting relation that is claimed to hold in many cases of scientific
change has not been explicated and that the validity of the principle
has not been established for those cases. Since then, some work has
been done in the philosophy of science to elucidate the relation and to
establish its existence. A more important objection, however, results
from Bunge's construing the Correspondence Principle as concerning
theories in a very comprehensive sense. For Bunge, the validity of
the principle would presuppose, for example, that the whole of classical mechanics is obtainable from the special theory of relativity and
from quantum mechanics. Thus Bunge seems to object to the validity
of the Correspondence Principle in such a comprehensive sense as
stated by Heisenberg, discussed above. Bunge's objection is clearly
justified. It may even be difficult to understand what such a comprehensive Correspondence Principle actually means. Heisenberg is not
very clear on the matter. As we shall see later on, the principle can
only be defended if it is taken in a sufficiently narrow sense, for

instance, as concerning the relation of individual laws rather than the relation of comprehensive disciplines.[9]

There are also logical problems. In many fundamental changes, especially in physics, some auxiliary hypotheses that would be needed to establish reduction relations are counterfactual, and, as sometimes thought, the theories involved in a reduction can be mutually incompatible. In such a case the derivation needed for an alleged reduction relation either does not make sense or is only valid approximately or in the limit, so it follows that there is no nontrivial logical connection between the theories of the kind required by Nagel. Thus, for example, Newton's gravitational theory is, strictly speaking, incompatible with Kepler's laws and, similarly, relativistic particle mechanics is incompatible with classical particle mechanics. So in these cases – in order to 'derive' in each case the latter theory from the former and thus to establish a reduction – we need counterfactual assumptions such as that the forces between the planets can be neglected or regarded as being infinitesimally small, and that the velocity of light approaches infinity.

Such difficulties have created uncertainty concerning the logical and explanatory status of intertheoretic relations, but they have also led to new attempts to study reduction. Several kinds of intertheoretic relation, which modify Nagel's model, have been subsequently suggested in the literature, as, for example, counterfactual correspondence,[10] nonlinguistic reduction,[11] factualization,[12] and approximate reduction.[13] While Nagel's reduction should lead to deductive-nomological explanation, the explanatory import of the other models is far from being clear. It is not evident, for example, that if a theory T is approximately reduced to another theory T', then T' would provide an explanation of T; for, even though it yielded an explanation of a theory which in some sense is an approximation of T – this is what the approximate reduction amounts to – it would not necessarily provide a conceptual relationship of T to T' of a kind that would be needed for an explanation.[14] Reductions of the other kinds mentioned above are problematic as well, this time since there does not seem to be any relevant why-questions to which they would provide answers. Hence, the explanatory import of the kinds of question to which these reductions provide answers is not quite evident, so we have to ask whether they play any role when one tries to explain and understand reduced theories by means of reducing ones.[15]

1.4. KUHN ON CONTINUITY

Let us now have a closer look at Kuhn's criticism of reduction and of the Correspondence Principle. From a point of view of counterfactual correspondence and counterfactual explanation, to be defined in chapters to come, Kuhn provides more interesting criticism than the other critics, since, as we shall see, he presents arguments whose logical structure is not clear, but which, on the other hand, seem to be substantially close to some ideas which in a natural way follow from the very notion of counterfactual explanation. There is, furthermore, some discrepancy between some of his arguments, and that discrepancy will also be scrutinized later.

Kuhn (1962) argues against the view that revolutionary change is cumulative. Thus, for example, Newtonian dynamics cannot really be derived from relativistic dynamics, in the sense required by the Correspondence Principle and by Nagel's notion of reduction. In order to be able to evaluate its logical structure and relate it to our forthcoming definitions, let us in some detail quote Kuhn's argument.[16] It is often said that from the laws of relativity theory together with additional statements, such as the one saying that the velocities of particles (i.e., their absolute values) are much smaller than the velocity of light, one can deduce by means of logic and mathematics a set of new statements which are identical in form with Newton's laws of motion, the law of gravity, etc. Therefore, says Kuhn, it seems that Newtonian dynamics has been derived from relativistic dynamics, enlarged by appropriate limiting conditions.

This is not really the case, Kuhn claims. The statements deduced are special cases of the laws of relativistic dynamics, but they are not Newton's laws despite their formal identity since the physical referents of the variables and parameters (that is, of variables and parameters which represent physical concepts) occurring in the set of the deduced statements are different from those occurring in the set of Newton's laws. In the former set, they still represent Einsteinian concepts, but in the latter, they represent Newtonian ones. The statements deduced are, therefore, reinterpretations of Newton's laws rather than those laws themselves, and such a reinterpretation would have been impossible until after Einstein's work. Hence the fact that statements that are identical in form with Newton's laws can be deduced from Einstein's laws together with the limit conditions does not show that the former laws are a limiting case of the latter. This is contrary to what used to be argued by what Kuhn calls positivist philosophers of science. The deduction only explains why Newton's laws seemed to work in cases where velocities are low.

This can be generalized so as to concern the main revolutionary changes in the sciences:

> Though an out-of-date theory can always be viewed as a special case of its up-to-date successor, it must be transformed for the purpose. And the transformation is one that can be undertaken only with the advantages of hindsight, the explicit guidance of the more recent theory.[17]

However, Kuhn goes on to say that the usefulness of the transformed theory is not very great even though the transformation were a legitimate device, since it could only restate what was already known.

1.5. KUHN ON TRANSLATION

Especially in Kuhn's later works the notion of translation becomes important. Let us first consider what he says of translation in the Postscript for the second edition of Kuhn (1962).[18] There he accepts the view that the participants in a communication breakdown can become successful translators in that each of them becomes able to learn how to translate the other's theory into his own language and, furthermore, to describe in his language the world to which the other's theory applies.[19] He seems to think, albeit somewhat hesitatingly, that this is what regularly takes place in the history of science. Translation in this sense is for Kuhn a tool for persuading one participant to accept the other's view, since it may allow him to experience merits and defects of that view, and therefore it provides motives for a conversion. It is not a technique that would guarantee the continuity of scientific developments. Even if it is associated with the transformation, that is, reinterpretation in the sense discussed in the previous section, as one might expect, it does not guarantee the existence of a limiting case correspondence between classical and relativistic particle mechanics.

In his later work, however, e.g., in Kuhn (1983), the notion of translation is different. Now it is very constrained, and less useful, and therefore somewhat controversial, particularly if compared with what we saw him saying earlier about translation. In this article, Kuhn makes a distinction between translation and interpretation and argues that some critics of his notion of incommensurability incorrectly equate the two notions. Actual translation involves two components or processes. One of them is called 'translation' in recent usage in philosophy, as he says. It is called by Hoyningen-Huene (1993) "translation in the narrow, technical sense." The other component is interpretative.[20]

Interpreting a culture, language, scientific theory or paradigm may, in turn, involve learning a new language or possibly an earlier version of the interpreter's own language. Learning a new language is not the same as translating it into the interpreter's own language. This much is obvious, of course. However, if the interpreter succeeds in translating, there will be no incommensurability. This notion means here the lack of an adequate translation: There is no language into which the two languages or two theories in question could be translated without residue. Incommensurability does not mean incomparability, however, since a comparison can be made to the extent to which there are terms, common to the two languages, whose meanings are preserved from the one language to the other.

Kuhn (1983) seems to vacillate between different criteria of adequacy concerning translation. For him, a translation of a text in one language into a text in another is a systematic replacement of words or strings of words of the former language by words or strings of words of the latter to produce an equivalent text in the latter language. Kuhn considers here two candidates for the notion of equivalence: that the two texts have the same extension or the same intension (or meaning). In the translation, the referring expressions of the former language must correspond to coreferential expressions of the latter in such a way that the structure of the world determined by the lexical structure of the former is preserved. What Kuhn thinks of the role of intensions remains somewhat obscure, but they seem to have something to do with criteria, though they are not equated with them, that are employed by speakers when picking out referents for the terms of their language.

Another feature that strongly constrains the notion of translation, in

Kuhn's newer sense of the word, is its holistic character. Because learning is holistic, also translation is holistic. Thus in classical mechanics, the notions of mass and force must be learned together with Newton's second law and this is why the Newtonian terms 'mass' and 'force' cannot be translated into the language of a physical theory, as, for instance, of relativistic mechanics, in which Newton's second law does not apply. Why Kuhn thinks that its holistic nature makes a translation in this specific case and elsewhere impossible, can be understood when it is recalled that, first, a translation means for Kuhn the preservation of meanings or extensions, in short, the equivalence of two texts, and, second, meanings and extensions are for him context dependent, wherefore there may not be any equivalence even when the same terms occur in different theories.

1.6. KUHN'S TENSION

If we look at Kuhn's criticism of the idea of limiting case correspondence (quoted in Section 1.4, above) more closely, we can see that it is, strictly speaking, logically fallacious in two respects. First, it is not the case, for example, that from the laws of relativity theory together with the statement saying that the velocities of particles are much smaller than the velocity of light, and with some other statements, one can deduce statements that are identical in form with Newton's laws. One can at most deduce statements which in some sense formally approximate Newton's laws.

Second, even if statements identical in form with Newton's laws could be so deduced, it would not follow that the variables and parameters occurring in the statements deduced would represent Einsteinian concepts and thus have Einsteinian and not Newtonian physical referents. It is not clear what Kuhn's claim that they do, in fact, represent in this fashion ultimately means. Assume that a law belonging to one paradigm is a deducible from another law, belonging to another paradigm, plus from some auxiliary statements. Consider any term that the two laws have in common such that its physical referent within the first paradigm is different from that within the second. Now it seems likely (or, at least, it is allowed by a consequence relation of the kind in question[21]) – if the term has different meanings within the two paradigms, as Kuhn proposes – that the auxiliary statements dispense with the physical referent within the first paradigm, that is, the referent is not preserved in the inference. This is because the very function of the auxiliary statements is to take account of the meaning changes from one paradigm to the other, that is, to switch the intended referent. It seems that only in some trivial cases a transfer from one paradigm to another is of the kind Kuhn is considering in his argument.

Therefore, it does not follow from the fact that there is such a deduction that the physical referent of the term, whatever it might be among its possible referents, is somehow automatically preserved in the deduction. Hence, no matter what the notion 'physical referent' means here, terms occurring in the statements that are identical in form with Newton's laws and deducible from the laws of relativity theory together with additional statements would logically admit of physical referents that are Newtonian rather than Einsteinian.

For these two reasons, it seems that Kuhn's criticism of limiting case correspondence is not logically accurate. But our first counterargument above can as well be applied to the philosophers whom Kuhn is here criticizing, since they are equally inaccurate in describing what limiting case correspondence is. Therefore, Kuhn's first objection is based on the fallacious starting point, derived from the philosophers he criticizes. As we shall see later on, the approach to the structure of limiting case correspondence must be radically changed, to be logically feasible.

What is more important here, however, and what even looks to me like a crucial point in Kuhn's reasoning, is the conclusion he draws, that is, what he is saying about transformation, or reinterpretation, of Newton's laws and, more generally, what is quoted above at the end of Section 1.4. The quotation seems to imply that if a correspondence relation is to produce more than a mere formal or syntactic analogy of an out-of-date theory, as, for instance, in an occurrence of limiting case correspondence, the theory must be transformed into, or, rather, reinterpreted in, the framework of its up-to-date successor. What he seems to suggest, then, is that to be more than a mere formal analogy, a correspondence relation has to take account of meaning change in some sense, in the form of a reinterpretation. As can be seen from the quotation, Kuhn is even saying that reinterpretations can always be worked out in the cases where limiting case correspondences are claimed to hold and, what is equally important, that they can only be worked out under the guidance of the up-to-date theory in question.

It does not become clear how Kuhn relates his earlier notion of translation to that of reinterpretation. If we try to understand, in a natural way, the relation between what he says of reinterpretation (of concepts), on one hand, and translation (of languages), on the other, it is difficult to avoid the idea that these two processes, one of which is semantic, the other syntactic, must somehow go hand in hand. The question of their interrelation will be studied later, especially in Chapters 2, 3, and 5, where we study the question from a slightly different point of view.

As it now seems evident to me, Kuhn's earlier discussion of translation and reinterpretation would provide a much more fertile basis for understanding revolutionary changes than does what he argues about translation some twenty years later, for example, in Kuhn (1983). Naturally, his earlier arguments are not directly incompatible with his later reasoning since he employs two different notions of translation, as noticed, e.g., by Hoyningen-Huene.[22] Nevertheless, it is somewhat confusing that the two processes which are of radically different kinds are both called translation. This ambiguity is later explained by Kuhn himself by telling us that earlier he used to be torn by competing senses concerning translation: that translation between an old theory and a new one was possible and that it was not possible; and he now recognizes that he was wrong to speak of translation in his earlier work.[23] It was language learning, not translation, he described, and these two processes are different.[24]

It seems, however, that Kuhn's new sense of translation is so rigid that it almost excludes any nontrivial, complete translation – thus almost excluding what he now calls commensurability – and he comes very close to saying it himself as he argues, for example, that different languages impose different structures on the world. If applied to theory change, his criteria would not only make two theories belonging to different paradigms mutually untranslatable but, as it seems to follow, also theories within the same paradigm. The requirements of equivalence together with the observation that translations are holistic make his new notion of translation, if not empty, at least impracticable, and, therefore, the range of his new notion of incommensurability very comprehensive.[25] As we shall see later, Kuhn is not alone with his concept of translation, however, since something similar seems to have been philosophers' standard view.

The criticism I have presented above is of a logical and methodological nature, and this kind of criticism has been rather common elsewhere in the literature.[26] It is therefore of some interest to see that Kuhn's notions of translation and commensurability have also been attacked from a point of view that is, so to speak, quite opposite. When commenting on Kuhn's and Feyerabend's notions, Hacking (1993) points out that their presentations of learning, translation, and incommensurability are "spectator-oriented." Hacking's view here is important, especially in that it relates to these notions an aspect that seems to be very much like a phenomenological one. In his view, if translation "is just replacing some words by others," as it is often understood in philosophy, and particularly by Kuhn and Feyerabend, it is not speaking or using words but merely displaying words instead of others: "The point is that one cannot speak the old science while using the projectible scientific-kind terms of present science." Even though one could learn to understand Paracelsus to the extent that one could write "a volume of pseudo-Paracelsus," one "couldn't live Paracelsus" nor project his terms, or if he could, he would drop out of his community.[27]

However, it is not quite clear to what extent Kuhn's earlier notion of translation or learning would be spectator-oriented since the notion itself is not clear. There can be some evidence for that. But it is obvious that in his earlier work incommensurability has to do with phenomenological aspects that are similar to what Hacking is after. For example, in his Afterword he argues that "[t]o translate a theory or world view into one's own language is not to make it one's own. For that one must go native, discover that one is thinking and working in, not simply translating out of, a language that was originally foreign." In brief, conversion means internalizing the new view.[28] There are similar passages in his more recent work, too. For example, Kuhn (1993) argues for the difficulty of transworld travel. If a modern physicist would enter the world of eighteenth-century physics or twentieth-century chemistry, ". . . that physicist could not practice in either of these other worlds without abandoning the one from which he or she came."[29] This is precisely what we just saw Hacking saying, but if he means Kuhn's notion of translation in (1983), he is obviously correct, for as we saw, it seems to be of a rather mechanical kind and at the same time defines commensurability.[30]

1.7. KUHN'S TENSION REMOVED

In spite of the logical flaws involved and the vagueness of presentation, Kuhn's criticism of the view that scientific progress is cumulative (and therefore of the Correspondence Principle), reviewed in Section 1.4, above, can be construed as stemming from an argument against the view that the notion of (limiting case) correspondence as a purely formal (syntactical) or mathematical relation would adequately describe the nature of revolutionary change. Whether or not the Correspondence Principle was thought of by physicists and earlier philosophers of science as being a mere formal principle, Kuhn, in any case, seems to claim that if it is not a purely formal principle, it is not valid. There is more to revolutionary scientific change, according to him; the meanings of familiar concepts change, and this change is not taken into account in the purely formal notion of correspondence.

In view of this construal, it is somewhat difficult to see why Kuhn (1970) did not follow the consequences of what he was arguing for when discussing reinterpretation, on one hand, and translation, on the other. Even if what he called translation in his earlier work is really to be taken as learning rather than translation, as he argues later on,[31] it is conceivable that it could have been studied together with what he called reinterpretation, and sometimes transformation, so as to produce an intertheoretic relation of some interest.[32] One reason why he did not elaborate such a relation might have been that translation and reinterpretation were considered by him as belonging to contexts of different kinds. The former is a tool for persuasion and conversion when a communication breakdown is experienced, but the latter pertains more directly to the problem of how the two theories or paradigms across a scientific revolution are related to each other. More importantly, perhaps, Kuhn did not seem to value the role of reinterpretation very high: "Furthermore, even if that transformation were a legitimate device to employ in interpreting the older theory, the result of its application would be a theory so restricted that it could only restate what was already known. Because of its economy, that restatement would have utility, but it could not suffice for the guid ance of research."[33]

It is no wonder, after all, that Kuhn later sticks to the strict notion of translation since this notion is more or less the standard one among philosophers, and moreover the common sense view — though he adopted it in the strict form that is almost useless.[34] For some reason, weaker forms of translation have not been of much interest to philosophers, though they are almost everyday practice for people working in science and in literature. It is well known, of course, and it is emphasized by Kuhn himself on several occasions, that one cannot often have translations which are strictly meaning-preserving or extension-preserving, or preserve the structure of the world, since so many things of one culture are ineffable in another. It is not well known among philosophers, on the other hand, that there is a device available which can be used to build a useful notion of translation — not presupposing that meanings or extensions, or structures, be preserved, in short, not requiring effability, but which nevertheless may enable one to move from one language into another in a controlled way. Somewhat surprisingly, perhaps, that device is analogous to one in mathematical logic, called there 'interpretation', but it can be applied in less formal contexts as well. The trick is to put appropriate syntactic and semantic transformations together in such a way that the semantic one reflects (hopefully, in a systematic way) the syntactic one. If the syntactic one does not preserve such semantic things as meaning or extension, the change can be seen by tracking the associated semantic transformation. As I remarked above, Kuhn (1970) can be construed as having components for something like a similar device, as well, but they were not integrated.

1.8. ON THE CONTINUITY OF CONCEPTUAL CHANGE

The original form of the Correspondence Principle is, as we saw, associated with the notion of limiting case correspondence. This notion in a strict sense is applicable to the kind of theory change that concerns theories whose principles can be mathematically formulated. As we shall see later, however, if the fundamental idea behind the Correspondence Principle is looked at from a sufficiently broad perspective, we might expect that something similar obtains, or should obtain, between conceptual and cultural frameworks more generally.

Conceptual developments are often possible only if an eye is kept on what is already familiar and if an alien culture is understood only in relation to one's own. This extension of the principle presupposes, naturally, that the relation of limiting case correspondence is replaced by a more general relation between conceptual, cultural, or linguistic systems, as the case may be.

Such a relation, to be called in this essay explanatory translation, will be studied and applied in the rest of this book.[35] I consider the question of whether this notion of translation would be more useful than philosophers' standard notion or Kuhn's strict notion in one's attempts to understand and explain the relation of two linguistic cultures in a communication breakdown. In the next chapter, it will be studied in the context of ordinary speech acts, and later on I shall use what we learn in the next chapter to investigate the same question in connection with literary works, works of art, and scientific theories. It seems to me that by consulting speech acts we can learn how syntactic and semantic components associated with translation are conjoined in practice. The conjunction is very much analogous to what is taught in the more formal context of logic. But what is explicit in logic is only implicit in speech acts and must be unearthed.

No matter how the notion of incommensurability is enunciated and generalized, it seems obvious by now that its core lies in something whose nature could be best captured in phenomenological terms, and which is referred to by Hacking when saying, as we saw, that the present-day scientist could not "live Paracelsus" nor project his generalizations, or, if he could, he could not survive with his present habits, and by Kuhn when speaking of the difficulty of transworld travel. If the roots of what is called incommensurability lie so deep, this feature definitely implies that there is no translation between two incommensurable cultures in Kuhn's strict sense of translation.[36]

However, it may still admit translation in the more relaxed sense to which I referred above, in which meaning variance can be taken into account. If it admits translation in this manner, then the existence of a translation in such a relaxed sense does not of course establish commensurability in Kuhn's sense. If its existence is thought of as establishing commensurability, as Pearce (1987) does, then it would be commensurability in a sense that is very different from what Kuhn (1983) means by it. As I have remarked before, and Pearce very con-

vincingly argues, Kuhn's notion of incommensurability is not adequate in so far as it is defined as something to which his strict notion
of translation does not apply. Why Kuhn does not think that it would
be enough to define the notion of commensurability as something
which admits cross-cultural understanding is not clear to me.[37] The
adequacy of a more relaxed notion of translation in contexts where incommensurability in Kuhn's sense applies has been pointed out in
earlier work by Pearce and me, and especially argued for by Pearce,
according to whom, for instance, ". . . one may uphold a view of
meaning variance similar to Kuhn's, as well as his assertion that no
theory-neutral language is available for expressing rival claims of
[classical particle mechanics and relativistic particle mechanics], without inferring that no ordinary logical comparison of the two theories
can be made."[38]

2

TRANSLATION

The standard view of translation contends that a translation must preserve meaning, or at least extension, as much as possible. As I noted
in the previous chapter, however, this requirement is in many cases
too strict and must be relaxed and replaced by the condition that there
be syntactic and semantic transformations that work together in a systematic way. As I also remarked, this generalized, nonstandard notion of translation has a predecessor in mathematical logic where it is
called 'interpretation'. Some reconstructed forms of reduction in the
philosophy of science also satisfy this condition. However, in the
present chapter I try to point out that the nonstandard notion of translation has its roots in everyday discourse, where it functions implicitly. We saw that what Kuhn (1970) says about revolutionary change
in science shows that he, too, possibly had elements of such a notion
of translation available but he did not put them together. Kuhn is not
very precise, however, and, furthermore, in his later work turns his
attention to the standard notion, so one cannot be certain about his intentions here.

It is more important to notice here, that if we study these implicit
sources of the nonstandard notion of translation, everyday communication and Kuhn's earlier work, we shall better understand how logical aspects of intertheoretic, intertextual, and intercultural relations are
associated with contextual, paradigmatic, and hermeneutic aspects.
Some aspects of contextual, paradigmatic, and hermeneutic kinds
must even be made explicit when such relations are modelled, whereas others rather function as guiding principles. This is particularly obvious in some cases of conceptual change in which communication
breakdowns occur and in cases in which the understanding of the
other's position presupposes a kind of explanation that does not
match any traditional notions of explanation. I start by studying translation and interpretation in ordinary discourse, and in later chapters

enlarge what is established here to more theoretical contexts.

2.1. PRINCIPLES OF INTERPRETATION

I shall consider interpretation and translation in terms of speech acts
since this may enable one to see more clearly crucial pragmatic and
intentional features of translation and, in particular, how these are associated with semantic and logical aspects. The notion of speech act is
here, and especially in the subsequent chapters, to be understood in a
generalized sense: the two parties of a speech act, a speaker and hearer, can be individuals or even (scientific or cultural) communities and
the speaker's utterances can be spoken or written texts – even theories and artworks. If, for instance, a theory is accepted by a scientific
community, that is, belongs to a scientific paradigm in something like
the Kuhnian sense of the word, it can be considered as an utterance in
a particular occasion where it, due to being an object of study, exemplifies some of its properties, to borrow this well-known notion
from Goodman (1968). I shall have more to say about speech acts in
connection with literary works and theories in Chapter 3.

Studying how interpretation emerges from reception makes it obvious that an interpretation on which a translation is based is more or
less implicitly guided by certain principles whose function is to increase a hearer's understanding of a speaker's position and this way
to close the gap between the positions of the two. The first of them is
well known, since it has been much discussed in the literature in the
context of Davidson's (1973) notion of radical interpretation. Therefore, I shall call it the *Davidsonian principle*. According to Wheeler
(1978), for example, a radical interpretation is such that the hearer interprets a speech act as being as much as possible like the speech act
he would make in similar circumstances. Hence, the hearer is trying
to maximize agreement, which, as it is said, is tantamount to maxi-
mizing the rationality of the other. At least on Wheeler's construal of
Davidson's theory, this is because the hearer assumes that the speaker's attitudes are the ones he would have in a similar situation, and in
like manner Davidson himself says that the hearer starts interpreting
by assuming general agreement on beliefs. General agreement means
here something that functions as a background which enables one to
engage interpretation but which makes disagreement in particular oc-

casions possible.

A similar position is assumed by some cognitive scientists. Dennett (1978), for instance, suggests that when we study the behavior of an intentional system, we start by assuming rationality. But this rationality is *our* notion of rationality. The presumption of rationality is so strongly established in our inference habits that we do not easily question the overall rationality of the system under study. This is, in other words, one's methodological starting point, but since exceptions are characteristic of intentional systems, intentionality does not always go together with ideal rationality, and one has to tolerate less than optimal performance, according to Dennett.

Important differences notwithstanding, it is evident that the philosophy which lies behind the positions of the kind which Davidson and Dennett represent, and which obviously comes from Aristotle, is the philosophy on which many literary theorists base their approaches to the interpretation of fiction. Ricoeur (1992), for example, argues that the significance of a narrative derives from the "intersection" of the world of the text and the world of the reader, that is, narrative understanding is anchored in living experience. This implies that the gap between fiction and life must be kept to a minimum when interpreting narratives. Lewis (1978), in turn, proposes that the body of beliefs to be used as the background for interpreting a story consists of the beliefs that are overt in the community of the origin of the story. All these positions cited above are much more sophisticated and complex than what I am able to quote here, but as a basic premise, they all regard the interpreter's or the hearer's side as the methodological starting point of understanding.

The second principle of interpretation, which I shall call the *minimization principle*, applies in contexts where a hearer starts the interpretation process by observing or believing that a speaker's utterances are (from the hearer's point of view) false or otherwise mistaken or inappropriate, or involve concepts not available and hence ineffable in the speaker's culture or ineffable in literal terms. However, even though an utterance is an expression of the speaker's attitudes, concepts, or truth conditions that are at least seemingly different from those of the hearer, the latter makes the gap as small as possible in order to understand, and possibly explain, the speaker's position. He may strive to understand the utterance by looking to the speaker's position. In other words, the minimization principle is functioning when the gap is minimized, not because the hearer assumes overall similarity of beliefs but *in spite of* assuming their dissimilarity. The direction of minimizing the gap is then opposite to the direction that results when the Davidsonian principle is applied. As we shall see from examples to follow, in some cases an effort to minimize the gap in this sense requires counterfactual thinking. Understanding is often successful only if the interpreter is ready to change, at least tentatively, his position so as to make assumptions or thought experiments that are contrary-to-fact by his lights. The minimization principle is appropriate in contexts where the notion of translation is taken in a nonstandard sense and where it is of explanatory import.

If it is the Davidsonian principle that enables one to engage in a process of interpretation in the first place, then it seems to follow that the minimization principle depends on it, and becomes applicable only after the former has done its work. While the former principle can be active independently of whether or not one is aware of it, that is, it functions in any case, it seems likely that to apply the latter presupposes a conscious effort. Be this as it may, it is natural to think of the principles as providing two basic components for a dialogue, even though their applications may not be separable in practice. Often they work as a starting point so as to enable one to engage in interpretation but become mixed during a dialogue, so that their applications provide phases in a hermeneutic process of interpreting. The principle saying that they are repeatedly applied and combined to give rise to a process in which the positions of the two parties approach each other, will be called here the *dialogue principle*. In a sense, then, this principle can be seen as a mixture of the first two principles, from which it follows that later on these principles play a dominant role.

If we add to the minimization principle appropriate proposals concerning the role of tradition in Gadamer's (1960) sense of tradition, as, for example, to the effect that assumptions an interpreter makes to make the gap smaller by means of the minimization principle tend to modify his tradition, that is, change the interpreter's position permanently, then the minimization principle in such an augmented sense has something to do with Gadamer's notion of agreement or consensus in interpretation. Then Gadamer's position can be seen as analogous to the combination of the Davidsonian and minimization principles, that is, the dialogue principle. But if the assumptions are just temporary counterfactual thought experiments and do not imply that the interpreter seriously reconsiders his previous position, his applications of the dialogue principle cannot be regarded as resulting in a dialogue in something like Gadamer's sense.

The three principles of minimizing the gap can also be characterized in terms of Hacking's (1993) theorizing on the spectator-oriented nature of translation, which was outlined in the preceding chapter. First, it seems obvious, in general, that the minimization principle provides a more spectator-oriented basis for translation than the others. It is often applied consciously as a methodological tool, and at least in cases where a translator evokes imaginary situations to make some counterfactual thought experiments, it is difficult to imagine that he could "live" such situations; or, to apply another expression of Hacking's, when the speaker belongs to a culture that is alien to the hearer, the latter agent cannot "speak" that culture while projecting the terms of his own culture.[1]

As we shall see later, in Chapter 5, the minimization principle, once it has been generalized so as to be relevant for speech acts in the generalized sense referred to above, is prominent in cases of limiting case correspondence. There we shall be dealing with translations between scientific theories, and therefore Hacking's criticism is directly applicable. However, its force here and elsewhere is dependent, e.g., upon the status of the old science in relation to the new one during the historical period one is talking about, and upon how radical the scientific change in question was. In many cases where a limiting case correspondence is currently claimed to obtain, even though an old science is superseded by a new one in the Kuhnian sense of scientific revolution, scientists are trained to speak the old and the new science equally well. In this sense, the relation of classical mechanics and relativistic mechanics, for example, is completely different from that of modern science and Paracelsus. In the former case, Hacking's criticism misses the point.

On the other hand, a translation based on the Davidsonian principle of interpretation is only slightly spectator-oriented since its application means that the hearer starts interpreting by taking, consciously or not, the speaker's speech act as being like a speech act he would make in similar circumstances, that is, by assuming general agreement on beliefs, or, in Ricoeur's terms, by anchoring understanding in living experience. If the third, Gadamer-type principle is a mixture of the first two, it places itself in between the others as far as spectator orienting is concerned. It is plausible, in any case, that Gadamer's notion of consensus admits the construal on which the tradition of the interpreter is changed in a dialogue and a new one emerges. On this construal, consensus means that the interpreter learns to project some of the speaker's notions, that is, (partially) to live the speaker's culture.

Another principle of interpretation, different in kind from those discussed above but sometimes effective, is what I shall call the *refinement principle*. It also seems to play an implicit role in speech acts, as we shall see in the next section, but its roots in a more explicit and precise sense can be found in some variants of the notion of scientific reduction. In some ordinary speech acts its functioning is a consequence of uncertainty in interpretation. When a hearer makes an effort to understand and explain an utterance, there may not be a unique way to do that, even contextually, as a result of which he will be forced to consider several interpretative alternatives. Considering different alternatives often means making finer distinctions, finer in comparison with distinctions made by the speaker, or by the hearer in the first instance of the process. The import of the principle will be seen in the rest of this chapter and later on when we study what I shall call explanatory translation, and, in particular, when discussing the correspondence relation of theories.

2.2. ON THE STANDARD VIEW OF TRANSLATION

We saw above that the later Kuhn maintains the standard view of translation. It is also held by Quine (1960), and it is the common sense conception as well. According to Quine, ". . . meaning, supposedly, is what a sentence shares with its translation . . .."[2] This is an ideal, however, which only rarely can be reached in actual linguistic practice; as we saw, this very fact was Kuhn's point when introducing the notion of incommensurability, and the fact is, of course, recognized by many others. According to Kuhn (1993), difficulties in translation arise from the failure of different languages to preserve structural relations among words, and, in science, among kind-terms. In literary texts, associations and overtones depend on such relations, and in science criteria for determining referents of scientific terms depend on them as well.[3]

Though the standard notions were not applicable in practice, one may argue that this is what we should strive for. But we shall see that often it is even desirable to *depart* from attempts to preserve meaning or reference. This may happen both in everyday communication and with intertheoretic relations. We shall point out that this is most prominently the case when translations can be used for explanatory purposes. Less clear-cut cases occur, for instance, in translations of literary texts, because there preserving aesthetic and stylistic values is often considered more important. Burge (1978) notices this deviation from the standard view when he says that good translations of literature are often such that they preserve "rhyme, rhythm, alliteration, or idiom, at the expense of 'literal meaning' and even reference." To see what properties translation should preserve, "[i]t is necessary to understand the wider context in which the sentence is used − the presuppositions and intentions of the user, and the character of the passage or argument in which the sentence is embedded."[4] This is close to what we saw Kuhn saying about the holistic nature of translation. On the other hand, at least in so far as natural languages and literary texts are concerned, aesthetic and stylistic features often contribute to meaning − they may contribute to what is sometimes called connotative or secondary meaning[5] − and therefore attempts to preserve such features can be thought of as attempts to preserve meaning. Contrary to what Kuhn argues, however, it is evident that the preservation of aesthetic and stylistic features does not necessarily depend upon preserving structural relations among words but, on the contrary, presupposes a deviation from them.

2.3. LOCAL TRANSLATION AND SPEECH ACTS

In the rest of this chapter, I shall consider ordinary speech acts, hence translations in a *local* sense, that is, translations of a single or a few utterances, rather than translations of languages or their comprehensive fragments or texts. *Global* speech acts are considered in forthcoming chapters, where I generalize what we learn here. Clearly local translations lack many of the most interesting and difficult aspects of translation that are typical for globally translating long texts or whole

languages, but a number of basic principles involved in any translation can already be learned from local cases.

I introduce now some concepts that are model-theoretic in spirit,
or, rather, pertain to possible worlds semantics. In local cases I shall,
however, speak of *situations* rather than possible worlds or models;
but the term 'possible world' can be thought of as a general name that
among other things may refer to situations and models. In possible
worlds semantics, situations are often referred to as 'small' possible
worlds. In everyday usage, a situation is an entity that is language
independent but in which a language, or part thereof, can be interpreted. Its model-theoretic counterpart would be the concept of structure rather than the concept of model, which is language dependent
including interpretation.[6] Whether situations can be language independent is of course a controversial issue in current philosophy. But I
shall here adopt a technical rather than everyday notion of situation in
the sense that I assume an interpretation of a language as belonging to
a situation, roughly as in model theory where an interpretation is a
component of a model. One reason for adopting this conception of
language-dependent situation is methodological in the sense that some
of the semantic relations involved in translations and principles of interpretation can then be treated by considering relations between situations, and therefore we need not consider relations between interpretations as detached from situations.

An utterance often has connotations presupposing possible worlds
that are, intuitively and nonformally speaking, more comprehensive
than what the word 'situation' suggests. I shall not consider them in
this section, but later in the context of theories and narratives. The
concepts to be introduced, as well as some other aspects below, cause
our theory to deviate slightly from ordinary speech act theories, but
they are necessary if we are to generalize the standard notion of translation.

I shall only consider those speech acts which have propositional
contents and, moreover, whose utterances can be *satisfied* in various
situations. Hence the notion 'conditions of satisfaction' is relativized
to a situation, but the manner in which it is relativized may depend on
context.[7] The notion will not, and possibly cannot, be given any exact characterization here. I follow Searle (1981) in that I assume that
by performing an illocutionary act with a propositional content a

speaker expresses an intentional state with the same content, but I generalize upon Searle by saying that the act, or the respective utterance or expression, is satisfied in a situation if and only if the expressed intentional state is satisfied in it.[8] This makes it possible to accept such a thing as what is often called the speaker's meaning and to say that a speech act is satisfied in some situations even though, for instance, the act has broken semantic or other linguistic conventions and rules or portrayed fictional characters or events. In other words, it is more or less up to the speaker what the situations are that are represented by his utterance, in which it is intended to be satisfied. Accordingly, when I in what follows speak of the *speaker's situations* associated with a given speech act, I refer to such situations, or possibly to a kind of situation. In like manner, I speak of a *hearer's situations*. They are situations in which the hearer's translation is satisfied and to which he intentionally refers by the translation, and they result from his interpretation and explanation of the speaker's utterance.

My view here is similar to Grund's (1988) theory of counterfactuals in that both admit that a speaker's utterance may be satisfied in a given situation according to his interpretation and conditions of satisfaction (or according to those of his culture) but not according to those of a hearer (or the hearer's culture). This would happen, for example, in cases of meaning variance. On the other hand, according to the present technical notion of situation, if the interpretations of the speaker's and hearer's utterances are incompatible, their situations are different. Thus such utterances could only be satisfied in a single situation if we had the language-independent notion of situation. What this possibility really means will be seen from a couple of examples in the next chapter. Those examples are supposed to illuminate the distinction between the two kinds of situation and its consequences.

Part of what I have said above may appear to employ controversial, and not very clear, assumptions. We have to be careful here, because there is no general way to distinctly characterize a speaker's situations, for what they are depends on context and on the kind of speech act. Their characterization is a matter of pragmatics and a matter of understanding the speech act. I shall not defend the terminology here by reviewing or commenting on the vast discussion in the literature that is related to similar notions. I mainly support it by trying to point out by means of examples, that it is natural in some contexts – those of common parlance – and apt for my nonstandard notion of translation.[9]

Assume, for example, that an uttered expression is a sentence, capable of having truth values in various situations.[10] Then it is true in each of the speaker's situations. For example, if the speaker by uttering the sentence 'The door is closed' makes the statement that the door is closed, his situations are such that the door in question is closed in them (under his interpretation of the language). No matter whether the statement is true by a hearer's lights (under his interpretation), that is, whether or not the hearer considers the door in question to be closed, it is true in the speaker's situations. That is to say – even if the speaker erroneously (erroneously according to the hearer) referred to a situation where the statement is considered false by the hearer, that is, where the hearer thinks the door is open – that the hearer's situations do not count as the speaker's situations in so far as the speaker made the statement that the door is closed. On the other hand, if he by uttering that very sentence (and for whatever reasons) makes the statement that the door is open, then in each of his situations the door is open. These results do not as such depend on whether or not the speaker is sincere, but the case where he is just lying when making a false statement does not seem to be very interesting from a point of view of translation. It does not call for explanation in the sense to be discussed below.

The issue is made still more complicated by the practical fact that a hearer cannot always find out what a speaker intended; and, as we soon can see, when interpretation and translation are used for explanatory purposes, that knowledge would be indispensable for a correct explanation of a speech act. Sometimes a speaker may not have distinct meaning intentions and they are often ambiguous, and, as we have seen, his intentional states may not be satisfied.[11] In such cases, a hearer can only do his best to find out the most likely or warranted interpretation, and appropriate situations (satisfying the act) accordingly. In what follows, I may occasionally yield to such epistemic difficulties and consider a speaker's situation in this relaxed sense. This will have no effect on the logical features of the forthcoming theory.

A translation need not be actually satisfied, by a hearer's lights, since sometimes the hearer's situations can be fictional from his point of view, that is, based on a thought experiment. Therefore, the hearer's situations may result from an application of the minimization principle that transforms some of the situations originally considered correct by the hearer, to be called *actual situations*, into situations considered by him imaginary or otherwise 'less correct'. How this works out will be seen from examples below. The term 'actual' does not necessarily mean here the same as in possible worlds semantics ordinarily, since here it is relativized to the hearer. In what follows, I may speak about situations of any of the kinds defined above in the singular or in the plural, depending on which way is more convenient and appropriate (and similarly for the worlds to be discussed later).

In this framework, each principle of interpretation whose function is to minimize the gap between a speaker's and hearer's positions, implies that the hearer's situations are assumed as being as close as possible to the speaker's ones. I shall not try to define closeness; the principles will be thought of here as intuitive and qualitative rather than explicit or measurable constraints, but the role played by closeness can be understood as analogous to that played by similarity or closenes in Lewis' (1973) theory of counterfactual conditionals.

The principles can also be distinguished from each other by characterizing the direction of convergence differently. A case for the first – Davidsonian principle – is such that though a hearer may not be able, in the beginning, to exactly identify a speaker's situations, he assumes them to approach actual situations, which result from his interpreting the speaker's utterance in the context in question and are "anchored in living experience." Therefore, these actual situations now play the dominant role in the interpretation. On the other hand, in an application of the minimization principle, the hearer's situations are not necessarily anchored in living experience, but, on the contrary, tentatively removed farther away, even detached, from the actual situations (for explanatory or whatever reasons) so as to approach the speaker's ones, which now play the dominant role.

2.4. EXPLANATORY AND CORRECTIVE TRANSLATION

As usual, the term 'translation' will be used ambiguously in what follows, and it refers both to a syntactic transformation and the resulting expression, and sometimes even to the whole process involving the syntactic and semantic components introduced above. This ambiguity will not cause any confusion since the context shows which of these alternatives are referred to. When trying to understand a speaker's utterance, a hearer must start considering the speaker's position from his own presumptions. This is what the Davidsonian principle of interpretation in fact says. Therefore, if the principle has already done its work in the sense that the hearer has a conception of the speaker's situations in relation to the actual ones, it seems natural to assume that the hearer would be able to transform the latter situations (or some of them) into those of the speaker. This *semantic transformation* may in general have two parts. As indicated above, the understanding of an utterance may involve an application of the minimization principle and hence a transformation of (some of the) actual situations into situations (i.e., the hearer's situations) that are closer to the speaker's situations, but which can sometimes be counterfactual with regard to what the hearer considers true. I shall call this first part of semantic transformation a *minimizing transformation*. Since the translation of the speaker's utterance is to be satisfied in the latter situations, how they are determined depends on how the hearer intends to translate the utterance – and as we shall see, for different aims different translations are relevant. The task of the second part is to transform (some of) the hearer's situations into those of the speaker, and it reflects part of the semantic change induced by the corresponding (syntactic) translation. This transformation will be briefly called a *correlation*.

The interplay between the various kinds of transformations of situations and, on the other hand, between them and translations of utterances (that is, syntactic transaformations) can be schematically and generally presented as in the diagram of Figure 2.4.1, in the below, in which (for typographical reasons) the fact is ignored that there can be several situations of each kind. In the next chapter, where we investigate some examples, diagrams of this kind are used to systematize various kinds of translation occurrences in the contexts of natural language and pictorial speech acts. The diagrams are presented in the hope that they would make the occurrences more understandable. When we discuss nonstandard translations between single utterances, it is not always very clear in what sense we are really entitled to say that they are translations. This is because our present, more or less technical, notion of translation is very different from what is usually understood by translation. Then the alleged purpose of the respective diagrams is to clarify the logical structure of the cooperation between the semantic and syntactic sides of the translations to be considered.

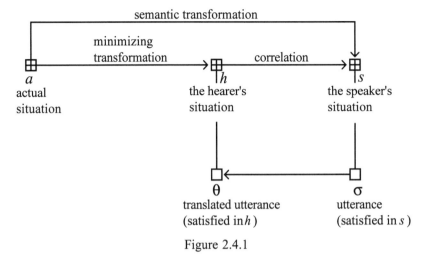

Figure 2.4.1

In what follows, we shall additionally denote

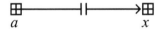

(where x is *h* or *s*) in the diagram if the effect of the minimizing transformation is counterfactual, that is, the transformation leads away from the actual situations, and

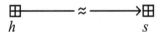

if *h* is very close to *s* (but not identical), where the meaning of 'closeness' depends on context (and can be either an intuitive or more formal notion).

As we shall see later, not any semantic transformation is good for explanation, since its both parts must be pragmatically grounded in such a way that they, when conjoined with the translation, help one to understand the speaker's utterance. It should be evident that pragmatic and hermeneutic features of translations cannot be seen from the respective diagrams which only display some logical elements of translations. If a semantic transformation is associated with a syntactic translation in such a way that they are of explanatory import, then we have what I shall call an *explanatory translation*. How the two parts of a semantic transformation are related depends on the case. As we shall see in Chapter 6, their relation is extremely complex when explanatory translation is the same as limiting case correspondence. In local cases, where the relation between translation and correlation is logically very simple, pragmatic elements play a specifically crucial role in attempts to avoid trivial translations. On the other hand, in global – especially formal – cases, translations have to have a systematic, often inductive, nature, and this feature plays a similar role.

The present account of translation is a generalization of the standard account. If we have a standard translation, then the associated correlation is identity, and therefore the speaker's and hearer's situations coincide. The hearer may still make up an appropriate minimizing transformation, as, for example, in a case where he considers the utterance incorrect, but nevertheless (by first applying the minimization principle) translates (or paraphrases) it so that its meaning is preserved. A standard, meaning-preserving translation leads to one of the following diagrams, depending on whether (as in Figure 2.4.2) or not (as in Figure 2.4.3) the hearer accepts the speaker's utterance, that is, whether or not the hearer thinks of the utterance as referring to an actual situation. Since a meaning cannot be represented by means of a single situation, the diagrams also depict the various cases where extensions are preserved. In the second diagram of Figure 2.4.2, in contrast to that in Figure 2.4.3, the arrow from an actual situation (that is, *a*) to one of the hearer's situations (*h*) is unbroken, meaning that the minimizing transformation does not lead outside the actual situations. The hearer uses the minimization principle to reach an actual situation that is among the speaker's and hearer's situations.

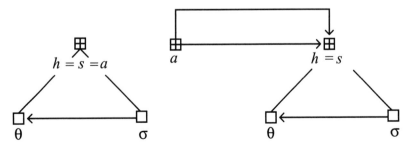

Figure 2.4.2

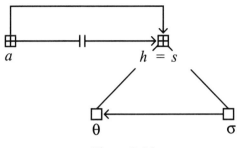

Figure 2.4.3

As noted above, a speaker's utterance and its translation by a hearer may be contingently satisfied in a single situation (in case they are compatible) or because they are equivalent. In more trivial special cases of these diagrams, $\theta = \sigma$, which means that no translation (nor paraphrase) has taken place. This would of course make the first diagram of Figure 2.4.2 completely trivial, but not the others since the minimization principle is applied there − which is indicated by the nonvacuous minimizing transformation.

The idea of generalizing the standard concept of translation by adding the notion of correlation is, as a matter of fact, not new. Similar notions, analogous to correlation, occur explicitly in some text-books of logic, accompanying such concepts as reduction and interpretation of theories,[12] and, I think, implicitly in the classical models of reduction and deductive-nomological explanation in the philosophy of science,[13] and, again, explicitly in what is often called structuralist philosophy of science.[14] It seems that in the philosophy of science correlation was for the first time formally associated with translation in a general, model-theoretic form by Pearce (1979) and (1982), and applied by Pearce and Rantala (1984a) to an actual theory change in the form to be discussed later in the present book.[15]

Obviously something like the notion of semantic transformation occurs implicitly in the processes of translating expressions of natural languages or literary texts, and the idea of minimizing transformation is analogous to, or a generalization of, a certain part of the notion of limiting case correspondence.[16] In what follows, I shall try to illuminate the importance of semantic transformation by considering its role in natural language translations, and later on it will be studied in the contexts of scientific and artistic change.

If a hearer considers a speaker's position incorrect − that is, the speaker's situations do not agree with the actual ones – or not being coherent relative to some other assumptions, as, e.g., assumptions pertaining to the speaker's situations, his translation may be *corrective*. Then he makes an effort not to find an equivalent translation of what is uttered, but to correct the speaker's cognitive representations of the world. For instance, if what is uttered is a sentence being capable of having truth values, then, even though the utterance is true in the speaker's situation, for the hearer it may appear to be false about the world in the sense that the speaker's cognitive representations on which the utterance is based do not seem to match the hearer's cognitive representations of the world, that is, the general cognitive strategies which the hearer employs in order to cope with the world. Therefore, the hearer tries to correct the utterance by looking for a translation that would be true of the world according to his representations, or at least true in situations that are closer to the actual ones.

We shall see later that explanatory translations are often based on the Davidsonian and minimization principles of interpretation in some form or other. These principles were argued for above, and it is obvious that they have consequences for translation. As we have seen, adopting the principles means that a hearer's situations are assumed to be as close as possible to those of a speaker's. The effect of the transformation accomplished by a correlation is then minimal. When applied to scientific change, they enable us to take a middle course in studying the role of translation in revolutionary scientific changes. In other words, they make it possible to try to moderate the implications of Kuhn's observation concerning the lack of perfect translations while preserving something of Nagel's classical view of reduction.

We have been looking at the notion of translation from a hearer's vantage point, assuming that a translation of an utterance is acceptable by the hearer's lights rather than by the speaker's.[17] The translation is based on the hearer's explanatory and possibly corrective interpretation of what is uttered. Therefore it seems indispensable that in many, if not most, applications of our generalized notion of translation the refinement principle is valid. In the present framework, this means that to each of the speaker's situations there correspond several situations of the hearer, that is, there are several situations of the latter kind that are in the associated correlation with, and hence transformable into, the former. There is, in other words, a correlation that is a many-one mapping from the hearer's situations to those of the speaker.[18] As we shall see, whenever the principle holds, it matches the requirement that translation be of explanatory import.

If the refinement principle is adopted in this form, it and my above argument for it agree with what is usually called the structuralist notion of reduction.[19] The principle is defended by philosophers of science by saying that the reducing theory is, typically, more fundamental than the reduced one in that it describes the world more successfully or more precisely, or is able to make more distinctions than the reduced theory.[20] The principle is also involved in the notion of interpretation as this notion is defined in mathematical logic, and its justification is the same.[21] As will be seen below, another kind of context in which the principle is valid is one in which a hearer's translation is corrective, but he has a number of alternative situations to refer to, due to his ignorance concerning the speaker's situation. In applications where the minimizing transformation is different from the correlation, the latter mapping (and hence the refinement principle) plays a role that is slightly different from the role played by its predecessors in the philosophy of science. We shall also see that minimizing transformations do not only yield applications of the minimization principle but often the material with which the refinement principle is associated.

EXAMPLES AND APPLICATIONS
OF LOCAL TRANSLATION

In the examples and applications of Sections 3.1-3 below, utterances are simple natural language expressions, but the conclusions are applicable to more complex utterances and to scientific contexts involving local translations. There are so many different types of translation contexts which may occur in ordinary discourse, that it may not be possible to consider them all. I try to explore, however, by means of an appropriate set of examples and in an orderly way, the relation between translation, correlation, and minimizing transformation, and to show that there is an intensive interplay between these elements. In particular, the translation chosen by a hearer from among all possible translations of an utterance usually determines the nature of the other elements. On the other hand, how the hearer understands the utterance and how he employs principles of interpretation determine the translation. Pragmatic and hermeneutic elements are crucial here, and they determine the logical structure of a translation. In the last section of this chapter, I try to indicate that the theory can be applied to visual art, too, if we accept the idea of pictorial utterances and speech acts and the idea that the distinction between syntax and semantics applies to pictorial representation.

3.1. LOCAL TRANSLATION AND EXPLANATION

It seems, for instance, that Danto's (1981) well-known analysis of the question of how the title of a work of art affects its interpretation provides us with an appropriate theoretical context in which to investigate problems of nonstandard (local) translation in a systematic way. For Danto, "[t]o interpret a work is to offer a theory as to what the work is about . . .," and on the other hand, the title is often intended to structure the work itself: it is a direction for interpretation.[1] Changing the title may change interpretation and this may transform the work into another work. As we shall see below, semantic transformations associated with translations can be nicely related to transformations of artworks.

As an example, Danto considers the painting *Landscape with the Fall of Icarus* by Pieter Breughel the Elder and examines the question of how its title (and the associated 'theory') may structure the painting, especially in comparison with the structuring that would result if the same picture possessed a different title. As soon as the title is acknowledged and the story of Icarus recalled, the small white spot that can be seen in the foreground of the picture may be identified as the legs of Icarus. The story gives the spectator a theoretical framework which organizes the picture around the white spot. Danto then asks us to imagine that the picture would have a different title, say *Industry on Land and Sea*. This would be appropriate since one can see ships and a plowman in the picture. Then the spot might be understood as representing the legs of a pearl diver or an oysterman; there is nothing in the picture as such with which this would be incompatible. On the contrary, this understanding of the meaning of legs is, I think, even more coherent with this title than to understand them as belonging to, for instance, someone swimming for fun. In any case, if this were the title, the painting would no longer be organized around the legs; thus, for instance, the fact that the plowman pays no attention to the legs is no more deeply significant, as it is in the case of Icarus.

This is just a small part of Danto's instructive discussion of the import of titles. Since it is theoretically oriented, it yields a nontrivial framework for examples that can be used to bring forth the fine-grained character of explanatory translation, i.e., to show how even small contextual and theoretical differences may influence the way in which the syntactic and semantic transformations cooperate.

Assume first that someone looking at Breughel's painting thinks that the white spot in the painting in front of him represents legs and then utters the following sentence:

(3.1.1)   The legs belong to a pearl diver or an oysterman.[2]

Assume furthermore, that by uttering this sentence the speaker asserts what he seems to assert, that is, that the legs in the painting in front of him belong to a pearl diver or an oysterman. Then, if a hearer wants to translate the utterance into French in the standard sense of translation, and if he is competent enough in French and English and believes that the speaker makes that assertion, his translation is likely to refer to the same (kind of) situation, and hence no semantic transformation is needed. The meaning is preserved. It can of course be argued that there nevertheless are tiny differences that, in fact, render a semantic transformation necessary; for instance, in the speaker's culture the way of understanding situations where paintings occur may be different from the way provided by the hearer's culture, or paintings in the two cultures may be perceived differently, or divers may have slightly different meanings in the two cultures. More generally, the role played by paintings in the speaker's culture is different from that in the hearer's. If there are cultural and connotative differences even in simple cases like this one, there hardly can be many opportunities for meaning-preserving translations. Be this as it may, in what follows I study examples that clearly, and for various reasons, call for semantic transformations.

It is sometimes argued that criteria for correct (meaning-preserving) translations between two languages are similar to criteria for the correctness of paraphrases within a given language.[3] It is equally obvious that what we have said about translations in the nonstandard sense is indifferent with respect to whether the expressions considered belong to two different languages or to a single language. In either case, it is essential to have an interplay between a (syntactic) translation and a semantic transformation in such a manner that a possible meaning or reference change becomes understood. Therefore, it will not make any difference if we in what follows mainly consider utterances belonging to a single (natural) language. Influential cultural differences may occur, moreover, between groups or persons representing a single natural language in the same way as such occur between representatives of different languages. To unify terminology, I shall therefore mostly use the word 'translation', even though I am going to consider paraphrases and other syntactic transformations within a single language.

We are now ready to discuss different kinds of explanatory translation to which an utterance of (3.1.1) may lead. If the hearer knows that the legs belong to Icarus, his most straightforward corrective and nonstandard translation of (3.1.1) would of course be

(3.1.2)   The legs belong to Icarus.

The manner in which the hearer's situation (that is, h), in which (3.1.2) is true, is associated with the speaker's one (s) is evidently sensitive to context, specifically to the way in which the hearer constructs his explanation. Let us suppose, for example, that the hearer knows that the falsity of the assertion is due to the speaker's ignorance of the story of Icarus, and, furthermore, that the hearer himself has never seen the painting, though he has been told that there is such a spot that represents the legs of Icarus. Then the hearer, in trying to explain the erroneous claim of the speaker, may choose a number of situations in which (3.1.2) is true, that is, he may choose a number of situations from his own situations that are as close as possible to, and hence easily transformable into, the speaker's situations in which the spots represent the legs of a pearl diver or an oysterman. They are, e.g., situations in which the legs in the painting look (to the hearer) like the legs of a pearl diver or an oysterman.

It can be directly seen that the minimization and refinement principles hold here. Since in the actual situations (that is, a) (3.1.2) is true as well,[4] they coincide with some of the hearer's situations, and the respective diagram is as in Figure 3.1.1, below, where σ is (3.1.1) and θ is (3.1.2). Therefore, an application of the minimization principle (i.e., the minimizing transformation) does not lead outside the actual situations, unlike in some other examples that follow. As I explained above, the plurality of situations (in particular, the hearer's situations), and hence the role of the refinement principle, is ignored here for typographical reasons:

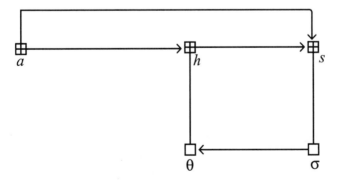

Figure 3.1.1

A translation is not always corrective in such a transparent sense, however, and its explanatory role may not be as obvious. Much depends on a speaker's attitudes and a hearer's aims, such as, for instance, the role the latter wants to give to his translation and semantic transformation. In order to see the point, let us continue considering Danto's thought experiment and assume next that the speaker does not see or understand the title – or for some other reason, as, for example, because of incorrectly thinking that the title is *Industry on Land and Sea* – and hence does not associate the picture with the story of Icarus. Assume furthermore, that he utters (3.1.1) since he thinks that the white spot represents legs that belong to a pearl diver or an oysterman, rather than to someone just swimming for fun. For the hearer, however, the title and representation are as they should be, and therefore he tries to understand the speaker's utterance by asking: "Under what conditions *would* the legs belong to a pearl diver or an oysterman?"[5] To answer the question, he may imagine, say, a counterfactual situation where the title is *Industry on Land and Sea*. By referring to this imaginary situation, he might then reason that:

(3.1.3)   If the title were *Industry on Land and Sea*, then the legs would belong to a pearl diver or an oysterman.

If the conditional is relevant to this context, its antecedent shows the hearer how the speaker's situation can be approached; by using it the hearer realizes the minimization principle, or even discovers the speaker's situation. In the former case, by means of the counterfactual assumption about the title, he departs from the actual situation and moves towards the speaker's situation to end up with what we have called the hearer's situation, which will be closer to the speaker's situation than the actual one is. In the latter case, he ends up with the speaker's situation itself.[6] In either case, conditional (3.1.3), if successful, yields a way to understand the speaker's position since it reveals how his situation can be approached. Therefore, it may be of some explanatory import.

Suppose now that the hearer's aim is to translate utterance (3.1.1) in the standard way, in order to save its general (or grammatical) meaning.[7] In other words, he does not intend to be corrective in the syntactic sense, as in our example above. If, in particular, the hearer's language is the same as the speaker's (as we assume here), this amounts to either paraphrasing (3.1.1) or saving its syntactic form. In the latter, more trivial case, the consequent of conditional (3.1.3) is its translation, which is then an identity. In the former case, the translation is not an identity (since it is a rewording), even though it preserves meaning as a paraphrase.

Now there are two alternatives concerning the semantic transformation. The first one occurs, e.g., if the speaker thinks, for some reason or other, that the title of the picture is *Industry on Land and Sea*, and the hearer knows that. Then the hearer's (contrary-to-fact) situation (h) is likely to be identical with the speaker's (s), so the correlation is also an identity. Therefore, if the hearer's language is different from that of the speaker, the translation is standard, as displayed by Figure 2.4.3; if not, and the translation (i.e., paraphrase) is vacuous (as is the correlation), we have a rather trivial case that can be taken as a 'limit' of more substantial instances. The hearer's semantic transformation is not trivial, however, but it is composed of the minimizing transformation alone, which now indicates how he has to (temporarily) change his original position, i.e., the actual situation (a), so as to reach the speaker's situation and thus to make the identity translation possible. Here, then, both σ and θ are (3.1.1):

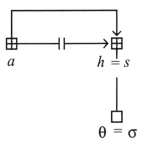

Figure 3.1.2

The second alternative occurs, e.g., if the following happens. The speaker does not see any title and yet claims (3.1.1) and the hearer responds by telling the speaker that if the title were, for instance, *Industry on Land and Sea*, then the legs would belong to a pearl diver or an oysterman, but not if there were no title at all. In the latter case the hearer then reasons that such an interpretation would not be appropriate. Then some situation (h) where the title is this is the hearer's (imaginary) situation (reached by means of the minimizing transformation, as indicated by the antecedent of the conditional), but not the speaker's (s), where no title exists. Since it does not agree with the speaker's situation, the respective correlation, transforming the situation where the title of the painting is *Industry on Land and Sea* into the situation in which it has no title, is not vacuous.

Here we have a case where a translation of an utterance is vacuous, that is, the utterance is not translated or paraphrased at all, but the hearer's (imaginary) contextual meaning of the utterance is different from that of the speaker's. Their respective situations, since they are different from each other, somehow manifest the difference, and the correlation indicates the respective meaning change. Both the minimizing transformation and correlation are effective here, together representing the semantic transfer with which the hearer is occupied in his attempt to understand the utterance. The hearer departs tentatively from a situation (a) which he thinks is correct and adopts a contrary-to-fact situation when realizing the minimization principle. In this imaginary situation the utterance is true, and the manner in which it is transformed into the speaker's situation, where the utterance is likewise true, would indicate how the contextual meaning gap, which still remains after the above application of the minimization principle, is bridged. This case is logically of the following kind, where σ and θ are (3.1.1):

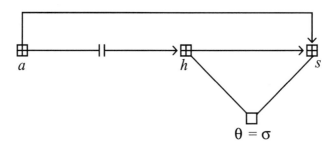

Figure 3.1.3

The counterfactual conditional has a similar effect if in the speaker's situation (s) the title is as indicated in the antecedent of (3.1.3) and the hearer knows that, and he wants to ground and explain the speaker's utterance by means of the conditional without giving up his corrective translation (3.1.2). Then, he sticks to his original actual situation (a) and uses the conditional to reach the speaker's situation, i.e., to obtain the correlation transformation. Now it seems that the minimization principle is not used. Rather, the Davidsonian principle is functioning here, if we adopt the Lewis (1973) type of semantics of counterfactuals,[8] which presupposes, intuitively, that the situations in which the consequent is true are as close (similar) as possible to the actual situation. In any case, the translation seems to be corrective and explanatory since some kind of 'counterfactual explanation' is involved. The hearer points out that actually the legs belong to Icarus, but if the title of the painting were *Industry on Land and Sea*, as assumed by the speaker, they would belong to a pearl diver or an oysterman. In the corresponding diagram, σ is now (3.1.1), as above, but θ is (3.1.2):

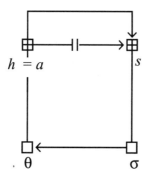

Figure 3.1.4

We must still modify these examples to obtain an application of counterfactual conditionals that would be more substantial in that none of the transformations is vacuous, and which would lead to a situation that is as close as possible (in a relevant sense) to the speaker's situation. Suppose, for instance, that the speaker wrongly suggests that we have here Breughel's work whose title is *Industry on Land and Sea*, but fails to coherently interpret the presence of the legs in the sense of not thinking of the one to whom the legs belong as being at work. Consequently, he argues:

(3.1.4)   The legs belong to someone swimming for fun.

The hearer, however, who is an admirer of Breughel's and therefore convinced that Breughel's artistic intentions could not be so incoherent, may correct the utterance by pointing out that if the title were as indicated, the meaning of the legs would be different from what the speaker proposes; they would belong to a pearl diver or an oyster-

man, as suggested by conditional (3.1.3), rather than to someone just
swimming for fun. Therefore we may assume that the speaker's and
hearer's situations differ from each other only in that in the latter the
legs belong to someone working (as a pearl diver or an oysterman),
whereas in the former they do not.

Now the conditional indicates how to convert the actual situation
(*a*, in which the title is *Landscape with the Fall of Icarus*) into an imaginary situation (that of the hearer) in which the title is as suggested
by the speaker but in which it is true that in the painting the legs belong to a pearl diver or an oysterman. This situation (*h*) is as close to
the speaker's situation (*s*) as the hearer can possibly go without giving up his conviction about Breughel's artistic coherence. Therefore,
the respective correlation, which transforms the hearer's situation into
the speaker's situation where the title is the same but (3.1.4) holds
true, shows a radical difference in the attitudes of the two parties reflecting on Breughel's artistic qualities. In this dialogue, (3.1.1) is the
corrective translation of (3.1.4). However, it is corrective only in a
relative sense, that is, it is not actually true. Whether it is explanatory
is not quite obvious, since this would depend on some additional
pragmatic considerations. The corresponding diagram below, Figure
3.1.5, exemplifies the general case, as does Figure 2.4.1, where all
the transformations are nonvacuous, but the minimizing transformation is indicated by a counterfactual conditional and *h* is very close to
s. Here σ is (3.1.4) and θ is (3.1.1):

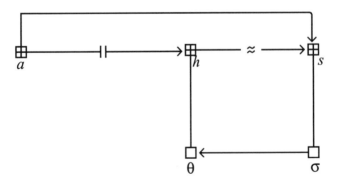

Figure 3.1.5

I believe that these examples present most of the principal types that
can occur in dialogical processes where local utterances are translated

in a nonstandard and explanatory way. Yet some other types of translation – which may not all be explanatory – will be presented later in this chapter. We can see now that differences between types of translation are often very fine-grained. The explanatory role of counterfactual conditionals, and the difficulties involved, will be further examined in the next section and later on in the context of scientific change, where it can be investigated in a somewhat more precise setting. What we have learned of the role of conditionals so far is that they show how semantic transformations function in certain cases; they indicate how hearers can approach speakers' positions.

As the final example of this section, let us consider a translation that apparently is not an instance of explanatory translation.[9] Let there be a cat on a mat by a speaker's light, and let the speaker utter:

(3.1.5)   There is a cat on the mat,

Assume that a hearer only considers the cat as being almost on the mat. Therefore, the obvious corrective translation of the speaker's utterance is something like:

(3.1.6)   There is a cat almost on the mat.

I have been assuming that situations are language dependent in that interpretations belong to them, but in the previous chapter I promised to illuminate the distinction between this concept of situation and the concept which supposes that interpretations do not belong to situations as components. This can be done in terms of the present dialogue. Consider the latter possibility first, and assume accordingly that the hearer refers to the speaker's situation, so that the hearer's and speaker's situations are identical. The situation consists of how the cat is physically located with respect to the mat. Then the semantic transformation is an identity mapping. The translation is now extension-preserving, but does it preserve meaning? There may not be any direct answer. If looked at from a point of view of the hearer's culture, the meanings of (3.1.5) and (3.1.6) are, of course, different, but a neutral observer may or may not argue that they have the same meaning relative to the culture of the observer. The problem here is that what meanings are is not fully understood.

On the other hand, it is quite clear that generally speaking the sentences are not equivalent. Moreover, if we accept what is often said

about meanings in possible worlds semantics, then we may see that
their meanings are different from the point of view of any culture or
any subject. Briefly stated, the meaning of an expression is defined,
e.g., by Hintikka (1969), as a function (possibly partial) from possible worlds to its extensions at those worlds. Thus, for example, the
meaning of a sentence, that is, the respective proposition, is a function from possible worlds to the truth-values of the sentence at the
worlds – or alternatively, the meaning is the collection of all worlds
at which the sentence is true. This characterization requires qualifications, for it is vague and simplified. In formal semantics, the necessary logical qualifications can be made, but they cannot be discussed
here in any detail. On the other hand, this is not necessary for our
present purposes. It is intuitively speaking evident that irrespective of
whether we consider possible worlds from the point of view of the
speaker's, hearer's, or observer's culture, the two sentences will not
be true exactly at the same worlds in so far as they are taken as sentences of ordinary English.

In any case, the translation is logically trivial, and it is not explanatory (though it is corrective), since there is nothing in the translation
that would help the hearer to understand the speaker's position. One
indication of this character is the fact that the hearer's interpretation
does not satisfy the minimization nor Davidsonian principle, since it
is simply supposed from the very beginning that all three situations a,
h, and s are identical, and hence no effort to approach the speaker's
position is needed. It follows that the first diagram of Figure 2.4.2,
where σ is (3.1.5) and θ is (3.1.6) – together with the idea that we
are here speaking about reference and not meaning – depicts the logical structure of the present case:

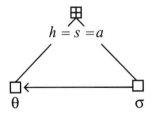

Figure 3.1.6

Now we also come across the problem whether we are really entitled to say here that the speaker and the hearer refer to the same situ-

ation. For example, if in the two cultures, represented by the speaker and the hearer, respectively, the proposition that something is on a mat is understood differently, it is possible that the two agents see different things when looking at the cat and mat. In such a case, their respective situations would be different. This would make our principal notion of situation more relevant, according to which interpretations of languages are parts of situations. In this case, Figure 3.1.6 does not exhibit the diagram corresponding to the translation. Instead, the following is now correct:

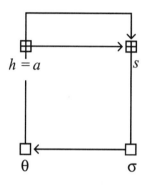

Figure 3.1.7

It now seems that the latter construal, exhibited by Figure 3.1.7, is less trivial than the former, in Figure 3.1.6, where no semantic transformation is involved. This is only apparent, however, since this example shows in the first place that our notion of nonstandard translation would not be felicitous at all if situations were considered as not containing interpretations as parts. Interpretations must be somewhere in any case, and therefore the former figure would give us a deficient picture of the dialogue in question, and it should be supplemented in such a way that it would somehow take account of the relevant interpretative transformation. Therefore, to avoid giving an excessively simplified picture of translations, interpretations should be handled separately if they were not supposed to belong to situations. This is one reason why in our technical notion of situation interpretations are supposed to be components of situations. Another is that on this construal the notion of situation is closer to that of model, which we shall need later on.

The examples of the present section serve as evidence of what was observed earlier about difficulties to provide an unambiguous treatment of the notion of a speaker's (or hearer's) situation and, more generally, of the semantic elements of nonstandard translation. As far as explanatory and corrective translations are concerned, there are indefinitely many pragmatic variations of speech act contexts (in the sense of everyday speech acts). There may not be any rules of how to classify them. In connection with scientific change and global translation, the number of different kinds of context is evidently more limited. Here I have tried to classify the different logical structures of translations, as represented by the respective diagrams.

3.2. METAPHORS

The present theory of nonstandard translation can be applied to the problem of metaphor.[10] I shall employ counterfactuals in the manner learned above to show that the semantic and pragmatic import of a metaphoric utterance can often be understood by means of explanatory translation which is not corrective. This account of metaphor explains two earlier theories, Tormey's (1983) construal of metaphors by means of counterfactuals and Levin's (1988) view that metaphors refer to metaphoric worlds at which they are literally true. Grund (1988) elaborates Tormey's idea by using a more exact theory of counterfactuals and possible worlds semantics. Therefore, Grund's account − which is independent of Levin's, though − can be seen as an advancement of the two theories. What I shall say below evidently complements some aspects of her account.

Tormey proposes to think of metaphors as elliptical counterfactuals. Accordingly, for example, when Romeo − the character is here considered a speaker − utters

(3.2.1)   Juliet is the sun,

this would yield something like

(3.2.2)   If Juliet were a celestial object, she would be the sun

(and not Mars or Jupiter, or any other celestial object). Thus metaphor (3.2.1) is recast as a counterfactual conditional whose antecedent is implausible or "not seriously possible," that is, cannot seriously be entertained. According to Tormey, (3.2.2) is not a paraphrase of

(3.2.1), since metaphors are not paraphrasable, but it indicates (something of) the meaning of (3.2.1). We may somewhat more precisely say – in view of the above theory – that (3.2.2) brings out the meaning of (3.2.1) by displaying a semantic transformation, that is, pointing out that the actual situation of the story, where Juliet is a human being, is to be transformed into a situation where she is a celestial object. In Goodman's (1968) terminology, the antecedent 'Juliet is a celestial object' is telling that the name 'Juliet' is detached from its home realm and applied in an alien realm, while Levin (1988) might say that the antecedent is telling how the name is transported across the conceptual space.

Levin proposes that metaphors should be taken literally in the sense that they literally describe deviant, metaphoric worlds.[11] In the present framework, this amounts to saying that when a speaker utters a metaphoric sentence, it is (literally) true in the speaker's situations. For example, (3.2.1) is (literally) true in Romeo's situations; in them, Juliet is the sun. Both Tormey's and Levin's proposals nicely fit our theory of explanatory translation, in terms of which they can be accounted for. It is obvious – and argued by Tormey and others – that metaphors, such as (3.2.1), are not translatable into literal expressions in the standard sense of translation. For example, Tormey says that the listing of analogical resemblances, or properties, only provides a grounding of a metaphor, not a translation. On the other hand, we may see that a metaphor may still be translatable in our *nonstandard* sense. Listing analogical properties that are shared by Juliet and the sun would clearly provide such a translation of (3.2.1):

(3.2.3)   Juliet is warm, central, brilliant, ...

It is likely that this sentence is true in a hearer's (reader's) situations. When the reader tries to understand what Romeo is after when uttering (3.2.1), it is likely that he accepts (3.2.3), though its meaning is not identical with the meaning of (3.2.1).

Conditional (3.2.2) shows how the reader's situations are correlated with, and transformed into Romeo's situations, which are now the metaphoric worlds in Levin's sense at which (3.2.1) is literally true. As we have seen earlier, if the conditional is held true, it also seems to indicate that the principles of interpretation, whose function is to minimize the gap between the speaker's and hearer's situations, are valid here. The existence of a gap does *not* mean this time that the hearer would consider the speaker's position incorrect – since the hearer accepts the metaphor – as in some earlier examples we have considered. The present translation is not corrective; it is only explanatory. If the reader accepts the metaphor, it is metaphorically true in his situations, but literally true in Romeo's. Moreover, something like (3.2.3) is (literally) true in Romeo's situations and in the reader's, and if the reader approves Tormey's construal, conditional (3.2.2) is also true in his situations. The diagram corresponding to the present case is therefore different from the earlier ones. Its truth-relations in the literal sense are denoted as earlier, but in the metaphoric sense by a broken line; σ is now (3.2.1) and θ is (3.2.3):

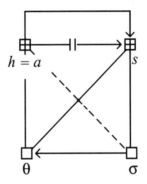

Figure 3.2.1

Even if the present theory of nonstandard translation illuminates the kind of meaning variance that is typical of metaphors, we have to ask whether it explains anything of why they can be emotionally and cognitively so powerful. Mac Cormac (1985) argues – as a tension theorist – that a semantic anomaly presented by a metaphor, which is caused by juxtaposing referents not normally associated, often results in emotional tension, and this tension may be cognitively effective. One aspect that may be cognitively relevant to this problem – at least if we consider the class of metaphors to which Tormey's construal is applicable – is the following. We have seen that in so far as a metaphor is acceptable or true, the respective nonstandard translation is not corrective. This is as it should. Nevertheless, as soon as a hearer accepts the metaphor, he may be struck with the observation that even though it does not make any sense to start looking for an expression that would correct it (as in the case of corrective translation), *yet* an appropriate counterfactual conditional indicating both a requisite semantic transformation and the functioning of the principles of interpretation would be in order for explanatory purposes (as is usually the case with corrective translations). Compare, for example, (3.1.3) with (3.2.2): the former points out under what conditions its consequent (3.1.1) would be true (in a hearer's situations) and why it in fact is not, whereas the consequent of the latter, (3.2.1), is unconditionally true (for a reader accepting the metaphor). In brief, the respective counterfactual conditionals (3.1.3) and (3.2.2) play different roles in these two cases of explanatory translation. The role of the latter could be expressed by saying that it yields another way to reveal the essence of metaphor (3.2.1), which – as we saw Mac Cormac claiming – amounts to an unusual juxtaposition of the familiar and the unfamiliar.[12] Furthermore, the antecedent of (3.2.2),[13] describing the outcome of the requisite semantic transformation, is highly implausible – unlike, perhaps, the antecedent of (3.1.3)[14] – as is the transformation itself. Yet it is a necessary condition for the truth of metaphor (3.2.1) itself, at least if Tormey's theory is acceptable. This association of implausibility and truth represents in its own way the tension for which the tension theorists are arguing.

3.3. PERCEPTION

In this section, I apply the notion of nonstandard translation to explore different views of how perception, especially seeing, and interpretation are connected with each other. The two views, and the distinction they lead to, that I shall mainly investigate are well known from the philosophical and psychological literature. Similar distinctions are much discussed in the histories of art and philosophy, but I shall only consider a certain recent discussion where the views are clearly analyzed. The idea that our eye is not 'innocent' is, I think, well established in current philosophy and psychology, both theoretically and empirically.[15] It means that one's perception, rather than being objective, is relative to context, one's background and culture, the theories one is holding, and similar things that may vary from one situation to another. This notion of perception (and observation) has been especially crucial in the latter part of the last century, in philosophers' attempts to disprove certain empiricist doctrines in epistemology and philosophy of science. But the views of what this relativity ultimately means seem to diverge from each other, and so do opinions concerning certain finer details within this doctrine of relativity. In what follows, I briefly quote one discussion, regarded important in current philosophy of science.

Consider first the so-called sense-data theory of perception, also called 'sensory core theory' by Suppe (1968), which implies that one can make objective or neutral observations, and therefore it complies with the empiricist doctrines of knowledge acquisition. It says, for example, that when two persons are watching something they *see* the same thing (that is, the same sense data), but their interpretations of what they see can be different. Thus when two persons, one living in Stockholm and the other in Helsinki, are watching a white ferry sailing on the sea in front of Helsinki, they see the same thing, the same sense data, that is, a white object moving on a green surface.[16] However, they may interpret what they see differently. The former may interpret the ferry as sailing to Stockholm (taking his friend back to her home town) and the latter interprets it as leaving Helsinki (taking his friend away). Opinions differ as to whether this interpretation is simultaneous with the perception of the sense data or comes after. The former view is more common: the perception and its interpretation are simultaneous but logically distinguishable from each other.

Such philosophers as Wittgenstein, Hanson, Kuhn, and their followers – who resist radical empiricism – and cognitive psychologists, hold an opposing view. Hanson (1958) argues that two persons holding radically different scientific theories may see different things when looking at the same object. Tycho Brahe and Johannes Kepler see different things when watching the dawn: Brahe sees the sun rising and Kepler the earth moving. More generally, Hanson argues that what one sees depends on context and ones's knowledge, experience, and theories. Brahe and Kepler are visually aware of the same object, but yet their experiences are differently organized conceptually: seeing is epistemic. Likewise, we should say that when watching the ferry on the sea the two persons may *directly* see different things: one sees the ferry going to Stockholm and the other leaving Helsinki. This is a matter of seeing and not of interpreting.

Another example that Wittgenstein, Hanson, and others often employ here is the well-known duck-rabbit picture. What one sees may change from a duck to a rabbit or conversely. According to the sense-data theory, two persons have the same visual image with a certain shape, but one interprets it as a duck and the other as a rabbit, but according to the opposing theory, they directly see one of the two different things, and this is not a matter of interpretation. Interpreting is a kind of thinking whereas seeing is an experiential state, and it is spontaneous. Both persons have the same visual image, but, again, they organize it differently and appreciate the elements of the same visual field differently.

An important and well-known aspect of Hanson's theory is that seeing is to see that something is the case, i.e., it amounts to *seeing that*. This is important, for example, because it would explain why we can infer new states of affairs from our observations (e.g., from what we see), and it would imply that there is a linguistic component in seeing. Since different persons may see that different states of affairs are the case, when they watch something, it is likely that they infer different things from their observations. Hence we may say that seeing is a cognitive process in a broad sense, but it is also important to emphasize here that it is also relative to emotions (which are part of our cognition), culture, and so on. Kuhn, for instance, stresses the latter feature when discussing his notion of paradigm.

If what one sees (or observes) involves an element of seeing that something is the case, as Hanson argues, it means that when one expresses linguistically what one sees is based on this propositional aspect of seeing. The expression is a linguistic formulation of the proposition in question, for instance, in the sense that one interprets the linguistic elements of the utterance according to the propositional aspects of what he sees. On the other hand, if the sense-data theory is correct, then to say that one expresses linguistically what one sees would be a controversial matter, as is well known from the discussion concerning the positivist doctrines of the so-called protocol language. On one proposal, protocol language is a sense-datum language that correctly characterizes one's sense experience. If this is the case, then two persons who see the same sense data and for whom the protocol language is available are able to express what they see in that language in the same way.

It may be difficult to know whether either one of these two alternatives of what seeing ultimately is can be conclusively defended or whether their distinction is conceptually clear. It is not my concern here, however, to ask whether one of them is correct, since my aim is to characterize the distinction in terms of nonstandard translation. The following kind of argument, often presented in favor of the Hansonian view, is important for the distinction. It is a linguistic argument purporting to show that our usage of the verb 'to see' implies that the language in which one talks about seeing is intensional rather than extensional. But it would be extensional if the sense-data theory were correct. This argument is of the same type as the more general arguments presented in the literature whose aim is to point out that the language we need when speaking about propositional attitudes, which are intentional states or acts, is not extensional.[17]

Let us now go to the characterization in terms of nonstandard translation and speech acts. Consider first the above Brahe-Kepler example in the light of the Hansonian theory, and assume that Brahe is a speaker and Kepler a hearer and that they express what they see when watching the dawn, and that their respective situations are what they see. Suppose, then, that Brahe utters, in accordance with Hanson's theory of seeing:

(3.3.1)   The sun is rising.

Kepler's utterance, to be considered here a corrective translation of Brahe's utterance, would then be something like:

(3.3.2)   The earth is moving.

Brahe's situation, s, is a situation in which the sun is rising, and Kepler's, h, is one in which the earth is moving, and the latter is also the actual situation a. Hence the corresponding diagram would be as in Figure 3.1.7, where, however, σ is now (3.3.1) and θ is (3.3.2). Whether this translation has any explanatory power seems to depend completely on the nature of the semantic transformation − which now amounts to the correlation, there being no attempt to use any minimizing transformation − since the syntactic translation, its being local and simple as such, does not involve any explaining elements. Obviously Kepler's transformation of h into s is of an explanatory nature since he knew the reason for Brahe's erroneous vision and the reason was part of the old cosmological paradigm. It would be an explanatory enterprise to try to transform a situation where the earth is moving (so that the sun can be seen) into a situation where the earth is not moving (but the sun comes into sight).

Now consider the sense-data theory. Then only on the assumption that the protocol language exists, we are entitled to say that Brahe and Kepler are able to describe what they see. Then – since according to the sense-data theory they see the same thing, namely the same sense data – their situations are identical, composed of those sense data plus the interpretation of the protocol language. Therefore, if Brahe does not make any robust mistake in his description, this amounts to saying that Kepler has nothing to correct, that is, their utterances would be identical and no translation would occur. This case of translation is as trivial as a translation can ever be, assuming that Brahe and Kepler use the same protocol language:

$$h = a = s$$

$$\theta = \sigma$$

Figure 3.3.1

Here σ and θ represent the common protocol language description in question. It seems obvious, in view of Figure 3.3.1, that the sense-data theory of perception is in some logical sense more trivial than the theory put forward by Hanson and the others mentioned above, and therefore the latter is at least logically more interesting, irrespective of whether it is correct or incorrect. If there is no protocol language – as convincingly argued by many, such as Hanson, ever since logical empiricist doctrines were abandoned – this case is not possible at all, since we cannot then say that Brahe and Kepler described what they saw – in so far as seeing means seeing sense data, as we assumed in this case. Hence it seems that there are two reasons to think that the above trivial case of translation may not occur in the context of perception at all. Due to their extensionality, both the sense-data theory and the assumption that there is a protocol language appear doubtful.

3.4. ON CONCEPTUAL CHANGE IN ART

As we saw in Chapter 1, in current philosophy of science questions of scientific progress have been extensively debated ever since Kuhn and other critics of the Received View presented their positions, in the early sixties. It may sound surprising at first sight that in the philosophy of art there are similar discussions about the nature of historical developments in art.[18] Part of this interest has been motivated by new and newly invented ideas of the character of perception, such as those discussed in the preceding section, so that here we can see the philosophies of science and art as having a common starting point for their criticism. I try to point out here that the notion of explanatory translation, especially schemata of the kind developed above, can be used (under appropriate pragmatic conditions) to distinguish between certain views of artistic developments. However, I shall restrict my comments to visual arts and their relation to perception, namely seeing, particularly considering Gombrich (1960) and what was said in the preceding section.

The problem to which Gombrich seeks an answer is why the visible world has been represented so differently in different ages and in different cultures − or, more succinctly, why art has a history. It has been suggested, for instance, that progress means a better imitation of nature in the sense of improved modes of seeing, and it has also been suggested that progress means a better imitation in the sense of improved skill. These are only partial answers, however, Gombrich argues, since artists must always take account of conventions: ". . . but every generation discovered that there were . . . strongholds of conventions which made artists apply forms they had learned rather than paint what they really saw."[19] Since conventions give rise to a style, which might be considered as meaning the same for art as a paradigm means for science, we should perhaps study translation in the global sense here, that is, as being something like a transformation between styles, or, rather, between some of their respective conventional elements. However, it seems that if there existed such a global translation − which is doubtful for many reasons − it might not be applicable to individual works of art, because of the contextual and intensional character of vision and skill.

Therefore, I shall now mainly indicate in general terms, relying on what was said of perception in the preceding section, how one could distinguish between different ideas concerning artistic progress in so far as progress means progress in imitating what one sees. Let us assume, then, that making a work of visual art can be thought of as a speech act in the generalized sense discussed earlier in such a way that we can distinguish its syntactic and semantic parts. It is not necessary now to get involved in the much-discussed problem of the ontological and conceptual nature of artworks, but I shall assume that an artist intends to represent, by means of the syntactic part of a picture or statue, what he sees. This assumption means in our framework of speech acts that (i) the artist, who is now the speaker, transforms what he sees into a work of art, (ii) the work will be the utterance of this speech act, and (iii) what he sees plays the role of what I have earlier called the speaker's situation, s – in which the artist considers the work (or, rather, its syntax) as being satisfied. Speaking about a possible situation and about satisfaction or truth in a situation is perfectly intelligible here, if the Hansonian view of perception is adopted, since for Hanson seeing means seeing that something is the case, that is, seeing is propositional.

Assume next that another artist, considered as being a hearer (in the context of our generalized speech act theory), who is possibly working within another set of conventions, responds to the first work by making a new one intended to represent the same, or possibly similar, subject, as he sees the subject. What is represented is the hearer's situation, h. Let us suppose – this is obviously a somewhat restricting assumption about art – that art aims to be progressive (as science does) in the sense mentioned above, or that the second artist thinks so. Then it is natural to say that the latter artist's, i.e., the hearer's, work – as an utterance of his speech act – is a corrective translation of the speaker's utterance in the framework of the conventions which the hearer is using.

It may be of some interest to see that here too we can make important distinctions already by means of the above schematic representation of translation. We can distinguish between the different opinions, mentioned above, concerning artistic progress. Consider first the case that both agents paint what they see, but the hearer's vision of the subject differs from that of the speaker. The logical structure of this case is presented by Figure 3.1.1 or Figure 3.1.4, depending on whether or not h is considered as being equal with the actual situation, a. This, in turn, depends on what the hearer regards as being the actual situation.[20] If, for instance, he thinks that he is able to see how the world 'really is', i.e., is able to capture the 'visual truth' − which he subsequently represents − then the latter diagram is the case for him.[21] If not, but he thinks that his representation is more advanced, then h is by his lights closer to a, that is, 'closer to the truth', than s is. Whether the translation is of any explanatory import depends on pragmatic and hermeneutic conditions and principles that are similar to those indicated earlier, but any explanation should include an account of how the modes of perception have changed. In our terminology, this change corresponds to the notion of semantic transformation. If the subject is identical with that of the first artist, then this case can only occur if seeing means seeing in the Hansonian sense.

The second possibility, that progress means the improvement of skill only, would be more trivial in terms of our schemata. If the two artists see the subject similarly, but the latter artist is more skillful in representing what he sees, the logical structure is as in Figure 3.1.6, which again implies that no semantic transformation is involved, and hence the explanatory force (in the sense discussed earlier) is weak.

On the other hand, Gombrich does not accept the view that historical developments of art would just consist of changes in representing what the artists see. They also consist of changes in conventions. If this amounts to what progress is and if this idea is combined with the idea of changing modes of perception, we are back to the first case (Figure 3.1.1 or 3.1.4), where, however, a possible explanation must also include a description of how conventions and styles change from one work to another.

4

GLOBAL TRANSLATION

In Chapter 3 nonstandard translation was studied in the local sense. When scientific theories and literary works are translated, possible situations and languages must be explored more globally. In the present chapter, I generalize the earlier ideas about local translation by studying a global version (concerning literary works) of what was in Chapter 2 called explanatory translation. The notion will cover, as earlier, both translations between two languages and paraphrases. The generalization will be rather immediate, and it will gradually lead us to the definition of correspondence relation, which is considered in Chapter 5, below. Recall, however, that what is here taken as the notion of translation is not strict enough to satisfy Kuhn (1983).

When we in previous chapters investigated satisfaction in various situations, we only needed the notion of satisfaction that is *internal* for speakers and hearers. Such a notion will also play an important role in global translations, as, for instance, in cases of fictional narratives, and — perhaps somewhat unexpectedly — sometimes with respect to theories. When a scientific theory is reconstructed in model theory, the model-theoretic notion of truth is employed, which in fact means internal truth. On the other hand, the sequential character of narratives, whether fictional or not, presents specific problems for the notions of satisfaction and truth. These problems will be explored here in some detail after a more general discussion about textual interpretation and translation.

4.1. TRANSLATION AND LITERARY WORKS

We proposed in Section 2.1 to understand the notion of speech act in a generalized sense, so that speakers and hearers can even be, besides individuals, (scientific or cultural) communities, and utterances can even be written texts. This generalization will be assumed in what follows, but I try to justify it only in the next section. Furthermore, it has been suggested by several students of literary theory that a natural way of analyzing a literary work is to think of it as referring to possible worlds, some or all of which can be fictional.[1] How the appropriate possible worlds are chosen by a speaker, on one hand, and a hearer, on the other, is in accordance with the general principles of interpretation, similar to those briefly considered in Chapter 2, above. There are additional difficulties in the present case; they will be discussed later in this chapter.

The possible worlds in question are those at which the work or its translation is satisfied, by the speaker's or hearer's lights, respectively. Hence we can here consider the *speaker's worlds* and the *hearer's worlds* as in the local case their situations; and *actual worlds* will be analogous to what were called actual situations in the local case. At this juncture, I speak of possible worlds rather than of situations, but otherwise the notions in question, and the problems and reservations thereupon, are similar to those discussed in the local case.[2] However, since situations are here considered as 'small' worlds, the concept of world covers both in what follows. Since literary texts are often complex and their contexts vague, the three collections of possible worlds are generally large (larger than in local cases), that is, a text cannot be thought of as describing a single world or a tiny collection of them even though its meaning were assumed to be unique. This will not jeopardize the applicability of our earlier theory, but below we take into account the fact that we are here investigating expressions, languages, and worlds in a more global sense.[3]

If the collections of a speaker's and hearer's worlds are disjoint, then, for reasons discussed in the local case, there has to be a *correlation* between them that harmonizes with a *translation*. The correlation is a mapping from a subcollection of the hearer's worlds into worlds of the speaker, and the translation maps into the hearer's language at least the expressions of which it is meaningful to say that they and their translations are or are not satisfied at the speaker's and hearer's worlds, respectively. Let these mappings be denoted by 'F' (correlation) and 'I' (translation). Then, on the ground of what we have discovered so far, we have to require at least that for all expressions φ in the domain of I, and all worlds w in the domain of F,

(4.1.1) φ is satisfied at $F(w)$ if and only if $I(\varphi)$ is satisfied at w.

What we are up to here when talking of meaningfulness should be intuitively obvious though the notion cannot be made precise in the present informal context. The expressions of the text are meaningful in the intended sense, since, by definition, they and their translations are satisfied at the speaker's and hearer's worlds, respectively. One could argue against the supposition adopted here that *all* expressions of the text should have a translation, even in our nonstandard sense. The supposition could of course be dispensed with, but this would cause more complications in the definition. Likewise, the domain of the translation is deliberately characterized in vague terms; it may depend on the case, but it is obvious that it is not always sufficient to consider just the expressions contained in the text. Furthermore, what it means for an expression to be satisfied (or, rather, for a sentence to be true) relative to a text and a world is unclear, and there is no common agreement upon the matter. In the next section I shall briefly discuss this question.[4]

The two mappings have to satisfy logical and pragmatic conditions of adequacy, the nature of which is dependent on the text that is to be translated and on linguistic, aesthetic, and contextual specialties. The mappings must be transformations in some formal or, as here, intuitive sense of the word, and they display how the respective changes are obtained. This means that they must establish a sufficiently informative conceptual relationship between the two texts. The most obvious condition (4.1.1) is logical, and it is explicitly included in the definition. The notion of semantic transformation will also be analogous to its namesake in the local case; as before, it is composed of a minimizing transformation — which transforms (some of the) actual worlds into a hearer's worlds, more precisely, into the worlds belonging to the domain of F — and of a correlation. The minimizing transformation is not explicitly mentioned in (4.1.1) for reasons that will become more evident in later chapters where global translation is studied in an exact framework. As we shall see, its role is more prominent on the pragmatic side than on the logical side of explanation.

This is only a first step, so to speak, and needs qualifications. Since a text is a sequence (sometimes even a sequence of sequences) of expressions, these are not satisfied at possible worlds individually but in relation to other expressions of the text. More generally, how a reader receives and interprets an expression is constrained by other expressions of the text. The meaning of the whole text is not a function of the meanings of its constituent parts, considered individually. At least in this sense the text is intensional; the principle of compositionality does not hold. This feature also implies, as a special case, that how the beginning of the text is to be interpreted is (or may be) dependent on its end and *vice versa*. Narratives describe histories rather than mutually independent situations. A semantic analysis of this feature and a discussion of logical difficulties caused by the sequential character of narratives, and an analysis of hermeneutic features involved, can be found in Section 4.4, below.[5]

Furthermore, when a speaker or a hearer is fabricating possible worlds to which he assumes the text referring, he – in the case of narratives, at least – is doing that by considering appropriate courses of events, that is, sequences of situations (sequences of 'small' possible worlds) which are dependent on each other. Hence it seems logical to enlarge our previous terminology and talk about the *speaker's* and *hearer's*, and *actual courses of events*, and assume that the text is satisfied at the speaker's course of events and its translation at the hearer's course of events.

Now the more comprehensive worlds in question – i.e., worlds that we mean when we talk about the world or worlds created by a story – are constructed not only by employing courses of events but also by means of connotations the text suggests and by means of other conditions pertaining to the background involved. I shall not here discuss the ontological difficulties philosophers have often found in the notion of possible world – though their queries are often misleading. Small worlds, situations, of which the speaker's and hearer's worlds are constructed, can be thought of as imaginary events that follow one another when the consecutive expressions of the text are received. This is a heuristic device that is relatively free from ontological problems (at this level of analysis), and it is in accordance with our disposition to visualize the events the story is describing. The speaker's and hearer's worlds are something more extensive, which can be visualized by means of the small worlds.[6]

The two dimensions or kinds of possible worlds mentioned above – small worlds, i.e., situations, and more comprehensive worlds – have been given different names in the literature, of which *denotative* and *connotative* dimensions, in the sense in which Barthes (1967) speaks of denotation and connotation, are the most suggestive. What an interpreted narrative text refers to in the first place, what it denotes, is courses of events.[7] What it connotes is more vague, in its construction the denotation plays a crucial role.

Evidently the principles of interpretation, as defined in Chapter 2, and the notion of semantic transformation apply in cases of global interpretation and translation in the same fashion as we saw them as functioning in the local case and they can be used for explanatory purposes in the same way. The principles can be associated with possible worlds in a similar way as earlier, so as to function in the connotative dimension. Their intuitive meanings in the global case are similar to what they mean locally, but what they exactly mean for narratives is less clear. As suggested by Ricoeur, something like what I have called the Davidsonian principle of interpretation is emphasized when fictional narratives are interpreted.[8] We shall see in Chapter 5 that for the study of intertheoretic relations the minimization principle becomes emphasized as well, particularly in cases of scientific change where theories are superseded by new ones. The meaning of the refinement principle is reflected in the above definition of correlation as a mapping from a hearer's worlds to a speaker's. Since new explanatory aspects come up with the correspondence relation, in particular with its formal version, I shall defer, until Chapter 6, the study of global versions of explanation connected with translation.

In the denotative dimension, on the other hand, a hearer usually attempts to keep the consecutive situations belonging to the course of events (suggested by the text) as close to each other as possible. This is done for the sake of coherence – in so far as consecutive situations described by the narrative are not separate but dependent on each other – and not so much for the sake of closing the gap between the two agents. So we can there talk about principles that are analogous to, but yet different from, our earlier principles of interpretation. This feature of receiving and interpreting narratives is discussed below, in Section 4.4.

Since the denotative dimension (that of sequences of situations) is logically prior to the connotative one (that of more comprehensive possible worlds), one may argue that instead of, or perhaps in addition to, considering single expressions and possible worlds as is done in (4.1.1), an analogous condition should be stated for whole texts and courses of events. So, if σ is any text and $I(\sigma)$ its translation such that it is meaningful to say of them that they are or are not satisfied at a speaker's and hearer's courses of events, respectively, and if F is a mapping from the hearer's (i.e., translator's) courses of events to the speaker's (i.e., author's), then for every c in the domain of F:

(4.1.2) σ is satisfied at $F(c)$ if and only if $I(\sigma)$ is satisfied at c.

It may not be too difficult to roughly understand what it means for a sequence of expressions to be satisfied at a course of events. It pertains to our everyday experience of reading. It is more difficult to see how the respective conditions of satisfaction could be defined and, in particular, how figurative expressions, metaphors and the like, and misuses of language should be taken into account in them. Similar problems are much discussed in the literature, and I consider them in some detail in Section 4.4; we shall see that conditions of satisfaction have to be holistic since compositionality does not hold.

4.2. STORIES AS UTTERANCES

Juhl (1980) argues for the conclusion that the meaning of a literary work is logically connected with its author's intentions; and, in particular, if the work conveys certain propositions, then the author has asserted these propositions. Thus the meaning of the work is very much like the meaning of a person's utterance in a speech act. On this view, writing a literary work can be considered as a speech act with the author as the speaker, and, consequently, the work has only one correct or acceptable interpretation, though it may have several possible interpretations. According to Juhl, furthermore, textual features constitute evidence of what the author intended to convey with the work. A somewhat earlier, and better known, proposal emphasizing authors' intentions is Hirsch's (1967) claim that only they can provide a discriminating norm for the meaning of the work.

But if the intended meaning cannot be seen from a text itself, there can be what Wimsatt and Beardsley (1978) call external evidence to the same effect.[9] In any case, if a translator, to be considered a hearer in the context of his translating, believes that he has found out (by whatever means) what the author intended to convey by means of the text, that is, what the intended interpretation is, then he may try to translate the text so as to preserve the assumed meaning, to the extent to which this is possible. But if cultural and linguistic differences may make a meaning-preserving or extension-preserving translation impracticable or aesthetically inferior, the principles of interpretation and the notion of semantic transformation, in something like the sense discussed in Chapter 2, become important.

It has been argued, on the other hand, that the intended interpretation cannot be known or it is not relevant. The author's intentions are private and not clear, or we may not even have indirect access to them, or the text may not bring out the author's intentions, wherefore its meaning may be different from what the author intended. According to Wimsatt and Beardsley (1978), the correct interpretation ought to be searched for by appealing mainly to internal evidence, that is, to linguistic, aesthetic, and textual features of the work, and to what is known about general features of languages and cultures. On this view – if we insist on the idea of speech act in this connection – what literary theorists call the implied author of a work, or, alternatively, the narrator or the speaker, can be considered a speaker in our sense. Another suggestion, made by Lewis (1978), is that the body of the beliefs that were manifest in the community in which a text was written determines its meaning. If we think so, then we regard the community in question, rather than the author, as the speaker. No matter which of the above suggestions is adopted, it becomes then possible to say that there is here a speaker and a hearer, and consequently the principles of interpretation become applicable. If there are several possible (or correct in some relevant sense) interpretations, the principles become applicable as soon as the translator has chosen a reading on which to work.[10]

4.3. TRUTH IN A STORY

We have tried to justify the idea of using the notions of speaker and hearer in connection with literary works, and we have given a survey of different views of how the meanings of works are established. Let us next take a sharper look at the question concerning how a reader can or should pick out worlds that we have been calling a hearer's worlds, which play a role, e.g., in condition (4.1.1) above. As suggested by many[11] – and noted above – when interpreting a story, one cannot restrict oneself to what is explicitly said in the text. One has to consider the text against an appropriate background that provides a framework within which to interpret the text. This means that it is not sufficient to consider just those possible worlds that the story defines, that is, the collection of all possible worlds at which each sentence occurring in the story is true. The background, and especially the beliefs belonging to it, do their work by supplementing the story and reducing the possibilities the story would otherwise admit of. This idea means, on the other hand, that additional sentences are admitted to be 'true', not just those sentences which explicitly occur in the text or are implied by it.[12]

The problem of appropriate background makes global speech act occurrences more involved than local ones, where backgrounds are usually more immediate since they are determined by more immediate contexts, and where speakers' intentions are more readily available. Therefore, we have to make a short digression to what it means to say that a sentence is true – or, more generally, an expression satisfied – according to a story.

If a speaker utters a single sentence – and the speech act is in this sense local – that is related to a story, a hearer is apt to explain and possibly correct it according to the background and possible worlds he considers as being relevant to the story. We have seen, however, that there are so many diverse proposals concerning a relevant background that we can hardly assume that there would exist a unique set of constraints for the story. There exist some detailed suggestions in this matter. Lewis (1978) and Wolterstorff (1980), for example, argue that it should consist of appropriate beliefs of the author and the intended audience. Lewis is quite explicit and detailed in his arguments, in one of which he says that the body of beliefs to be used as a background consists of the beliefs that are overt in the community of the origin of the story.[13] Consider, for example, the question whether it is true about *Hamlet* that Hamlet has an Oedipus complex. If the background contains Freud's theory, this truth can be argued for. But according to Lewis' and Wolterstorff's suggestion, Freud cannot provide a proper background for *Hamlet*. Rather, Hamlet's melancholy should be explained by means of the humours theory, since the theory belongs to the set of beliefs of the author and his audience at that time.[14]

It was proposed in Rantala and Wiesenthal (1989), on the other hand, that Lewis' analysis should be regarded as a special case of a more general pattern. It is conceivable – and counterexamples can be provided to Lewis' proposal[15] – that other sets of beliefs and assumptions should be sometimes used as a background. For one thing, it is often the case that understanding a story requires that the reader intentionally deviates from a background of the kind suggested by Lewis. Secondly, an author's ideas may not agree with the beliefs of the intended audience, for some reason or other. Therefore, it seems, despite many arguments to the contrary, that there may in general exist several potential backgrounds against which a given sentence related to a story can or should be justifiably studied, but all of them must, naturally, be compatible with the story itself. In any case, even though it is more difficult to identify speakers' intentions in global cases than in local ones,[16] they play also in many global cases an important role, in principle at least, in so far as we want translations to be of some explanatory import in the sense discussed in previous chapters.

4.4. NARRATIVE TEXTS AND SATISFACTION

It is a conspicuous feature of literary texts that a reader must, as it were, move back and forth in three dimensions: between a text and possible situations, which the reader is going to associate with the consecutive expressions of the text, between these expressions, and, finally, between the possible situations in question. This feature is a consequence of the fact that any interpretation of an expression occurring in the text is constrained by the linguistic meanings of other expressions in the text and their actual interpretations. Furthermore, the meaning of the whole text is not a function of the meanings of its constituent parts, considered in isolation, so that the principle of compositionality does not hold in any immediate sense here. These facts are obvious and familiar to us, but it is less obvious how they should be taken into account in semantics. One consequence is clear, however, as remarked above. If possible worlds semantics is here used, one has to consider courses of events, that is, sequences of events or situations rather than mere sets of them. In this section, I briefly investigate the question of how to do this, but I am not going to develop any full-blown semantics.[17] We try to see how the reader's courses of events, like *c* occurring in condition (4.1.2), above, are chosen.

In order to gain some concrete insights into the contextual features of interpreting narrative texts, let us consider the following short passage from 'Death and the Compass' by Jorge Borges:[18]

> The train stopped at a silent loading station. Lönnrot got off. It was one of those deserted afternoons that seem like dawns. The air of the turbid, puddled plain was damp and cold. Lönnrot began walking along the countryside.

Since the passage is a short part of a longer text, it should not be detached from the whole story. Nevertheless, we can study certain aspects of the passage, in particular, mutual relations between its sentences and the question of how consecutive events should be chosen to interpret the passage correctly.

Let the five sentences of the passage be denoted by the letters α, β, γ, δ, and ε, respectively, and hence the entire passage by the concatenation $\alpha\beta\gamma\delta\varepsilon$. The manner in which the sentences constrain each other is rather evident. Thus, for instance, how β is to be interpreted is constrained by the fact that it comes after α: Lönnrot got off the specific train that stopped at a silent loading station. Likewise, γ and δ pose some restrictions on the interpretation of ε, that is, conditions under which Lönnrot started walking along the countryside. α influences the interpretation of δ, since it implies together with δ that the air around the loading station was damp and cold.

Constraints may also be directed from the right to the left, that is, a sentence may constrain sentences that are preceding it. Thus β implies that Lönnrot was in the train that stopped at the silent loading station. Similarly, γ and δ state the same restrictions on the interpretations of α and β as on the interpretation of ε, and so forth.

I shall call the constraints of the kind discussed above – that is, constraints that are due to the fact that when interpreting sentences of a text, one has to consider its other sentences – *textual constraints*. The presence of other sentences in the neighborhood of a given one makes the meaning of the latter more specific by cutting off possibilities which it might otherwise admit of. I also extend this notion of textual constraint to concern sentences (and utterances) not in the text but otherwise related to it. We have to improve, therefore, what we have said earlier about the background when one is interpreting sentences related to a story, by taking textual constraints into account. The improved account does not only concern sentences explicitly occurring in the text, but all sentences related to it.

Let us now use the above text to illustrate the role that courses of events play in general, by thinking of them as imaginary consecutive situations that a reader is 'picking' or 'choosing' one after another when interpreting consecutive sentences of a narrative. We may even think of them as mental images or mental scenes of the reader. The reader's actual images may not be very definite, but we can use the notion of mental image as a heuristic device that is relatively free from ontological problems. This usage accords with our disposition to visualize events stories describe. When using the metaphor of picking or choosing events, I am ignoring the obvious fact that one usually reads a text too fast to be able to consider individual sentences, thus to visualize the events they describe, *one by one*. I am here thinking of something like an *ideal reader* who is able to do so. Naturally, an ordinary reader can also do so by reading the text very slowly.

Let us now go back to our example text. When a reader (ideal reader) picks a situation v which he thinks of as corresponding to the sentence β occurring in the text $\alpha\beta\gamma\delta\varepsilon$, i.e., in which β is true, then – according to what we have said about textual constraints concerning β – v must be such that in it Lönnrot got off the very same train that belongs to the situation, say u, which the reader chose in order to interpret the sentence α. On the other hand, in v there must be a deserted afternoon that seems like a dawn, and the air must be damp and cold. If the reader picks the situation v for β so that this is not the case, then it does not provide a correct interpretation for the occurrence of β in the text $\alpha\beta\gamma\delta\varepsilon$. Therefore, after having read γ (and δ) the reader has to go back, as it were, and choose a more appropriate situation for β, that is, a situation that satisfies the conditions set forth by γ (and δ).

It is obvious that such back-and-forth adjustments concerning interpretations of individual sentences of a text can be performed only locally, and, on the other hand, not all adjustments are important. Incoherent, incomplete, and even wrong moves from one image to another are often made, and this results in wrong interpretations. An ideal reader, however, would always proceed in the correct manner, and not only locally – and an ordinary reader may gradually approach it by reading the text repeatedly.[19]

Textual constraints derive from what is explicitly said in the text. Common knowledge and beliefs and social rules about language – or whatever is regarded as forming an appropriate background – are influential here as well. A reader of αβγδε is not able to conclude that Lönnrot got off the train that stopped at a silent loading station, if he does not understand how the language is generally used in narratives or does not know the function of trains. Textual constraints, in the sense I understand them, have to do with readers' abilities at some common and conventional social, linguistic, and possibly historical levels – and therefore with the idealizing assumption that texts possess conventional meanings at a general linguistic level and with relevant backgrounds. Such meanings may not be unique, however, and some texts may not even possess independent meanings at all. Consider, e.g., ambiguous sentences, metaphors, complicated theoretical expressions, and works of art. Such expressions may not even mean the same thing for two readers belonging to the same linguistic community and being equally competent language users. Moreover, two readers may have different cognitive and aesthetic views, abilities, and skills, different experiences in literary criticism, or different intentions concerning how a given text should be read. A reader may even approach a given text differently on distinct 'literary occasions'.

Therefore, it seems appropriate to say that there are also personal constraints on which interpretations depend. They are constraints that pertain to more personal levels of a reader's background and are less dependent on the text under consideration or common rules. They may, naturally, provide different frameworks for different texts – that is, the reader's background may be differently 'oriented' for different purposes, intentionally or nonintentionally – but they are not primarily due to the syntactic structure of any given text. Thus I shall call them *pragmatic constraints*.

There are indefinitely many personal factors – some of which are even contingent – and principles that may influence local features of one's interpretation. Such local features are not always cognitively or aesthetically significant, such as possible details concerning the station at which Lönnrot got off the train. Any reader probably has mental images of some such details, but they are not helping him in the process of interpretation – and they are not mentioned by the author. On the other hand, theoretical pragmatic constraints, that is, various theoretical principles and attitudes of a reader, are often global governing the whole interpretation or a great part thereof. If, for instance, the reader assumes that environmental conditions have causal effects on how people feel or if he assumes that the author whose text he is reading thinks so, then he cannot assume, to be coherent in his interpretation, that Lönnrot felt very happy when he got off the train and started walking along the countryside. In terms of possible situations or worlds, this would mean that since Lönnrot did not feel very happy in the situation x where the air of the turbid, puddled plain was damp and cold – and which was chosen by the reader to interpret the sentence δ – he has to choose such situations for the sentences β and ϵ that Lönnrot was not very happy in them either. The same conclusion follows, of course, if the reader does not assume any general causal connection between the environment and mood, but only assumes that in 'Death and the Compass' there is a similar connection between the environment and Lönnrot's mood.

From these examples we can see, furthermore, that pragmatic constraints can be influential backwards as well as forwards, in the same manner as textual constraints.

When a reader is picking new situations while reading, he has to do it so that it is in conformity with the pragmatic constraints. In many cases it may not even possible to choose otherwise. For instance, he cannot read the text so that it is beyond his abilities or knowledge, although the general linguistic meaning of the text left room for such readings. Pragmatic and textual constraints are, however, interwoven in actual interpretations, but in a pragmatic sense the former precede the latter, though logically and linguistically the latter are primary. We might say, roughly speaking, that the reader has to proceed so that relevant pragmatic constraints will be satisfied, but he attempts to do it so that it corresponds to the text. Pragmatic constraints indicate how he is permitted to go on when picking new situations for consecutive sentences of the text. As we saw above, what he is permitted to do at each step may depend on his earlier choices, and sometimes on his choices to come.

In a sense, then, pragmatic constraints indicate the situations that are accessible to the reader at each stage of an interpretation, if the

reader assumes these constraints. It follows that they also have a pre-
dictive function in the following sense. Let us again consider the text
αβγδε. As soon as the reader has reached a situation x for the sentence δ such that in x Lönnrot is in the loading station in the middle of
the damp and cold air of the turbid, puddled plain, the reader has *expectations* concerning the next step of the story. The expectations are
dependent on his interpretative assumptions and on the situations he
has chosen before x. Thus, for example, on the assumption concerning the influence of environmental conditions on Lönnrot's mood the
reader certainly expects that Lönnrot will not feel very happy at the
next stage of the story, if he is still around. This amounts to the fact
that all situations which are accessible from the situation x are such
that in them Lönnrot does not feel very happy, if he is still somewhere in the neighbourhood of the loading station. On the other hand,
the reader may not expect, for example, that Lönnrot began walking
along the countryside.

   For the sake of the argument, let us assume now that later on in the
text the author indicates – he does not actually do so – that Lönnrot
felt happy, after all, when he started walking along the countryside.
In this case, the reader's expectations in this respect will not be satisfied. Then, on our assumptions concerning x, the reader cannot con-
sistently continue from x, and so – if we are talking about ideal reading – he has to return to some earlier situation, say w, for the sentence γ in order to reach another situation x' for δ which is accessible
from w and such that there exists a situation for ε which is accessible
from x' and where Lönnrot felt happy.[20] If such an x' cannot be
found – this would be the case, for instance, if the reader's assumption concerning the relation between environmental conditions and
mood is global and thus intended to cover the whole interpretation –
pragmatic constraints are not compatible with the general linguistic
meaning of the text, that is, with textual constraints, and must therefore be changed.

   Another modality that the notion of pragmatic constraint suggests is
possibility. As we have seen, expectations amount to a kind of necessity that may emerge at some stages of reading. Possibility and expectation, as related to the same pragmatic constraints, are therefore dual
notions. Reading a text in the framework that is provided by an appropriate system of pragmatic constraints may give rise to expectations

concerning forthcoming events and, on the other hand, indicate possibilities the reader can speculate about. Both notions have cognitive, aesthetic, and conceptual connotations since the notion of pragmatic constraint has, but, as it seems to me, it is the relation between what is predictable and what is possible in the story that is of greater cognitive and aesthetic significance than these modalities considered individually. If too much can be successfully predicted about the story, the text is conceptually narrow in the sense that it does not leave very much room for the reader to approach it creatively. If, on the other hand, the text admits too many possibilities, it is conceptually loose, lacking cognitive and emotive intensity. Only if these dual features are in harmony – no matter what this exactly means – the story exhibits cognitive and aesthetic tension and complexity.

—— PART TWO ——

THE LOGIC AND PRAGMATICS

OF

SCIENTIFIC CHANGE

THE CORRESPONDENCE RELATION

In the rest of this book I modify the notion of explanatory translation in order to define a general correspondence relation, and I apply the results to some actual cases of scientific change. I shall not consider such Kuhnian questions as how scientific changes have taken place historically, how communication breakdowns have been handled and scientists persuaded by means of translation or learning, or how scientists have experienced the changes. Much of what I shall do is, however, relevant to the second question. There is nothing in principle that would prevent us from saying that translations of local utterances and literary texts are correspondence relations, but for historical reasons I reserve this term for certain relations between scientific theories — theories of the human sciences and aesthetics included.

Most of what follows is partly reconstructive in nature. On the other hand, if I have managed at all to realistically present the character of explanatory translation in earlier applications, and if I shall be able to persuade the reader in this and later chapters that pragmatic and hermeneutic considerations are closely associated with logical ones when the notion of correspondence relation is applied, the reader will also see the realistic side of my forthcoming reconstructions. While Kuhn emphasizes the role of pragmatics and hermeneutics in theory choice, I try to point out that they are also important for problems concerning intertheoretic relations and intertheoretic explanations, that is, in the context of justification.

The most intricate kind of correspondence relation is the so-called counterfactual correspondence — particularly its special form, limiting case correspondence — since it involves contrary-to-fact assumptions, similar to the ones we investigated earlier in some examples of local translation. As we shall see in later chapters, this notion plays an important role in many actual cases of scientific change that Kuhn calls revolutionary. The correspondence relation is investigated in this chapter with respect to both nonformal and formalized theories.

5.1. THE CORRESPONDENCE OF THEORIES

Even if the languages of two theories have terms in common, they call for a translation whenever the terms are given different meanings within the two theories. It is often an explanatory, even corrective, global translation that is needed,[1] which, as noticed earlier, does not in general meet Kuhn's criteria of adequacy, and therefore it would not be a translation by his standards. It follows that the existence of an explanatory translation does not suggest commensurability in his sense. On the other hand, Pearce (1987) argues that Kuhn's notion of commensurability can be challenged and replaced by weaker notions that admit of the cognitive comparability of theories in a sufficiently strict sense. I shall point out later on that the notions of corrective and explanatory translation in a global sense are the first step toward a type of cognitive comparability that is different from the kinds discussed earlier in the literature, but which is in many ways congenial with Kuhn's (1970) insights.[2] Therefore, I start by applying, *mutatis mutandis*, what we have so far learned to scientific theories, in order to define the notion of correspondence relation. It will appear, however, that we need something more than merely this notion to understand revolutionary conceptual change. What it is that is needed depends largely on the kind of change one is studying.

If a theory is formal or can be formalized, the questions of the composition of its language and of the type of possible worlds it describes are assumed to be solved.[3] Otherwise there may not be any exact conventions as to what belongs to its language – and this is true, in particular, about theories belonging to sciences that are not exact. A similar uncertainty concerns then the structure of possible worlds that the theory is describing. It also concerns the question of which of those worlds are correlated with worlds described by another theory. It turns out that something like Kuhn's notion of paradigm is what is here needed. If paradigms or scientific communities play such a crucial role in interpretation as he argues, they determine how these and similar questions should be answered.

Though the import of notions related to speech acts may here become even less obvious and less important than, for instance, in the case of fictional texts, it is natural to say that (in a given context) a scientific community (belonging to an appropriate paradigm), or even an individual researcher trained in that community, is a speaker and another community or individual (possibly of another paradigm) a hearer. Consequently, our earlier principles and concepts concerning interpretation, translation, and their cooperation with a speaker's, hearer's, and actual worlds become meaningful and applicable to scientific theories, whether formal or not. Furthermore, since the correspondence relation can be taken as a type of explanatory translation, it means that a theory which is in a correspondence relation to another, that is, whose language is translated into the language of the other, will be explained and possibly corrected from a point of view of the paradigm associated with the latter theory rather than of some other paradigm or framework.

In so far as the relevant collections of linguistic expressions and possible worlds for the theories in question can be delineated, the correspondence relation can be defined as follows. Consider first the collection of all sentences belonging to the language of a theory T, a *secondary theory*. If the collection is vague, it can nevertheless happen that only the sentences belonging to some subcollection S, which can be more strictly delineated, need to be translated into sentences of another theory T', a *primary theory*.[4] Next, let K' be a collection of possible worlds that are described by T' and possibly some special conditions, that is, at which the principles and laws and axioms of T' and the special conditions are true. K' may, for instance, consist of worlds that are considered relevant to an explanation of T, in some sense of explanation (to be specified in each case). It need not contain all the worlds described by T'. In terms of speech acts, the worlds in K' are what we have been calling a hearer's worlds. K' may sometimes result from an application of a relevant minimizing transformation, in the sense discussed earlier, to some of the worlds that T' is supposed to describe in the first place. These intended worlds, which in the philosophy of science are sometimes called 'intended applications' of the theory, play here the role of actual worlds. As before, if the minimization principle is used in a radical, counterfactual way, the respective minimizing transformation may lead to worlds that differ radically from the actual worlds.[5]

Now let I be a *translation* mapping which converts the sentences belonging to S into sentences of the language of T', and let F be a *correlation*, which is a mapping from K' to possible worlds described by T (plus possibly by other conditions), i.e., to a speaker's worlds. If condition (4.1.1) of Section 4.1 holds for all sentences in S and all worlds in K', then the pair $\langle F, I \rangle$ is called a *correspondence* of T to T'.[6] We may add here the requirement that the collection K' be definable, or more loosely characterizable, in the language of the theory T', but this requirement is as sensitive to the effects of vagueness as the assumptions concerning K' itself and the two mappings.[7]

Since Aristotle, philosophers have kept saying that the research of history differs from natural science in that its task is not to search for general laws but describe particular events, but this view is opposed by some modern scholars, as, for instance, Hempel (1965) who tries to point out that general laws are an important instrument in historical research. They can be used in historical explanations. If this is true and such laws can be found for theories of history, then our notion of correspondence would be applicable to such theories – presupposing, furthermore, that the above conditions concerning the delineation of the languages and possible worlds of the theories in question are satisfied. Hempel goes on to say that since laws, and other relevant conditions, in history are in most cases only vaguely indicated, we should in such cases talk about explanation sketches rather than explanations in a proper sense of the word. To turn it into an explanation, an explanation sketch would need to be completed by inserting further details. The same holds for other fields of research in the human and social sciences. Now it is straightforward to carry Hempel's idea over to the notion of correspondence and say that in cases where laws and the relevant collections needed in its definition are only vaguely indicated, and no correspondence relation can be defined in any precise sense, we may still have a *correspondence sketch*. Thus we may continue to talk about correspondence even on pain of being inaccurate if what we mean are correspondence sketches.

5.2. NARRATIVE CORRESPONDENCE

The view that historical research presents narratives, that its descriptions of past events are organized into temporal sequences of sentences, is evidently more popular, currently, than the view that they present universal laws. According to Danto (1968), for example, historical research explains changes by using narratives; its task is to organize the past into temporal wholes. However, many criticize the idea that history describes the past as temporal sequences. The reasons are similar to those which we indicated in the previous chapter. Thus Solomon and Higgins (1993) ask whether any historical story is ever really over since history is constantly rewritten, and 'history' is ambiguous meaning both the sequence of events and the ordering of this sequence. A narrative text is an integrated whole, and therefore it is evident that if the text is continued by descriptions of new historical facts, the meaning of the text may change.

In any case, if theories of history are narrative sequences in the first place, and not sets of independent laws or lawlike sentences, then our definition of correspondence is not applicable to them. We have to consider condition (4.1.2), Section 4.1, in order to define the correspondence relation for theories of history. From our discussion concerning global translations of literary narratives we may see that there is similar vagueness here (and the same holds, of course, for local translations of speech acts), similar to what Hempel is talking about, wherefore we should rather talk about correspondence sketches here, too. Whether sketches or more precisely definable, I shall in what follows distinguish the notion of correspondence as applied to narrative theories by simply calling it *narrative correspondence*. Since historical narratives also have the connotative dimension in the sense discussed above, that is, the dimension consisting of possible worlds constructed by means of narratives, it is conceivable that the notion of correspondence (sketch) in our original sense, as defined in the previous section, is relevant here as well. This is at least the case if historical narratives suggest connotations that have sufficiently great universal importance, even though they could not be regarded as laws or lawlike statements.

Now the kind of criticism Solomon and Higgins present against Danto's view of historical theories as temporal narratives suggests in fact an opportunity to apply narrative correspondence to historical theories. As soon as a given narrative is rewritten or continued, we have two theories of history, of which the new dislodges the original one, and it is then natural to pose questions concerning their relationship, similar to questions usually posed about scientific change more gener-

ally. We can ask, for example, whether there is a narrative correspondence relation between the two theories. In view of vagueness,
one can then try, at best, to establish a sketch. On the other hand, it is
almost everyday routine in history (and in social science, for that matter) to change theories in a radical way – at least it seems to take place
more often than in natural science – and therefore, for example, the
question of incommensurability, in something like the Kuhnian sense,
could there be raised more frequently.[8]

In view of (4.1.2), the notion of correspondence relation, which
has so far been associated with scientific theories only, could be extended, in the form of narrative correspondence sketch, so as to apply
to fictional narratives too. If this is done, we can say that a minimum
condition a translation has to satisfy is that there be a correspondence
relation between the two narratives. As remarked above, however, I
shall continue to use the term 'correspondence' in the context of scientific change, rather than just 'translation', and use the latter term,
rather than the former, in connection with fictions, where the term is
usually used. This terminology is motivated, furthermore, by the fact
that the point of translating texts of fiction is that it makes them available in different natural languages, whereas in history rewriting and
paraphrasing are more crucial tasks.

5.3. CORRESPONDENCE AND FORMAL THEORIES

In model theory there exist general and formal notions of reduction
establishing both syntactic and semantic relations between theories
that together can be thought of as reflecting the respective meaning
changes.[9] The version of correspondence to be defined in this section
can be considered a general, but nontrivial, formal notion of reduction. It can be shown that even on some very general conditions for
the notion, if a correspondence relation obtains between two theories,
the translation of the secondary theory (or, rather, of the conjunction
of its axioms) is a logical consequence of the primary theory plus additional hypotheses. Whether correspondence relation can be considered a generalized reduction in something like Nagel's (1961) sense
of reduction depends on its logical properties and on pragmatic conditions one can assign to it, and such properties and conditions are even
more crucial if the consequence relation is to establish a deduc-

tive-nomological explanation of the secondary theory.[10]

Actual scientific theories, or their laws, and their reductions and explanations are, of course, normally expressed by means of nonformal scientific or mathematical languages. However, to be able to discuss correspondence in explicit terms I shall assume in this section that they are formally representable in appropriate logics.[11] This implies that the approach to correspondence which follows is model-theoretic, and therefore I shall here speak of models rather than of situations or possible worlds. Each theory is assumed to be model-theoretically reconstructed and to determine a class of models (of a given type) that is definable in an appropriate logic. There are always several logics in which a theory can be defined, if it can be defined in some logic, and which of them is chosen depends on methodological, pragmatic, and purely logical criteria.

We can immediately see that the following definition agrees with our earlier conditions and principles for global translation and correspondence, and more or less with the notion of interpretation in logic. In the formal treatment the import of speech act theory and pragmatic considerations is suppressed to a minimum, but not lacking. Evidently we can speak here of a *speaker's* and *hearer's*, and *actual models* in the same manner as we spoke of their counterparts earlier in this chapter. In particular, by actual models I shall mean what are often called *intended models* of a theory. We shall take up these and other pragmatic things in Section 5.5 and especially in the next chapter. Such formal things as the structure of a language needed, that of appropriate models, the definitions of relevant collections of models, and so on, are characterizable in detail as soon as it is known what the logical framework to be used is.

Let T and T' be two theories such that their models (the models of their axioms) are of types τ and τ', respectively. Let H and H', respectively, be the classes of all models of T and T'. Let a logic L be assigned to T and L' to T', whence H is $L(\tau)$-definable and H' is $L'(\tau')$-definable. Assume there is a correlation F, i.e., a mapping from a (nonempty) $L'(\tau')$-definable subclass K' of H' onto an $L(\tau)$-definable subclass K of H, and a translation, i.e., a mapping I from the set of all $L(\tau)$-sentences to the set of $L'(\tau')$-sentences.[12] Then the pair $\langle F, I \rangle$ is a *correspondence* of T to T', *relative to* $\langle L, L' \rangle$ (*relative to L if L'* is *L*), if the following condition holds for all models $\mathfrak{M}$ in

K' and all $L(\tau)$-sentences φ:

(5.3.1) $F(\mathfrak{M}) \models_L \varphi \Leftrightarrow \mathfrak{M} \models_{L'} I(\varphi)$.

The two mappings have to satisfy logical and pragmatic conditions of adequacy, the nature of which seems to be dependent on the theories in question and on paradigmatic and contextual features. The mappings must establish sufficiently informative conceptual relation between the two theories; the most important condition (5.3.1) is model-theoretic, and thus it is embedded in the definition. As we have already seen, what is here taken as a translation is too general to be called translation by Kuhn; condition (5.3.1) does not in general meet his criteria of adequacy, and therefore the existence of a translation in the present sense does not imply commensurability in his sense.[13] I shall discuss the cognitive comparability that results from correspondence in the next chapter.

The translation could also be defined as a *partial* mapping; then its domain should usually comprise all the axioms of T. In translating the language of T, one is usually interested in the axioms, and as we shall see later, some results concerning explanation depend on this matter. On the other hand, if the translation is defined recursively according to the recursive definition of $L(\tau)$-formulas (if there is one), the requirement that the axioms must be included in the translation often excludes partiality. This is so because in view of the economical character of axiomatization all nonlogical and logical symbols (variables perhaps excepted) of $L(\tau)$ occur in the axioms and thus participate in the translation, making it total.

As noted above, a general correspondence relation can be thought of as a weak reduction. Stronger forms are obtained as special cases. If K is H, the class of all models of T, $\langle F, I \rangle$ is a *reduction* in a more usual sense of the word; whereas if K' is H', the class of all models of T', we have an *interpretation*, roughly in the sense in which logicians speak about interpretation of one theory in another. Although weak, the general notion of correspondence has interesting logical and philosophical consequences, to be studied in chapters to follow.

5.4. THE CORRESPONDENCE OF LOGICS

The formal definition of correspondence, as presented above, can be modified so as to provide a general notion of correspondence, or re-duction, of one *logic* to another.[14] There has been some discussion of special cases of such a correspondence, as, for example, in the frame-work of abstract logic, where the question of whether one logic is in-cluded in another, that is, whether the latter has a better expressive power than the former, is a crucial problem. Inclusion in this sense, which presupposes that the two logics which are compared have the same notion of model, is a special case of the forthcoming notion of correspondence. The question of expressive power also has a close connection with the problem of explanation.

In order to compare two logics having different kinds of model, the following notion of correspondence is needed. We say that a logic L is in a *correspondence* relation (or is *weakly reducible*) to a logic L' if for all types τ of L, there exist a type τ' of L', an L'$(\tau$')-definable class K' of L'$(\tau$')-models, a mapping F from K' onto the class of all $L(\tau)$-models, and a mapping I from the set of all $L(\tau)$-sentences to the set of all L'$(\tau$')-sentences, such that (5.3.1) holds for all models $\mathfrak{M}$ in K' and all $L(\tau)$-sentences φ.

5.5. COUNTERFACTUAL AND LIMITING CASE CORRESPONDENCE

We saw in Chapters 2-3 that sometimes an utterance is explained or corrected by conjoining its (nonstandard) translation with a semantic transformation that involves a counterfactual condition. In an example modifying one of Danto's examples, in Section 3.1, we investigated a speech act context in which a hearer tried to understand a speaker's false assertion that the legs presented in Breughel's painting *Land-scape with the Fall of Icarus* belong to a pearl diver or an oysterman by imagining a counterfactual situation where the title of the painting is different from what it actually was, that is, *Industry on Land and Sea*.[15] This assumption justified, by the hearer's lights, the condi-tional saying that if the title of the painting were this, then the legs in the painting would belong to a pearl diver or an oysterman.

Similar cases may occur in the framework of global translation, and, in particular, in applications of the correspondence relation, as will be seen in Chapters 6 and 7. Consider a correspondence $\langle F, I \rangle$ of T to T', where F is a mapping from a class K' of possible worlds (or situations) described by T' to those described by T and I is a translation. If K' consists of possible worlds that are regarded as being contrary-to-fact (in an appropriate sense) by the scientific community holding the theory T', then the correspondence relation is called a *counterfactual correspondence*. It follows that if K' can be defined in the language of T', then some of the defining sentences are considered false – even radically false – by the community; that is, they are false in such worlds described by T' that are regarded actual, that is, intended.

If T and T' are formal theories, we can be more precise, since we required that the relevant classes of models are definable in the respective logics. The question remains, however, whether the actual or intended models of T' or T can be defined in an equally precise way, since the problem of the nature of such models is often pragmatic and scientific rather than model-theoretic. We can see from the case studies to follow, however, that in many cases of counterfactual correspondence it is enough to have pragmatic ideas of *why* some conditions are to be considered counterfactual, and then to express the conditions and define K' in L'(τ'), without any need to define the community-relative notion of actual model.

In the most important special type of counterfactual correspondence, which is closest to the original and the most controversial idea of correspondence, that is, in *limiting case correspondence*, some of the sentences defining K' express appropriate limit conditions in such a way that the possible worlds or situations in K' are 'very close' to their images (in the correlation mapping) in K. The expression 'very close' will be defined so as to be in conformity with the idea that the laws of the theory T are obtainable from those of T' by means of an appropriate limit procedure. In one of our former examples of local translation, which we again discussed a few lines ago in the present section, the condition saying that the title of the painting was different from its actual title is counterfactual in something like the sense meant here. Here, too, the minimization principle is evident since the minimizing transformation yields K', which is defined by using counterfactual assumptions.

We have to distinguish between two kinds of limiting case correspondence. They are analogous to certain types discussed in Section 3.1. If it is assumed that the laws of T themselves (or, rather, certain worlds at which the laws are true) are reached from the laws (worlds) of T' by a limit procedure and, hence, that K' is identical with K, we have a limiting case correspondence in the original sense advocated by Niels Bohr and many others after him, and criticized by Kuhn.[16] This is the standard way to understand limiting case correspondence, but it is trivial as an explanatory translation, since translation and correlation are identity mappings. Hence the respective semantic transformation, which is now composed of the limit procedure alone, is identical with the minimizing transformation. This kind of triviality may well have been in Kuhn's mind when he criticized the notion of limiting case correspondence saying that it lacks a transformation.[17]

In the second, nontrivial type, 'very close' does not mean 'identical', so that K' is different from K, even though it is as close as it can possibly be without being identical with it. Thus, for instance, while in the first, standard formulation of the idea that classical particle mechanics is a limiting case of relativistic particle mechanics the velocity of light is assumed to *tend to* infinity – meaning, among other things, that Newton's second law approaches Minkowski's force law – in this second type we shall say that if the velocity of light is assumed to *be* infinite, then the former law is infinitesimally close to the latter, and similarly for appropriate worlds. Again, the idea of limiting case correspondence in this second sense can be more exactly expressed for formal theories, so we turn now to its model-theoretic representation. In this more exact reconstruction of the idea, notions of nonstandard analysis will be used.[18]

By using terminology of nonstandard analysis, to be explained in Sections 7.3 and 8.6, and Chapter 9, below,[19] the limit conditions can be expressed as follows. Let $\langle F, I \rangle$ be a correspondence of T to T', as above. Then it is a limiting case correspondence if (i) K is a class of standard models of T and K' a class of nonstandard models of T' such that $F(\mathfrak{M})$ is the standard approximation of $\mathfrak{M}$, for all $\mathfrak{M}$ in K', and (ii) the correlation F determines the translation I in such a way that what the definition of I essentially amounts to is to trace the process of forming the standard approximations of the models in K'. The latter condition means that the terms occurring in a formula of the language of T are forced to refer to standard approximations of individuals in the models contained in K', and similarly for quantifiers. How this can be exactly done will be exemplified in Section 9.1, but what happens in the translation is, roughly speaking, similar to the following imaginary and simplified example.

Example. Assume that **f** and **g** are two function symbols in both τ and τ' that are interpreted in $F(\mathfrak{M})$ as real-valued functions and in $\mathfrak{M}$ as functions having standard and finite nonstandard reals as values. Consider an $L(\tau)$-sentence φ that is of the form

$$\forall \mathbf{x}(\mathbf{f}(\mathbf{x}) = \mathbf{g}(\mathbf{x})),$$

where **x** is an individual variable of an appropriate sort. Then in the translation $I(\varphi)$ the atomic formula $\mathbf{f}(\mathbf{x}) = \mathbf{g}(\mathbf{x})$ is replaced by a formula saying that in $\mathfrak{M}$ the equation holds between the standard parts of the interpretations of $\mathbf{f}(\mathbf{x})$ and $\mathbf{g}(\mathbf{x})$. (In $\mathfrak{M}$, these terms may refer to nonstandard elements.) Similarly, the quantifier occurring in φ is replaced by the respective bounded quantifier ranging over the standard individuals of $\mathfrak{M}$, denote them by A (which means in this case that it ranges over the standard real numbers). If the interpretations of **f** and **g** in $\mathfrak{M}$ are briefly denoted by f and g, then what $I(\varphi)$ expresses in $\mathfrak{M}$ is the following, where st means a 'standard part':

$$\forall x \in A(\text{st}(f(x)) = \text{st}(g(x))).$$

Then it is obvious, since $F(\mathfrak{M})$ is the standard approximation of $\mathfrak{M}$, that φ is true in the former model if and only if $I(\varphi)$ is true in the latter. Furthermore, since $\text{st}(f(x))$ and $\text{st}(g(x))$ are infinitesimally close to $f(x)$ and $g(x)$, respectively, we can immediately see that $f(x)$ and $g(x)$ are infinitesimally close to each other, that is:

$$\forall x \in A(f(x) \approx g(x)).$$

This result can be informally expressed by saying that

φ *almost* holds

in the models in K', where 'almost' means infinitesimal accuracy. This example is simplified but illuminates the general idea. An actual example will be presented in Chapters 7 and 9 where some case studies are described.

5.6. APPROXIMATE CORRESPONDENCE

It seems natural to say that a given relation of limiting case correspondence is, in a sense, a limit of relations of *approximate correspondence*. A relation of the latter kind is of course relative to the degree of approximation. In so far as it is possible to define what it means to say that an $L'(\tau')$-formula is an approximation of an $L(\tau)$-formula, and similarly for models, the respective translation and correlation must syntactically and semantically reflect the degree of approximation in question. I shall not study problems of approximate correspondence in detail — since the notion does not seem to be conceptually as interesting as the notions of counterfactual and limiting case correspondence — but only illuminate the notion by means of the following example, analogous to the one in the previous section.

Example. Assume again that **f** and **g** are function symbols in both τ and τ', and φ is the $L(\tau)$-sentence

$$\forall \mathbf{x}(\mathbf{f}(\mathbf{x}) = \mathbf{g}(\mathbf{x})),$$

as in the previous example. Now assume that given a positive real number ε, the absolute value of the difference $f(x) - g(x)$ is (for all values of x) smaller than ε in all models $\mathfrak{M}$ of T' in which appropriate conditions expressible in the language of T' (that is, by means of $L'(\tau')$-sentences) are satisfied. Then in the translation $I(\varphi)$ we let the atomic formula $\mathbf{f}(\mathbf{x}) = \mathbf{g}(\mathbf{x})$ be replaced by a formula essentially saying that the absolute value of the difference is smaller than ε.

In the transformation of $\mathfrak{M}$ into $F(\mathfrak{M})$ we may let f stay as it is but g transform into f, whence f is the interpretation of both **f** and **g** in $F(\mathfrak{M})$. What else must be done depends, of course, on the nature of $\mathfrak{M}$, T, and T', and similarly for how the translation is defined. This is not very interesting in itself, but we can see now in what sense a limiting case correspondence can be thought of as being a limiting case of an approximate correspondence.

5.7. CORRESPONDENCE AND SYMMETRIES

Though it is obvious that all scientific theories are dependent on there being invariant features present in the phenomena studied, symmetries have been mainly discussed by physicists and philosophers of physics. Wigner (1967) gives the principles of invariance a status of superprinciples by saying that laws of nature could not exist without them. There is a hierarchy in our knowledge of the world: our knowledge of events is expressed in laws and our knowledge of laws is expressed as invariance principles. The latter are in the similar relation to the laws of nature as the laws are to the events themselves.

Sometimes invariance principles are even thought to be more fundamental than laws, and therefore they have a justificatory function in scientific discovery: a new law will be accepted only if it satisfies appropriate invariance restrictions, considered universal. It follows that they may also have some heuristic power, since they enable a scientist to eliminate inappropriate law candidates. They may even be useful in attempts to find new laws. Thus, for example, Einstein (1905) and Wigner write that they are of great heuristic value in this sense. This view of the role of symmetries, the view according to which invariance principles, considered more fundamental and evident than laws, can be used to derive laws of nature, is called by Redhead (1975) the *a priori* approach. The empirical approach is the opposite view, according to which invariance principles are considered properties of laws and hence derivable from laws of nature. Both approaches are exemplified in the history of science.

Let us consider first Redhead's definition of a symmetry. According to him, the *symmetry* of a physical system or situation is a pair $\langle \text{Inv}, S \rangle$, where Inv is the collection of invariants, that is, the features of the system or situation that remain unchanged, and S is the collection of transformations that express those changes for which the invariants remain fixed. If this definition is applied to a law, the invariant feature is the form of the law and the symmetry transformations are operations affecting the terms by means of which the law is formulated. The operations leave the form of the law unchanged. For example, laws of nature − or, rather, what are laws of nature according to scientific theories − are usually assumed to be invariant with respect to operations transforming the space and time coordinates of the systems satisfying the laws.

Redhead's characterization can be immediately modified so as to be applicable within a model-theoretic treatment of theories. Since in that framework models represent systems about which a theory, say T, is talking about, the symmetry of T can be defined in terms of appropriate models and their transformations.[20] If the models of T are of type τ, assume that $\text{Mod}(\tau)$ is the class of all models of type τ, or possibly some relevant subclass of it, and H the class of models of T, i.e., the class representing the axioms, or laws, of T.[21] Then we can say that a *symmetry transformation* of T is any operation on models in $\text{Mod}(\tau)$

under which H is closed, that is, any operation P such that for all models $\mathfrak{M}$, $\mathfrak{N} \in \text{Mod}(\tau)$:

If $\mathfrak{M} \in H$ and P: $\mathfrak{M} \mapsto \mathfrak{N}$, then $\mathfrak{N} \in H$.

Not all such operations are important and, hence, when studying an actual theory one should restrict oneself to those which are usually studied with the theory, that is, which are intended transformations, or characteristic of the theory.[22]

Modifying in this way Redhead's characterization of symmetry, let us say that *the symmetry of the theory* T is the pair

$$\mathbf{H} = \langle H, S \rangle$$

where H is as above and S is the class of symmetry transformations of T (i.e., transformations that are characteristic of T). In Chapter 9, *subsymmetries* of T will also be considered. They are of the form

$$\mathbf{K} = \langle K, R \rangle,$$

where K is a subclass of H and R consists of all transformations in S under which K is closed. A symmetry can be thought of as a *category* in the sense of category theory, to be explained in Section 8.5, below.

Let us now study the claim that symmetries play a role in scientific change. There exist some theoretical discussions in the literature that support this claim, and some of the forthcoming case studies support it, too. Their methodological role in connection with the Correspondence Principle was recognized by Post (1971) and a little later studied by Redhead (1975). Post raises the question of how the symmetries of two theories are related if there obtains a correspondence relation between the theories, and Redhead studies their relation in somewhat more exact terms. Post and Redhead start with a principle, which Chalmers (1970) calls Curie's Principle, after Pierre Curie (1894). Its original formulation seems to be:

> *Curie's Principle* (Curie). When certain causes produce certain effects, the elements of symmetry of the causes must be found in the effects produced.

Redhead formulates the principle differently, and his formulation appears to be somewhat stronger:

Curie's Principle (Redhead). Symmetries are always transmitted from a cause to its effect, that is, the cause cannot be more symmetric than its effect.

They ask whether an analogous principle can be found that applies to the correspondence relation, that is, which would be valid when instead of speaking of cause and effect we speak of two theories T and T' such that T is in a correspondence relation to T'. If there is such an analogy, we can think of T' as being analogous to a 'cause' only in the sense of explaining the successful part of T. That the analogy holds means then that the symmetries of T' are somehow transmitted to symmetries of T. This analogy is suggestive, but Redhead's useful explication of the analogy is not entirely satisfactory. He uses a notion of correspondence that is purely structural and Post's and his analogy of Curie's Principle, which he calls the Curie-Post Principle, lacks sufficient generality:

The Curie-Post Principle (Post, Redhead). The primary theory T' cannot be more symmetric than the successful part of the secondary theory T in the case where T and T' stand in a relation of consistent correspondence.

As stated here, the principle excludes a kind of correspondence that is studied in the present book, namely counterfactual, especially limiting case correspondence, which Post and Redhead classify as an inconsistent correspondence. Such is, for example, the relation between classical particle mechanics and relativistic particle mechanics, which, according to Post and Redhead, would provide a counterexample without the restriction to consistent correspondence, since Lorentz transformations are not transmitted to classical particle mechanics. However, it is obvious that the original formulation of Curie's Principle, unlike Redhead's formulation, admits an interpretation according to which appropriate structural elements of a symmetry are transmitted in a correspondence relation, rather than the symmetry itself. If Curie's Principle is interpreted this way, it appears to be in harmony with the notion of correspondence as defined here.

Let us now restate the Curie-Post principle accordingly, within the framework of this chapter and of Section 8.5.[23] Assume that there is a correspondence $\langle F, I \rangle$ of T to T', where F is a mapping from a class K' onto a class K, as in Section 5.3, above, that is, the respective correlation. These subclasses of the classes of models of T and T' determine, respectively, subsymmetries $\mathbf{K} = \langle K, R \rangle$ and $\mathbf{K'} = \langle K', R' \rangle$ of T and T'. They are subcategories of the symmetries of the theories T and T', respectively. Now, if we want the correspondence relation in question to be in conformity with the subsymmetries, it is natural to require that the correlation F can be extended so as to become a *functor* between the two (sub)symmetry categories, that is, so as to correlate the morphisms in R' to those in R in such a way that the categorial structure of $\mathbf{K'}$ is preserved:

> *Curie's Principle for Correspondence* (the Curie-Post Principle restated) Assume a mapping F from K' onto K is the correlation in a correspondence of T to T'. Let $\mathbf{K}$ and $\mathbf{K'}$ be the subsymmetries of T and T' determined by K and K', respectively. Then F can be extended to a functor from $\mathbf{K'}$ to $\mathbf{K}$.

If G is the mapping extending F, that it is a functor implies that it correlates the morphisms (symmetry transformations) in R' and R,

$$G: \mathrm{P'} \mapsto \mathrm{P}\ (\mathrm{P} \in R, \mathrm{P'} \in R')$$

in such a way that the following diagram commutes:[24]

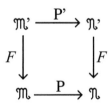

The principle, as stated here, does not require that a symmetry itself be preserved in a correspondence in some sense, but rather, that its *categorial structure* be preserved. If it can be said that the existence of a correspondence relation between two theories is a sign of a degree of continuity in the respective scientific change, then we can also say that if Curie's Principle for Correspondence is valid, it is a sign of continuity in the context of invariance principles. It is readily seen, however, that the principle, as stated here, is very strong, and, therefore, not easily satisfied. Its validity will only be studied in Sections 7.1 and 9.1, in connection with particle mechanics. There we can see, in any case, that the principle is relevant for actual theories.[25]

5.8. CORRESPONDENCE AND DEFINABILITY

Various kinds of definability provide special types of correspondence relation, where, roughly, translation amounts to elimination and correlation to model-theoretic expansion. In this sense definability as a correspondence differs radically from many correspondence relations to be studied later in Chapter 7, below. A survey of the concepts, results, and notation concerning definability to be needed in this section – and a discussion about the methodological import of definitions – can be found in the Appendix, which the reader should consult if not already familiar with the questions of definability. In this section, we shall mainly restrict ourselves to the model theory of elementary logic $L_{\omega\omega}$. Roughly speaking, the existence of the correspondence relations follows from the eliminability results, as presented in the Appendix, concerning the different kinds of definability.

As in the Appendix, we shall only study cases where the symbol whose definability we are considering is a unary predicate symbol, say **P**. It makes notation simpler but does not simplify the discussion too much since the results below can be generalized for an arbitrary predicate or function symbol. As there, we consider now a similarity type τ (which does not contain **P**) and its expansion $\tau \cup \{\mathbf{P}\}$,[26] and study the definability of **P** in terms of τ. Let us briefly write 'L' for $L_{\omega\omega}(\tau)$ and '$L(\mathbf{s}_1, \ldots)$' for $L_{\omega\omega}(\tau \cup \{\mathbf{s}_1, \ldots\})$, where $\mathbf{s}_1, \ldots$ are symbols (nonlogical constants) not in τ. T(**P**) will be a theory that is formulated by means of $L(\mathbf{P})$-sentences, that is, sentences formulated by means of nonlogical constants in $\tau \cup \{\mathbf{P}\}$. Let T be the set of all L-sentences deducible from T(**P**).

Let us start with explicit definability, which provides the most obvious case for correspondence, and assume that **P** is explicitly definable in T(**P**) (in terms of τ), i.e., an explicit definition deducible:

$$\mathsf{T}(\mathbf{P}) \vdash \forall \mathbf{x}(\mathbf{P}(\mathbf{x}) \leftrightarrow \varphi(\mathbf{x})).$$

Then it follows from the results concerning explicit definability, as presented in the Appendix, that T(**P**) is reducible to T. The correlation F that is needed assigns to each model $\mathfrak{M}$ of T its unique expansion $(\mathfrak{M}, P)$ that is a model of T(**P**).[27] The translation I assigns to each $L(\mathbf{P})$-sentence ψ the unique L-sentence that is obtained by eliminating **P** from ψ, that is, by replacing each subformula of the form $\mathbf{P}(\mathbf{t})$

(where **t** is an arbitrary term, i.e., variable or individual constant) by $\varphi(\mathbf{t})$. Then it is obvious by (A.6.1) of the Appendix that $\langle F, I \rangle$ is the required reduction, and at the same time it is also what was in Section 5.3 called an interpretation.

This result is obvious and well known, but we can show, moreover, that some weaker kinds of definability induce weaker, and more interesting, kinds of correspondence, that is, logically weaker than reduction and interpretation. Suppose first that **P** is piecewise definable in $\mathsf{T}(\mathbf{P})$ (in terms of τ):

$$\mathsf{T}(\mathbf{P}) \vdash \bigvee_{1 \le i \le n} \forall \mathbf{x}(\mathbf{P}(\mathbf{x}) \leftrightarrow \varphi_i(\mathbf{x})).$$

According to Theorem A.3.4 of the Appendix, this is equivalent to the condition that **P** is explicitly definable in every complete consistent extension of $\mathsf{T}(\mathbf{P})$. Take then any such extension, say $\mathsf{S}_i(\mathbf{P})$. Then **P** is explicitly definable as follows:

$$\mathsf{S}_i(\mathbf{P}) \vdash \forall \mathbf{x}(\mathbf{P}(\mathbf{x}) \leftrightarrow \varphi_i(\mathbf{x})).$$

Let K be the class of all models of that extension and let K' be the class of all restrictions to τ of the models in K. If we continue as above and define F as a mapping from K' onto K and define I by replacing $\mathbf{P}(\mathbf{t})$ with the formula $\varphi_i(\mathbf{t})$, we may see that $\langle F, I \rangle$ is a correspondence of $\mathsf{T}(\mathbf{P})$ to T. It can be seen, furthermore, that for each complete consistent extension of $\mathsf{T}(\mathbf{P})$, a correspondence, even reduction, can be obtained.

Restricted definability in $\mathsf{T}(\mathbf{P})$ implies, in turn, that there is a correspondence relation between simple extensions of $\mathsf{T}(\mathbf{P})$ and T. If

$$\mathsf{T}(\mathbf{P}) \vdash \bigvee_{1 \le i \le n} \exists \vec{\mathbf{x}} \forall \mathbf{y}(\mathbf{P}(\mathbf{y}) \leftrightarrow \varphi_i(\vec{\mathbf{x}}, \mathbf{y})),$$

then, as indicated in the proof of Theorem A.6.4 in the Appendix,

$$(\mathfrak{M}, \vec{a}) \vDash \forall \mathbf{y}(\mathbf{P}(\mathbf{y}) \leftrightarrow \varphi_i(\vec{\mathbf{a}}, \mathbf{y})),$$

where $\mathfrak{M}$ is an appropriate model of $\mathsf{T}(\mathbf{P})$, $\vec{a}$ are elements (i.e., a sequence of elements) of the domain of $\mathfrak{M}$, and $\vec{\mathbf{a}}$ are new individual constants. This means that **P** is explicitly definable in the complete consistent theory of the form $\mathsf{S}(\mathbf{P}, \vec{\mathbf{a}}) = \mathrm{Th}((\mathfrak{M}, \vec{a}))$, from which together with the results above it follows that there is a correspondence relation of an extension of $\mathsf{T}(\mathbf{P})$ formulated by means of $L(\mathbf{P}, \vec{\mathbf{a}})$-sentences to an extension of T formulated in $L(\vec{\mathbf{a}})$.

One correspondence relation that can be associated with conditional definability is obvious, in view of what we learned above about explicit definability. Let us assume, according to the Appendix, that

$$T(\mathbf{P}) \vdash \forall \mathbf{x}(\sigma(\mathbf{x}) \rightarrow (\mathbf{P}(\mathbf{x}) \leftrightarrow \varphi(\mathbf{x}))).$$

If $S(\mathbf{P}) = T(\mathbf{P}) \cup \{\forall \mathbf{x}\sigma(\mathbf{x})\}$ is consistent, it can be used to define in $L(\mathbf{P})$ the (nonempty) range of the correlation of the required correspondence relation, and since $\mathbf{P}$ is explicitly definable in $S(\mathbf{P})$, that is,

$$S(\mathbf{P}) \vdash \forall \mathbf{x}(\mathbf{P}(\mathbf{x}) \leftrightarrow \varphi(\mathbf{x})),$$

$T(\mathbf{P})$ is in correspondence to T, which can be shown analogously to the earlier results.

Finally, let P be finitely definable in $T(\mathbf{P})$, as in Theorem A.3.2 of the Appendix. If we assume again that $S(\mathbf{P}) = T(\mathbf{P}) \cup \{\forall \mathbf{x}\sigma(\mathbf{x})\}$ is consistent, then by combining the two preceding results, concerning restricted and conditional definability, we may observe that an extension of $T(\mathbf{P})$ formulated in $L(\mathbf{P}, \vec{\mathbf{a}})$ is in a correspondence relation to an extension of T formulated in $L(\vec{\mathbf{a}})$. We can now see that all the different formal definability notions discussed in the Appendix are closely related to our concept of correspondence.

5.9. CORRESPONDENCE VS. CLASSICAL REDUCTION

Consider next a correspondence $\langle F, I \rangle$ of T to T', as defined and denoted in Section 5.3. Assume the class H is defined by a sentence θ, H' by θ', and K' by sentences θ', $\sigma_0, \ldots, \sigma_n$. We may assume here that the sentences θ and θ' represent the axioms or laws of the theories T and T', respectively, and $\sigma_0, \ldots, \sigma_n$ are the requisite auxiliary assumptions (special conditions). θ is an $L(\tau)$-sentence and the rest are $L'(\tau')$-sentences. Since θ', $\sigma_0, \ldots, \sigma_n$ define K', which in this formal framework corresponds to what we have called a hearer's models, they indicate in appropriate cases[28] how the respective minimizing transformation has to work.

Now the following result is obvious. If $\mathfrak{M}$ is any model (of type τ') in which the sentences θ', $\sigma_0, \ldots, \sigma_n$ are true, then in its image $F(\mathfrak{M})$, θ is true (by the definition of F), whence $\mathfrak{M}$ satisfies $I(\theta)$.[29] It follows that the translation of θ, $I(\theta)$, is a logical consequence (in the Tarskian sense) in L' of the other sentences:

(5.9.1) θ', σ_0, . . . , $\sigma_n \models_{L'} I(\theta)$.

If L' is axiomatizable and complete, then (5.9.1) amounts to a deduction in L'. I shall simplify notation, and ignore the difference between logical consequence and deduction, by rewriting (5.9.1) in the form

(5.9.2) θ', σ_0, . . . , σ_n / $I(\theta)$.

Thus, if there is a correspondence of T to T', then the (axiom of the) latter theory and the auxiliary, connecting assumptions logically imply the translation of the (axiom of the) former theory. Argument (5.9.2) does not directly connect the primary theory with the secondary one itself but with its translation into the language of the primary theory. This feature is a consequence of the above model-theoretic definition of correspondence, which involves both syntactic and semantic connections. We may say about the role of the translation and correlation that they together with the minimizing transformation (leading to K') indicate possible changes in meanings, even though I have not tried to define what we exactly mean by meanings in this framework.

Argument (5.9.2) is analogous to, but clearly different from, the formal conditions for the classical notion of reduction as they are stated by Nagel (1961). The present notion of correspondence (and reduction) is more general in the sense that Nagel restricts his notion to cases where θ represents an experimental law of the secondary theory (or science) containing a term,[30] say **s**, that does not occur in the theoretical assumptions of the primary theory (science). More importantly, the present approach, unlike Nagel's, admits counterfactual premises, as we have seen. On the other hand, the present notion is stronger in the sense that it presupposes a global translation of the secondary science into the primary one, whereas Nagel just assumes that there is a postulate, say δ, establishing a link between **s** and theoretical terms of the primary science. He calls this requirement the 'condition of connectability'. Furthermore, Nagel requires a formal 'condition of derivability' to the effect that

(5.9.3) θ', σ_0, . . . , σ_n, δ / θ,

where σ_0, . . . , σ_n are auxiliary assumptions.

It is evident that pragmatic conditions for the relevance of the two notions of reduction must be somewhat different since they are differently conditioned in the methodological and formal sense. They are not generally equivalent, but on suitable pragmatic and logical conditions they are. Assume, like Nagel, that a symbol **s** occurs in the language of T, that is, is in the type τ but not in τ', and suppose furthermore, as presumably does Nagel, that each logic L and L' in the definition of correspondence is elementary first order logic $L_{\omega\omega}$. Then, it directly follows from the preceding section and the definition of correspondence that if (5.9.3) holds and δ is of the form of an explicit definition or piecewise definition or conditional definition, T is in a correspondence relation to T'. It is not quite clear, however, and not indicated by Nagel, what kinds of sentences except explicit and conditional definitions would be scientifically appropriate as postulates connecting the two languages. Possibly this depends on the theories.

Conversely, let T be in a correspondence relation to T', so (5.9.2) holds, and let the translation I be obtained by means of explicitly defining **s** in terms of τ'. Let the explicit definition in question be δ. This means, in particular, that $I(\theta)$ is obtained from θ by replacing every occurrence of **s** by the definiens in δ, and then δ implies that θ and $I(\theta)$ are equivalent, that is,

$$\delta \vdash \theta \leftrightarrow I(\theta).$$

If δ is now added to the premises of (5.9.2), then (5.9.3) follows, whence there is a reduction in Nagel's sense. The other kinds of definability studied in the Appendix and in the previous section are not relevant here, since they do not yield a unique translation, that is, a mapping.

5.10. TRANSLATION AND EMERGENCE

In the philosophical literature, there exist some much-discussed notions that seem to be to some extent analogous to the notion of correspondence. Such are, at least, *supervenience, idealization,* and *emergence,* of which only emergence is discussed here.[31] For the sake of clarity, I shall assume below without any special mention that in so far as the notion of emergence is compared with correspondence, it must be assumed that the notion, to be comparable, is lifted onto the theoretical or linguistic level where our notion of correspondence is, whenever needed.

Let us first quote what has been said about emergence in cultural entities since philosophers often characterize the meaning of 'cultural' by using this notion. For example, Margolis (1980) argues that cultural entities are something that are emergent but embodied, in some form or other, in material things. Thus works of art are embodied in material things, that is, in physical bodies or marks or movements, but the works themselves are not material things but something more, since they have emergent properties that material entities do not have. They have cultural, in the case of art artistic or aesthetic, properties. He says, furthermore, that these two aspects, emergence and embodiment are the ontological core of cultural entities. What Margolis is mostly up to is to characterize the notion of artwork, but for him the cultural entities are theoretically similar in kind in the sense just mentioned. 'Cultural' means for Margolis any property in virtue of which cultural entities or systems exist apart from their physical existence. Such a system is rule-governed in that it is intelligible only in terms of rule-like regularities, pertaining to traditions, customs, practices, and institutions. Here, of course, 'rule' is to be understood in a broad sense. Cultural properties are representational, symbolic, expressive, intentional, causal, functional, historical, and institutional. Therefore, the relation between cultural entities and physical bodies is not dualistic nor reductive, he says and calls his position 'nonreductive materialism'. It is obvious that his position is similar to certain earlier ideas, as, for example, those of Wittgenstein and Popper.

Margolis (1980) and Danto (1981) compare cultural entities with persons in Strawson's (1959) sense, which does not accept dualism nor reductionism, and so they suggest that Strawson's pattern can be applied to works of art and to other cultural objects. For Strawson, physical bodies are not proper parts of persons nor identical with them. Persons can be distinguished from physical bodies in that different attributes can be ascribed to them; to persons one can ascribe two kinds of predicate, to bodies only one. Thus we can say 'a tall person' and 'a merry person', and 'a tall body', but not, of course, 'a merry body'. What the relationship between persons and bodies really is remains obscure, however, but in any case a person is a holistic thing. An important consequence is, as Margolis remarks, that persons and physical bodies cannot be distinguished from each other by perceptual means.

Apart from the philosophy of culture and art, emergence has become especially important in discussions about the relations between
consciousness and the brain, between different levels of energy and
matter, between life and matter, between different levels of knowledge, and so on. Though these fields are different from each other,
the notions of emergence as used in these fields seem to be, intuitively, similar to each other. Thus consciousness is said to emerge from
brain processes, it is embodied in the brain, and it has properties the
brain does not have as a material entity, and similarly for the rest of
the pairs indicated.

However, unlike Margolis and the other philosophers I mentioned
above, physicists and cognitive scientists may speak about emergence
as something that is compatible with (ontological) reduction.[32] Thus,
for instance, they may say that macroscopic phenomena and bodies,
such as tables and chairs and their relations and properties, emerge in
some reductionist sense from more fine-grained phenomena and bodies, such as elementary particles and forces between them. According
to this reductionist view, the behavior of tables and chairs, or rather
the theory thereof, is reducible to the microscopic phenomena in an
appropriate calculable or algorithmic manner, which is like the classical kind of reduction. It is well known that scientists may even say
that there is a reduction in the eliminative sense between psychological theories of the mind and theories of the brain, but whether this is
thought of as being compatible with emergence is not clear to me. As
we saw, such philosophers as Margolis often argue that reduction and
emergence exclude each other, but it seems to be the classical notion
of reduction they mean, though it is usually unclear what they mean
by reduction. In view of this controversy, it is necessary to study
more closely the relationship.

On the other hand, there are well-known views of the notion of
emergence that compare it with reduction, prediction, and similar concepts of the philosophy of science. Hempel (1965) criticizes the simple view that a phenomenon or property is emergent if it is not explainable or its occurrence predictable. This view concerns, in particular, properties of a system as a whole that are not explainable or predictable from the information about its constituent parts. A standard
example presented here is the claim that such properties of water − in
the well-known circumstances − as transparency and liquidity are

emergent since they are not predictable from the information about oxygen and hydrogen. Hempel argues that this notion of emergence must be relativized in three respects to be nonvacuous. Firstly, one should specify which of the several meanings of 'part' is referred to. Secondly, it is trivial to say that a statement about an occurrence of a property or phenomenon of the whole system cannot be inferred from properties of constituent parts unless it is specified what the properties are that are referred to in the premises of the inference characterizing the constituent parts. That is, emergence is relative to some properties. If any properties are accepted, then one may claim, for example, that the transparency of water can be trivially inferred from the property of hydrogen indicating that when suitably combined with oxygen, a transparent liquid results. Thirdly, whether a property of the whole system can be predicted or explained depends on theories or laws that are available for the task, that is, emergence is relative to a theory about the parts of the system.

A property or phenomenon may in this sense be emergent relative to the theories available today, but become explainable or predictable in the future. Hempel quite correctly notes, however, that the fact that some biological and psychological phenomena are emergent relative to physico-chemical theories in the above sense is trivial since their descriptions may contain terms that are not contained in the languages of physics and chemistry. Therefore the descriptions cannot be inferred from the latter theories. To make this question of emergence less trivial, one must require that the alleged explanatory theories contain 'presumptive laws' that connect the languages of the two levels in question. It is obvious that this requirement is similar to Nagel's condition of connectability.

Hempel's conception of emergence is embarrassing in that its definition is stated in negative terms, saying that emergence means nonpredictability or nonexplainability, and therefore it cannot help much if one wants to understand what it really means that something emerges, that is, what its theoretical structure might be, if there is any. More should be said about pragmatic constraints and adequacy conditions restricting, e.g., the kinds of theories to which emergence is assumed to be related. It seems that Margolis' theory of cultural entities, being illuminating and stated in positive terms, does a little better intuitively, but it also leaves the structure open. Both approaches maintain the view that emergence and reduction exclude each other. Let us next briefly investigate Nagel's view to see if it would be of any help.

Nagel also emphasizes nonpredictability as a characteristic of emergence and its relative nature. A property of a system is emergent relative to some theory if an appropriate statement describing the property cannot be deduced from the theory and relevant statements describing the properties of its constituent parts. Thus, for example, the statement 'Water is transparent' cannot be deduced from any statements about hydrogen and oxygen that do not contain the terms 'water' and 'transparent', and therefore transparency is emergent relative to such statements concerning oxygen and hydrogen. In this sense a property can be emergent relative to one theory but nonemergent relative to another.

This agrees with what we saw Hempel saying, but what Nagel argues about the relationship between statistical mechanics and thermodynamics is very telling. He draws attention to his arguments concerning reduction — e.g., to the reduction of thermodynamics to statistical mechanics — and claims that a certain gas law of thermodynamics involving the term 'temperature' can only be deduced from statements of statistical mechanics if one adds a postulate that relates this term to the language of the latter discipline, namely to 'mean kinetic energy of molecules'. It cannot be deduced from the assumptions of statistical mechanics alone since they do not contain the term 'temperature'. Nagel argues that such an *independent* postulate must be added to statistical mechanics to deduce the law, and this fact "... illustrates what is perhaps the central thesis in the doctrine of emergence as we have been interpreting it."[33]

If I correctly understand what he argues here, he seems to be saying that a certain property described by the gas law is emergent relative to statistical mechanics — where only statements containing mechanical terms are accepted — and yet the law is reducible to statistical mechanics in his sense of reduction.[34] If made exact, the independent assumption linking the two terms is either of the form of explicit definition or conditional definition, depending on the formulation. It follows, as we have seen earlier, in Section 5.9, that the alleged reduction would imply that the law is in a correspondence relation to statistical mechanics.

Thus we may observe that for Nagel emergence and reduction are not incompatible notions, and this is because a reduction must satisfy what he calls the condition of connectability requiring that there be an assumption as a premise that links appropriate terms of the two scientific languages. It is not clear to me how crucial it is to say that the linking assumption is independent and what independence is assumed to mean here. Nagel himself is unable to decide whether the assumption in the specific case of thermodynamics must be thought of as a convention or else as a physical hypothesis having indirect physical evidence (that is, in contexts of other physical theories) for the relation of the meanings of 'temperature' and 'mean kinetic energy of molecules'. This distinction is roughly analogous to the distinction often made between nominal and real definitions, to be discussed in the Appendix, below. The latter possibility is, however, problematic if meanings are theory-relative, and the former because what seem to be pure conventions are often results of some intuitive or empirical analysis of ideas, rather than *ad hoc*.[35] It seems, however, that it is sufficient for the purposes of the theory of emergence to say that independence means that the connecting link is independent of the primary, reducing theory – in the present case, statistical mechanics – in the sense, to say the least, that it does not follow from it logically or somehow by means of empirical evidence. In view of the latter characteristic it seems possible to say, if something like Nagel's theory is accepted, that a property that is emergent with respect to a theory may cease to be so when appropriate empirical evidence is found. We return to this question later in this section and in Chapter 7, when considering other examples.

Some physicists, such as Primas (1998), do not accept Nagel's theories of reduction and emergence. One reason is, as he says, that he has not found a single, nontrivial and physically well-founded example of reduction in the sense of Nagel and the other philosophers of science holding similar views of reduction. Philosophers' examples and ideas are simplified, and they are not mathematically or conceptually sufficiently rigorous. However, my main concern here is another argument by Primas. He claims, for example, that the link between thermodynamics and mechanics is not logical, since it is not correct to say that the concept of temperature can be reduced to statistical mechanics with the aid of a mere definition relating it to the mechanical concept of mean kinetic energy. Philosophers who accept that show ". . . an incredible ignorance of the most elementary concepts of physics," says Primas.[36] He justifies this opinion by saying that temperature is not a mechanical or molecular concept, but defined by a fundamental principle of thermodynamics, not derivable from the principles of mechanics.

So it seems that Primas does not accept that translations, at least definitions, linking terms that belong to two different disciplines, and hence indicating meaning changes, can be used in reductions, or that they even establish reductions. But if one takes seriously the conceptual import of interpretation and the role of logic in our extended sense of logic, then — in view of what has been said earlier in this book — the relevance of this argument is somewhat difficult to understand. Primas' doubts concerning translation may, however, help us to see better why many people in such cases of scientific change as the one just discussed are apt to speak about emergence.

It may be of some interest to study Primas' (1998) own approach to emergence since it seems to be related to our notion of limiting case correspondence. Since Primas defines emergence in purely topological terms and does not use nonstandard analysis, it is not easy to see whether this is really the case and why one is intuitively entitled to use there the term 'emergence'. But some special examples — such as the ones where what he calls asymptotic expansions are used — and case studies where limiting cases of theories are considered, suggest the analogy. I shall not study his approach here, but we may in passing notice that it is obvious — considering, for instance, some case studies of the next chapter — that when we go to a limit of a theory or function (in an appropriate sense of limit), "qualitatively new," as Primas says, properties or theories may come up. In some such cases one might be intuitively entitled to speak about emergence.

Let us turn back to Margolis' arguments concerning cultural entities. We may now see better why the idea entertained by some that emergent cultural properties are reducible to physical ones is problematic. If the former are intentional, symbolic, aesthetic, or whatever, it does not seem intuitively speaking appropriate to say so. This is, of course, as problematic as to say that the mind is reducible to the brain. If this reduction is accepted, there might be a roundabout way to defend the thesis of the reducibility or eliminability of cultural properties more generally. It is said sometimes that since such cultural properties are (in a sense) man-made, they can be reducible to something that is going on in the brain in so far as mental phenomena are so reducible. This argument can only be defended if the reducibility of the mind can be defended, but as we know, this is an open question in the sense that no one has yet presented anything like such a reduction. I shall consider this question in Chapter 7.

Recall, however, what Margolis and Danto say about the nature of artworks. Though we are not considering scientific theories now, we may say that if a work and its artistic and aesthetic properties are rule-governed, as Margolis proposes, we can of course speak of a theory or theories on which its properties as an artwork depend. This is what Danto argues when he says that "[t]o interpret a work is to offer a theory as to what the work is about, what it subject is . . . . To interpret Breugel's painting as simply about Icarus involves at best identifying the legs and the relationship between their owners and the sun, implying a narrative structure, a story the painting not so much tells as presupposes in order to integrate the elements."[37] Danto says much more in this book and other writings than the passage indicates, but the central idea is that interpretations and therefore the identities of artworks are dependent on theories of some sort and without interpretations they would be mere material things.

These ideas of the importance of such theories and conventions, concerning, e.g., styles and means of representation, seem to lead to the consequence that in many cases they can be used to reduce – after all! – symbolic properties of works by means of these theories and conventions, by using, furthermore, special assumptions concerning the physical configurations occurring in the work and, possibly, intentions. So it seems that in view of what Margolis and Danto argue it is not too brave to say that in many cases something like Nagel's condition of connectability is at work in interpretation, because interpreting requires assumptions that link certain physical features to certain symbolic and possibly other artistic and aesthetic properties. The story presupposed by *Landscape with the Fall of Icarus* and possible conventions of representation provide assumptions relating the white patch in the picture to the legs of Icarus, and therefore they connect a language talking about physical properties with a language of artistic properties.

It is certainly premature to infer from this observation that there is a correspondence sketch or reduction sketch, or something similar, relating the symbolic properties of this work to its physical properties. One should, among other things, answer the crucial question concerning whether the linking assumptions are independent – in the sense in which we saw Nagel using the word – of possible theories or other statements about physical properties of the picture. The assumptions may not be considered independent if they are based on rules that are justified, for example, by empirical or whatever knowledge that would show how people of the respective culture necessarily interpret such things as the white patch as soon as they are familiar with relevant background stories and theories. In such cases, we could perhaps justifiably, but hesitatingly, claim that the symbolic properties in question are reducible rather than emergent. If this is true, it seems to follow that it is more likely that, for instance, abstract and avantgarde paintings, and the like, would exhibit *emergent* symbolic properties better than classical and familiar ones.

6

INTERTHEORETIC EXPLANATION

In the previous chapter we defined what it means for a secondary theory to be in a correspondence relation to a primary one. We can now investigate whether such a relation is of any explanatory import, that is, whether or in what sense we could say that the former theory can be explained by means of the latter. I shall not make any survey of the complex and intriguing history of the philosophical research of the notion of scientific explanation.[1] Instead, I try to point out that the notion of correspondence relation can be used as a first step toward a type of cognitive comparability which is different from the types discussed earlier in the literature, but which is in many ways congenial with Kuhn's (1970) insights.

I present a kind of explanation that would be applicable to many cases of revolutionary scientific change – where the more conventional kinds of explanation turn out to be inadequate, as we shall see. I call these explanations counterfactual since they involve contrary-to-fact hypotheses and the notion of counterfactual conditional. On appropriate pragmatic and logical conditions such an explanation can be derived if there obtains a counterfactual correspondence between two theories, in the sense defined in the preceding chapter. Some of the pragmatic features involved seem to be close in spirit to what we have seen Kuhn arguing when criticizing earlier views of intertheoretic explanation. However, since I am employing a notion of translation that is more liberal than his later notion, and, furthermore, integrate logical and pragmatic considerations, it seems to me that we are able to go beyond many of his arguments and suggestions. It turns out that by means of such methods it is possible to see some problematic and much discussed intertheoretic relations in a new light. In what follows, I shall discuss problems of explanation mainly in the model-theoretic framework, but in nonformal cases – where we could only have correspondence sketches available – the conclusions we reach below can be understood less formally and modified accordingly.[2]

6.1. THE ROLE OF PRAGMATIC CONSTRAINTS

If one theory, a secondary theory, is to be explained by means of another, primary theory, various social, cognitive, and intentional presuppositions must be satisfied. For example, the nature of an explanation is dependent on the individuals who give the explanation according to some rules and procedures accepted by a relevant community. It also depends the individuals whose understanding, concerning the intertheoretic relation in question, is intended to be increased by the explanation. If the explanation involves extensional,
logical or mathematical, procedures and notions, they must be properly interpreted — to become scientifically meaningful in this explanatory context and to be of some explanatory import. Let us say briefly
that the notion of explanation is *pragmatically constrained*. Any feasible model of intertheoretic explanation has to provide an account of
the relevant pragmatic constraints.[3]

The deductive-nomological model of explanation is intended to be a
theory of ideal, 'objective' explanations that minimizes the role of
extralogical and nonextensional features, i.e., the role of pragmatic
constraints. This feature has been best realized in some noncontroversial applications of the model, but such easy applications are rare
in the history of science. On the other hand, the more problematic its
applicability is, the more prominent seems to be the role of pragmatic
constraints — and the less warranted it is to talk about explaining
theories by means of the model. Consider, for instance, the notion of
approximate deductive explanation (involving no counterfactual special assumptions) as applied to a scientific change where a theory is
superseded or replaced by a new one. It is customary to say that since
the new theory can be approximated by means of the superseded one,
this does not only explain why the latter is approximately true when it
is. It explains why scientists used to hold such a 'false' theory and
why it still is successful as a calculating device. On the other hand, if
the deductive-nomological model is used here, the role of extensional
logic is emphasized, and the model, as it stands, does not involve any
account of the behavior of scientists or of other pragmatic constraints.
Such an account would be external to the model itself.

There have been attempts to apply the notion of deductive explanation, in some form or other, to problematic cases of scientific change, cases which are said to represent scientific revolution, incommensurability, limiting case correspondence, and so forth, but these applications have appeared unsuccessful and extremely controversial. It is above all the extensional, paradigm independent character of the model of deductive explanation that has been problematic in such applications, since, if Kuhn (1962) is correct, such scientific changes are socially loaded: they cannot be described without giving an account of paradigmatic factors. What the relevant pragmatic constraints would exactly be and how they should be formulated in the models of intertheoretic explanation have not been very clearly stated in the literature. In the rest of the present chapter, some constraints are discussed, constraints that arise in the context of explanation derivable from applications of the correspondence relation.

6.2. CORRESPONDENCE AND EXPLANATION

In Chapter 2, we observed that in a local translation the respective semantic transformation, considered as the conjunction of a minimizing transformation and correlation, is of explanatory import if it helps a hearer understand how a speaker's situations can be reached from actual situations. It is associated with interpretation and its principles, whose purpose is to cope with the speaker's meaning, and its role is to show in what way his utterance is wrong if it is wrong (by the hearer's lights). Particularly the import of the minimization principle was seen to be emphasized in cases where the hearer has to imagine counterfactual situations in order to approach the speaker's situations. If the translation is corrective, the transformation might tacitly deal with explanatory questions concerning the speaker's intentions, such as 'Why the speaker uttered this sentence?', or questions seeking conditions which would make the utterance correct or understandable, such as 'How the actual situation should be changed to make the utterance true?'. This theory of local, explanatory translation can be generalized so as to apply in the global context of correspondence relation. For example, the generalization can be seen as providing similar ways of explaining the position of the scientific community holding an out-of-date theory from a standpoint of the community holding the more up-to-date theory.

In the model-theoretic framework an additional and exact result can be derived from the above conditions for correspondence. It shows that on appropriate logical and pragmatic conditions correspondence is related to deductive-nomological explanation. However, this kind of explanation is not my main concern here. As I already mentioned, I am interested in counterfactual explanation, a new kind of explanation that can be associated with the notion of counterfactual correspondence. If there is only a correspondence sketch of one (nonformal) theory to another, in the sense discussed in Section 5.1, then the results to be obtained below about the relation of correspondence and explanation must be paraphrased in a more sketchy way, but the main ideas remain the same.

Assume $\langle F, I \rangle$ is a correspondence of a theory T to a theory T', relative to $\langle L, L' \rangle$, and let the class H be defined by a sentence θ, H' by θ', and K' by sentences θ', σ_0, ..., σ_n. Let us consider again argument (5.9.2), discussed in Chapter 5:

(6.2.1) θ', σ_0, ..., σ_n / $I(\theta)$.

As shown here, if there is a correspondence of T to T', then the latter theory in conjunction with the auxiliary assumptions logically implies the translation of the (axiom of the) former theory. This can be generalized as follows. Let ψ be an $L(\tau)$-sentence that is true in every model belonging to the class K, the range of F.[4] Suppose that φ is an $L(\tau)$-sentence such that

(6.2.2) θ, ψ / φ,

where '/' is now a shorthand for deducibility or logical consequence in L. Then it is easy to see that (in L')

(6.2.3) θ', σ_0, ..., σ_n / $I(\varphi)$.

Now one can say, for instance, that if T explains φ in the D-N sense of explanation (and ψ satisfies the above constraint[5]) and appropriate pragmatic conditions (concerning σ_0, ..., σ_n and I) are satisfied, T' explains the translation of φ. Then, if there is a correspondence relation of one theory to another, explanations that the former theory provides can be transformed into explanations in terms of the latter (if appropriate conditions hold).[6] In what follows, explanation will be discussed in terms of (6.2.1), rather than in terms of this generalization, which, however, will be referred to in the next chapter.

In order to discuss these matters conveniently, I may occasionally use the following terminology in the rest of this section. Given a theory T, a *transformed theory* $I(T)$ is a theory whose axiom is represented by a translation of the form $I(\theta)$, as indicated above.[7] Furthermore, I shall say that to explain a theory is to explain its axiom, and I may also speak about correspondence between laws or axioms in the same sense as between theories.

Can an argument of the form (6.2.1) be of any explanatory consequence? Such an argument may qualify as a covering-law explanation of a transformed theory $I(T)$, in the sense of some of the deductive models proposed in the literature, if it satisfies appropriate theoretical and pragmatic conditions. It is not, strictly speaking, such an explanation of the theory T tself. But if the translation I is of explanatory importance, (6.2.1) provides a roundabout kind of explanation of the theory. Thus, on appropriate conditions (6.2.1) can be conceived as an explanation of the theory in an indirect sense since it provides an explanation of the transformed theory. Some of the conditions concern the translation. For instance, it must be workable and instructive and the result $I(\theta)$ must be empirically and conceptually meaningful. Then, *if* the requisite conditions of adequacy are satisfied, we may say that an indirect explanation of T by T' consists of the following steps:

(6.2.4)   (i) Interpreting T in the conceptual framework of T' by using the translation I;

(ii) Explaining the transformed theory $I(T)$ by means of a covering-law explanation of the form (6.2.1).

We have seen, however, that semantic transformations may also be of some explanatory importance, especially when considered together with respective translations. Therefore, we can further improve the concept of explanation by keeping an eye on the correspondence relation as a whole, and thus also on the notions of correlation and minimizing transformations (whenever functioning).[8] Given two theories T and T', and their respective axioms θ and θ', T may be explainable by T' by means of the following, more complex process, which now replaces (6.2.4):

(6.2.5) (i) Constructing a correspondence $\langle F, I \rangle$ of T to T' (relative to an appropriate pair of logics $\langle L, L' \rangle$);

(ii) Explaining the transformed theory $I(\mathsf{T})$ by means of a covering-law explanation of the form (6.2.1).

Even though the minimizing transformation is not explicitly mentioned in (6.2.5), its role is also important here to the extent to which it has been used to determine the class K', the domain of F.

It seems that all classical, syntactic approaches to intertheoretic explanation ignore the following general features that are crucial in the present notion of explanation. First, since the class of models of T, K, into which models of T' are transformed by the correlation F, does not in general contain all possible applications of the theory T but is only a subclass of H, we have good reasons to say that T' does not explain T *per se*. It will become evident from case studies in Chapter 7 that the class K can usually be considered as representing the successful part of T, that is, the phenomena to which the theory has been most successfully applied. Second, when a law is explained by subsuming it under a more general law, the former is looked at from a point of view of the latter. The earlier covering-law approaches to reduction, discussed in Chapter 1, above, which have been criticized by Kuhn and others, do not acknowledge this feature in its full generality. For instance, suggestions made by Nagel concerning how the terms of a secondary theory should be connected with those of a primary one are not general enough to really take care of the most intricate scientific changes, specifically those that involve idealizing or limiting or other counterfactual assumptions. Such cases will be studied in the rest of this section and in the next two chapters.

6.3. GLYMOUR ON COUNTERFACTUAL EXPLANATION

Glymour (1970) suggests that intertheoretic explanation is an exercise in the presentation of counterfactuals rather than Nagelian reduction.[9] One does not explain a theory from another by showing why it is true. Instead of why-questions one has to ask quite different questions, since often counterfactual special assumptions and limiting procedures are needed in the transition from a primary to secondary theory. Therefore, one has to ask: 'Under what conditions would the secondary theory hold?'. According to Glymour, the theory is then explained by means of the following steps:

(6.3.1) (a) Showing under what conditions it would hold;

(b) Contrasting those conditions with the conditions that actually obtain.

Notice, however, that Glymour has a deductive picture of explanation in mind here. His idea is, roughly, that the primary theory together with additional, special assumptions, some of which are counterfactual, and with definitions connecting the languages of the two theories, and possibly with limiting procedures, entails the secondary theory. Such an inference does the job of (6.3.1a). (6.3.1b) is accomplished by showing that without the counterfactual conditions the secondary theory is not in general entailed by the primary one.

Let us consider Glymour's idea more closely. I restrict myself to (6.3.1a). (6.3.1b) yields an answer of the form 'Since such-and-such conditions do not hold' to the question 'Why does the theory not hold?', and that may increase our understanding of the theory, but what its explanatory import is is not very clear. Assuming, for simplicity, that only one counterfactual special assumption, say σ, is needed, (6.3.1a) amounts to showing that a conditional of the following form is true:

(6.3.2) If it were the case that σ, then it would be the case that θ,

where θ is the law of the secondary theory, as in the preceding section. Hence, what Glymour seems to suggest is that by means of a relevant deduction, which is formally in conformity with the deductive-nomological model, one can establish (6.3.2). If this is correct, it means that Bonevac (1982) is not completely accurate when he argues that in the formal sense counterfactual assumptions are the only features that distinguish Glymour's analysis from a straightforwardly deductive account of explanation. On the other hand, Glymour's suggestion is not warranted, either. As we know from the analysis of counterfactual conditionals, the truth of such a conditional is established by a deduction only on certain conditions concerning the members of that deduction.[10] In the next section, Glymour's proposal is elaborated and improved in terms of correspondence.[11]

6.4. COUNTERFACTUAL EXPLANATION AND CORRESPONDENCE

As before, assume that $\langle F, I \rangle$ is a correspondence of T to T', and θ and θ' are the axioms of T and T', respectively. Let $\sigma_0, \ldots, \sigma_n$ be special assumptions, and let K and K' be the respective classes of models as specified in the definition of correspondence, in Section 5.3. If some of the special assumptions are counterfactual (contrary-to-fact) in the sense explained in Section 5.5, $\langle F, I \rangle$ is a counterfactual correspondence. Assume that σ_0, which will be more briefly denoted by σ, is the only counterfactual assumption. It is then the single counterfactual sentence among the sentences that define the class K'. If, for example, $\langle F, I \rangle$ is a limiting case correspondence, σ expresses formally the respective limit condition, such as, for instance, 'the velocity of light is infinite', and, as we stipulated in Section 5.5, $I(\theta)$ will be read as 'θ almost holds'.[12] The special assumption σ can of course be considered possible in a relevant sense even if it is not realizable in any actual models. It follows from (6.2.1) that in the case of limiting case correspondence θ almost holds in the models where θ' plus the special conditions hold. It does not follow that θ would almost hold in the actual models. Similar considerations concern other types of counterfactuality.

Since a counterfactual correspondence involves a contrary-to-fact assumption, its relation to scientific explanation is not obvious, and we even have to ask whether it is of any explanatory importance in the traditional sense of explanation. If argument (6.2.1) satisfies appropriate formal and pragmatic conditions of adequacy, we might perhaps say that it provides a deductive-nomological explanation of $I(\theta)$. But what would then be a relevant question to which this explanation would provide an answer? If we again consider a limiting case correspondence as an example, it is obvious that we cannot simply ask: 'Why does θ almost hold?' since θ does not almost hold in any actual models; and even if we could, (6.2.1) would not provide any adequate answer because one of the special assumptions is counterfactual. Rather, a relevant why-question would be in such a case: 'Why does θ almost hold in (possible) models where it does so?.' But since one of the special assumptions is contrary-to-fact, of what explanatory value would it be to correctly answer that θ almost holds in those models *because* θ' holds generally and the special assumptions hold in the same models and they logically imply $I(\theta)$? The only kind of explanatory import one can imagine an argument like (6.2.1) to yield, in general, is conceptual: By studying the special conditions and the translation $I(\theta)$ one learns something about the conceptual relation of the two laws. But if one examines the correspondence relation $\langle F, I \rangle$ itself, instead of its consequence (6.2.1), one in fact learns more about it. Therefore, we must preserve (i) but delete (ii) of (6.2.5).

Even if (6.2.5) would provide an instance of deductive-nomological explanation when σ is considered false – which I now think is not the case – its explanatory role *as* such an explanation would be very weak.[13] Therefore we have to shift the emphasis from explanation to understanding, and a natural way to do so is to elaborate Glymour's proposal, discussed in the preceding section, concerning the explanatory import of counterfactual conditions. To embed Glymour's idea in the framework of counterfactual correspondence, the following, small but essential, correction must be made in his model in the first place. Instead of including possible translations in the auxiliary assumptions, the conclusion θ is replaced by its translation $I(\theta)$, interpreted in the instances of limiting case correspondence as 'it is almost the case that θ'. Instead of conditional (6.3.2) proposed by Glymour, which expresses the condition under which the law θ would hold, we must then consider the following conditional:

(6.4.1)   If it were the case that σ, then it would be the case that $I(\theta)$,

which expresses the condition under which the translation of θ would hold. We write (6.4.1) more formally as follows:

(6.4.2) $\sigma \,\square\!\!\rightarrow I(\theta)$.

Consider again argument (6.2.1) and write it into the form

(6.4.3) θ', $\sigma_1, \ldots, \sigma_n, \sigma \,/\, I(\theta)$.

If what Glymour argues is corrected in this way, we can interpret him as essentially maintaining that an argument of the form (6.4.3) *makes* the conditional of the form (6.4.2) *true*. However, whether it makes it true is not obvious.[14] In other words, contrary to what Glymour suggests, the truth of the conditional does not follow from the validity of the argument. This means that their relationship must be studied more closely, and then we see that investigating the conditions under which the latter establishes the truth of the former provides additional pragmatic constraints in the sense discussed in Section 6.1, above.

We employ here Lewis' (1973) semantics of counterfactual conditionals, to be surveyed in Section 8.4. As to the general import of nonextensionality and pragmatic constraints of explanation in principle, it would not make much difference if another (nonformal and intensional) semantics were employed here. The question of whether a counterfactual conditional is true embraces, in any case, pragmatic considerations of the kind that are foreign to deductive explanation and to Glymour's model as well.

We shall assume from now on in this section that we consider a fixed possible world of a fixed Lewis model to which all semantic notions we need are relativized. Hence we may omit reference to the world and to the Lewis model. The fixed world is now an actual or intended model of T', but the exact nature of the Lewis model need not be characterized for our purposes here, which are philosophical and pragmatic rather than purely logical. How it should be exactly defined is subordinate to philosophical and pragmatic conditions, which I shall discuss later. According to Lewis' semantics, counterfactual (6.4.2) is *trivially* true if the antecedent σ is not possible, in which case the conditional is of no explanatory importance. If σ is possible, the conditional is true if there is a valid argument of an appropriate form which makes it true, that is, which is *backing* the conditional. But we are not in fact asking here whether there exists an argument that makes it true, or whether it is true on some other truth condition, but whether (6.4.3) itself makes it true. Thus we ask whether this very argument is backing the conditional, which is the case if each of the premises θ', σ_1, . . . , σ_n is *cotenable* with the antecedent σ. Therefore, we must ultimately study the question of whether the following holds:

(6.4.4)   The premises θ', σ_1, . . . , σ_n are cotenable with σ.

Whether (6.4.4) holds is relative to the notions of possibility and necessity and to the notion of similarity of possible worlds; that is, the answer depends on how these notions are construed and how the actual world is chosen in this context. In particular, if a premise is necessary, it is trivially cotenable with the antecedent σ. Though the antecedent is assumed to be contrary-to-fact, that is, false at the actual world chosen, it nevertheless is compatible with the premises θ', σ_1, ..., σ_n if they are cotenable with it, no matter how 'implausible' it may look, if only possible. Recall, furthermore, that if σ is impossible, then conditional (6.4.2) is trivially true, but the specific argument (6.4.3) makes it true only if θ', σ_1, ..., σ_n are all considered necessary, which, however, is never the case when actual theories are at stake.

In the light of Lewis' theory it is thus apparent, contrary to what Glymour seems to suggest, that an argument of the form (6.4.3) does not automatically make the respective conditional (6.4.2) true.[15] Instead, as it is easy to see now, investigating whether the conditions hold under which the truth of the conditional follows from the validity of the argument is also of explanatory importance. Therefore, the following more complex model of counterfactual explanation emerges when Glymour's model is improved by exploiting the model-theoretic notion of counterfactual correspondence and the Lewis type of semantics. A *counterfactual explanation* of T by T', in our new sense, consists of the following steps:

(6.4.5)   (i) Constructing a counterfactual correspondence $\langle F, I \rangle$ of T to T';

   (ii) Deriving an argument of the form (6.4.3) from $\langle F, I \rangle$;

   (iii) Showing that the resulting argument makes counterfactual (6.4.2) true.

The third step is relative to a given Lewis model and the actual world in it, that is, it is relative to context, especially because cotenability and backing depend on similarity and possibility. The first two steps involve the logics in which the theories are represented. Therefore, at least the following *pragmatic constraints* are embedded in this model of counterfactual explanation, either explicitly or implicitly:

(6.4.6)   (a) $\langle F, I \rangle$ must be informative and constructible under the guidance of what is known about the laws, paradigms, and logics involved; in particular, the respective syntactic and semantic transformations must be of explanatory relevance;

(b) Explanation is given from a point of view of an appropriately chosen Lewis model and world;

(c) σ must be possible and $\sigma_1, \ldots, \sigma_n$ contingent at the world chosen;

(d) θ', $\sigma_1, \ldots, \sigma_n$ must be cotenable with σ, hence also true, at the world.

Constraints (6.4.6b-d) mean that the explanation is only good relative to a properly chosen context. (c)-(d) are relative to the individual or scientific community who or which is giving the explanation and to the adopted scientific principles. (c) excludes trivial cases, as we saw above, and thus explanatorily empty counterfactuals. If the explainer holds the primary theory T' rather than T — which seems likely in case T' has superseded T — then those scientific principles are determined by T' and by the respective paradigmatic assumptions, rather than T and its paradigm. Thus it seems likely that the law of the primary theory, θ', is considered cotenable.[16] If it is, then the role of special assumptions becomes more conspicuous. In any case, there are no objective notions of possibility, cotenability, and truth that would be available here. Constraint (a) is intended to guarantee that the translation is explanatory in the sense discussed earlier in this chapter and Chapter 2; not very much is explained by the mere knowledge of the existence of the correspondence $\langle F, I \rangle$.

The most striking conclusion seems to be that an explanation of the above kind can hardly be objective, not at least as objective as deductive-nomological explanations are assumed to be. The pragmatic constraints show that one has to look at the secondary theory T, or rather its transformation $I(T)$, from the point of view of the new paradigm, and this is what we saw Kuhn arguing. Furthermore, there may even be individual disagreement, concerning the question of what counterfactual conditions can be considered possible. For example, when the limiting correspondence of classical particle mechanics to relativistic particle mechanics is studied, one has to consider the question whether it is in some sense possible that the velocity of light is infinite. It is known that there are philosophers and physicists according to whom this assumption is so much contrary-to-fact that it cannot even be considered possible, and others who consider it possible. For the former, the intertheoretic relation in question is of no explanatory value, not at least in the sense discussed here. There are interpretative elements involved here, and also methodological relativity is present, because, as we have seen, an explanation of the above kind is dependent on the methodological and logical tools employed.

6.5. TWO EXAMPLES

Example 1. In Chapters 7 and 9, below, some counterfactual correspondence relations between actual scientific theories are worked out. In the present section I shall only illustrate step (iii) of schema (6.4.5) by a simple example of counterfactual explanation. Let us consider two minitheories, T and T', whose laws are, respectively, the classical law θ saying that mass is independent of velocity and the corresponding relativistic law θ'. Here I present them informally (i.e., with no formalization) and discuss the correspondence informally by assuming that the relevant parts of the discussion and the relevant mathematical expressions can be formalized:

θ: $m = m_0$;

θ': $m = m_0(1-v^2/c^2)^{-1/2}$ $(v < c)$.

It is clear that mathematically the law θ is a limiting case of θ', i.e., in the limit when c approaches infinity the latter becomes the former. But one can even show that there is a limiting case correspondence $\langle F, I \rangle$ of T to T' in the formal sense of Section 5.3, if the theories are appropriately reconstructed. Instead of showing in detail that there is such a correspondence,[17] I shall make some informal remarks on the resulting argument of the form (6.4.3).

The following informal notation will be used here.[18] If an entity a is infinitesimally close to b, denote '$a \approx b$'. Write '$a = \infty$' or '$a < \infty$' according to whether a is a positive infinite or nonnegative finite real number.

The sentences in the argument of the form (6.4.3), which is yielded by the correspondence $\langle F, I \rangle$, are θ' and the following (in a nonformalized form):

σ: $c = \infty$;

σ_1: $v < \infty$;

$\sigma_2, \ldots, \sigma_n$: required theorems of analysis, etc.;

$I(\theta)$: $m \approx m_0$.

Assume that the actual world to which truth, possibility, cotenability, and backing are related is a standard model of T', in the sense explained earlier, or even an intended model of T'; wherefore in that model, θ', $\sigma_1, \ldots, \sigma_n$ are true and $c < \infty$. It is a world which is in agreement with relativity theory, whence it might be considered as a piece of our real world – if relativity theory is considered true. We ask whether the sentences θ', $\sigma_1, \ldots, \sigma_n$ are cotenable with σ (at the world in question), and hence, whether argument (6.4.3) makes the conditional

(6.5.1) $c = \infty \;\square\!\!\longrightarrow\; m \approx m_0$

true (at the world). If they are, we can say with good reason something like this:

(6.5.2) If the velocity of light were infinite, then mass would almost be velocity-independent,

where 'almost' means infinitesimal accuracy.

If the antecedent is considered impossible, the conditional is trivially true. If one thinks that it is not possible in any important conceptual sense, that is, in any sense that would be theoretically relevant, then one presumably holds that schema (6.4.5), or, more loosely, the fact that θ is a limiting case of θ', is of no explanatory value. Usually, however, it is thought that the assumption that the velocity of light is infinite (or approahes infinity), even though contrary-to-fact, provides an explanatory link between classical and relativistic theories, thus being conceptually relevant. Let us assume, then, that the antecedent is possible. If accepted laws tend to be cotenable, as Lewis (1973) argues, then it is justifiable to consider θ' cotenable with the antecedent. If θ' is true at a world u, then no world at which it is not true can be more similar to the actual world than u is.

The view that accepted laws tend to be cotenable ascribes a special importance to them, and it emphasizes the fact that T is being explained *from* T'. One way to argue against the view would be to claim that appropriate structural similarity criteria would provide more objective notions of cotenability, that is, criteria which more directly pertain to the mathematical structure of worlds. It is not quite evident, however, in what sense the result would be more objective if the structure of the actual world is dependent on accepted theories.

The assumptions $\sigma_1, \ldots, \sigma_n$ are also more or less like laws, physical or mathematical, whence they can be considered cotenable on similar grounds as θ'. If they are not laws, they are at least important truths; but this is, however, due to the fact that the present example is simple enough. In many of the forthcoming case studies, some of the assumptions needed are conditions which are more special in nature. Thus it seems that we are, with some reservations, entitled to maintain that the above counterfactual conditional is true. We can also see that an investigation of the conditions under which its truth follows from the argument of the form (6.4.3) is of explanatory importance since it pertains to many questions of adequacy and presuppositions involved.

Example 2. To further illustrate the explanatory role of step (iii) in (6.4.5), consider the same theories T and T', with the axioms θ, θ', but with a different correspondence. Write now '$a \geq b$' if $a - b$ is a finite positive, but not infinitesimal, real number or zero. Let us assume, just for the sake of the illustration, that there is a correspondence $\langle F, I \rangle$ of T to T' such that $\sigma, \sigma_2, \ldots, \sigma_n$ are as above, but σ_1 and $I(\theta)$ are as follows:

$\sigma_1:$ $v/c \geq 0$;

$I(\theta):$ $m \geq m_0.$.

Then it is likely that steps (i) and (ii) of (6.4.5) can be taken, but not (iii). It seems obvious that σ_1 cannot be considered cotenable with the antecedent σ since it would not be true at the actul world if σ were (for v is finite in the actual world).[19] Therefore, this example illuminates the negative fact, which we emphasized above with reference to Glymour's proposal, that the truth of a counterfactual cannot be inferred from a seemingly relevant deductive inference without any further consideration. On the other hand, though argument (6.4.3) is valid, it does not seem to provide any adequate explanation for the relation of the effective mass and the rest mass, as indicated by $I(\theta)$.

This follows from the fact the special assumptions σ and σ_1 are not adequate.[20] Thus this example seems to sustain my view that step (iii), if successful, provides an adequacy condition for step (ii), that is, for the respective 'deductive explanation'.

6.6. EXPLAINING AND UNDERSTANDING SCIENTIFIC CHANGE

We have seen that the familiar notions of explanation are not directly applicable to the most interesting cases of scientific change since the latter involve counterfactuals. It does not provide a model of intertheoretic explanation in cases where the theories in question are, in the Kuhnian sense, incommensurable. Therefore, we attempted to define a model of explanation that would apply to theories between which a relation of counterfactual correspondence obtains.

It appeared then that this model cannot be defined in an extensional framework, not even the aspect of the model that is obtainable from Glymour's (1970) suggestion – contrary to what both Glymour and Bonevac (1982) seem to maintain when they argue that Glymour's model is derivational. Its intensional features and pragmatic conditions bring in elements that are not objective in any usual sense of the word. Whether the counterfactual model can be successfully employed in a given case is in part a matter of interpretation. Furthermore, the kinds of question to which the counterfactual model may provide answers are different from those explanation-seeking questions to which we are accustomed. This being the case, is it, after all, justified to call a counterfactual explanation an *explanation* of something? This is what we already asked in the above. There may not be any clear-cut answer, but according to Scriven (1962), for instance, it might.[21] For Scriven, an explanation is something that ". . . fills in a particular gap in the understanding of the person or people to whom the explanation is directed." It is a description rather than an argument, and it may even be nonlinguistic, Scriven writes. More recently, Glynn (1993) argues that the thoroughgoing dichotomy between explanation and understanding cannot be maintained. To explain something is to interpret it and to make it understandable – from which it follows that even in the natural and mathematical sciences explaining involves interpretative or hermeneutic methods.

Whether or not a given counterfactual explanation is an explanation proper, it is rather evident that it may increase one's understanding of the scientific change to which it is applied. This increase is many-sided. Some aspects of it have been discussed above, but let us consider now another aspect, its holistic nature. The counterfactual model of explanation represents – in an exact manner – the complexity that is often characteristic of the conceptual relation of laws. Take, for example, the much-debated relation between Newton's second law and Minkowski's force law (which is discussed in more detail in the next two chapters). The mere facts that the former gives good approximations of the latter with respect to particles whose velocities are sufficiently low and that the former is a limiting case of the latter if the 'velocity of light' approaches infinity (and certain other conditions are satisfied) do not increase our knowledge of the complex relationship between the meanings of the two laws, since those facts as such are purely computational.

If, instead, the counterfactual model is applied, we learn how intimately syntactic, semantic, and contextual aspects of the relation are hanging together in a process that is hermeneutic.[22] It is obvious that a counterfactual explanation is a hermeneutic enterprise in a proper sense of the word, since it consists of several steps, each of which adds to our understanding of the subject matter and, furthermore, some of those steps clearly involve pragmatic and intentional processes. As we saw above, two of the steps are complicated processes. The first step, that of establishing a correspondence relation between two theories, may not be easily understood logically,[23] and, therefore, one is engaged in painstaking investigations of logical, methodological, and even substantial foundations of the theories. Furthermore, the enterprise is guided by the principles of interpretation that we have discussed in earlier sections. For example, we assumed that the investigator (i.e., the hearer) looks at the secondary theory from a point of view of the primary one and its paradigm. So it follows that the former theory must be understood in the light of the latter. Therefore, the Davidsonian principle is effective in the first place, enabling the investigator to start the interpretation. The roles of the minimization and refinement principles are more clear-cut in this context, and they were discussed earlier. The second step is a direct consequence of the existence of a correspondence relation, but the third one is again involved, as we saw from the examples above. Its complexity is not due to logical difficulties in the same sense as that of the first step, but is, rather, a consequence of one's uncertainty concerning interpretational and pragmatic matters.[24]

The positive role that translation plays in applications exemplifies the holistic nature of intertheoretic relations. The fact that translation plays such a positive role is somewhat contrary to what we have seen Kuhn maintaining.[25] The point is, however, that the present notion of translation is different from Kuhn's (later) notion. It is the notion that we called in Chapter 2 explanatory, even corrective, translation whose criteria of adequacy are different from Kuhn's and whose special case Kuhn's strict notion of translation is.[26] Detailed case studies below seem to call into question the adequacy of some of Kuhn's arguments concerning the relation of translation and learning.[27] Kuhn correctly claims that, for instance, when Newtonian concepts of mass and force are learned, they must be learned together. Newton's second law plays a role in their acquisition, that is, without recourse to the law one cannot learn to understand Newtonian mass and force. But then Kuhn draws the conclusion that this holistic nature of learning makes it impossible to translate Newtonian 'mass' and 'force' into the language of another physical theory in which Newton's second law is not valid, as, for instance, into the language of relativistic particle mechanics. Since they must be learned together, they cannot be translated individually.

However, in so far as Kuhn requires that a translation preserves meaning or extension – as we have seen him requiring – it does not make any difference whether or not learning is holistic, since in either case such a translation of 'mass' and 'force' is not possible. More importantly, a case study of the relation of these laws will make it clear that the holistic nature of how we understand mass and force, whether Newtonian or relativistic, is a presupposition of a corrective translation, rather than an obstacle. As we shall see, a corrective translation (needed for a correspondence relation) of the Newtonian language into the relativistic one is guided by a holistic understanding of the relation of Newton's and Minkowski's laws. As our condition (5.3.1), Section 5.3, shows, a translation in this sense presupposes that certain syntactic and semantic requirements fit into each other, but in a sense that is less strict than what Kuhn demands. When the condition is applied to the case under discussion, those requirements are determined by the force laws in the first place, in addition to the methodological and logical frameworks employed.

Why Kuhn thinks that the holistic nature of learning makes a translation impracticable can perhaps be understood in the light of his strict notion of translation. As we have seen, Kuhn requires that a translation should preserve reference, intension, and the structure of the world. We know, however, that this requirement is too rigorous, even in principle, and, as noticed by Pearce (1987), a translation that would not bring out the conceptual disparities of the central terms of the two theories would reproduce the syntactical inconsistency between the theories. Therefore, it seems evident that Kuhn's notion of translation is fruitless, at least in the sense in which the present one is not, since it cannot be positively applied to increase our understanding of intertheoretic relations.

6.7. UNDERSTANDING AND THE CONTEXT OF DISCOVERY

In the preceding section the understanding of a scientific change was discussed from a point of view of the primary theory, say T', and its paradigm – that is, the context of justification was emphasized – but let us now tentatively touch on the change from a point of view of the secondary theory, T. Scientists are historically familiar with that theory in the first place, that is, T exists first and T' is discovered later.[28] Thus to comprehend the nature of the change in its historical context, in the context of discovery, one has to look at T' from the perspective of T and its paradigm. Both perspectives are needed in order to better understand the nature of scientific change – and to appreciate philosophers' disagreement about its conceptual nature.[29]

The models of intertheoretic explanation and reduction discussed in the literature are usually applied so as to provide a way of understanding earlier or less advanced theories or laws in the light of later or more advanced ones. Thus it is often asked whether relativistic particle mechanics explains classical particle mechanics in some sense and whether it increases or changes our understanding of the latter. But looking at the new theory from the standpoint of the old paradigm may result in a completely different picture of the relation of the two theories. This is the more natural and more original direction of understanding such cases of scientific change where the old theory was the starting point when the new one was invented – cases where the Correspondence Principle was applied as a heuristic device, in the sense discussed in Section 1.1, above.

Consider, for instance, the case of quantum mechanics. The mathematical part of this theory was being developed by using some principles and concepts of classical physics as a starting point, whereby classical terms were carried over to quantum mechanics. It is evident, on the other hand, that Bohr's Complementarity Principle was motivated by the view that quantum-theoretical measurements must be described by means of classical terms despite the fact that quantum phenomena cannot be explained by means of classical physics. Bohr defends this view by arguing that by the word 'experiment' scientists refer to a situation about which they can tell other people what they have done and learned, and therefore experiments and their results must be reported by means of an unambiguous language in which classical terminology is being used. Bohr argues, however, that classical concepts can only be used in a restricted way. Hence one cannot obtain a single complete picture of the phenomenon in question, and one must be satisfied with two complementary pictures. It follows, according to Bohr, that in quantum mechanics there is a tension between the classical and nonclassical terminology.[30]

Similar problems may occur in other fields of inquiry, too. Let us imagine, for example, a radical, paradigmatic change in which some crucial principles of classical logic and mathematics are replaced by other principles, say intuitionistic ones. Then we have to ask to what extent we are capable of learning to carry out our logical and mathematical thinking in the new way and for how long it would take to learn it – or whether a complete change in our thinking is at all possible. Because classical terms are part of the intuitionistic language, there would be a tension, and if the change in question is 'universal' – as it would be in this case – so as to extend to the metatheoretical level, how could one learn to live with the tension?

Since a change in our basic mathematics and logic would also comprise the metatheory, it seems that the accompanying problems would be even more fundamental than in the change from classical to quantum mechanics, or in any changes of physical paradigms. Nevertheless, in physics and other empirical sciences it may often be equally difficult to fully adopt a new way of structuring the world because a classical term suggests a classical visualization. The great number of different quantum-theoretical interpretations is evidently one consequence of the difficulty. Since old terms, such as 'wave', 'particle', and 'mass', must be employed when scientific knowledge is improved or extended to a new area, these terms are then used metaphorically, and the manner in which they are understood is based on the manner in which they were understood classically. Thus there is some tension as always in cases of metaphor – at least if the tension theory of metaphors is correct – but to what extent this tension is cognitively fertile in science would deserve a closer study.

6.8. EXPLANATION AND REDUCTION IN MATHEMATICS

Do similar problems occur with respect to theories of pure mathematics, i.e., mathematics considered a nonempirical discipline? Reasons to be interested in reduction and explanation are partly the same in metamathematics as in metascience. Philosophers of science have been at pains to ask whether scientific change is progressive, and reduction has been one tool for answering such a question. Philosophers of mathematics have been interested in the problem of progress, but it seems, on the other hand, that they have not extensively studied the role of explanation in mathematical developments.

A prominent exception is Kitcher (1983), who distinguishes three types of mathematical explanation that are connected with different kinds of mathematical progress. Firstly, ". . . we can sometimes explain mathematical theorems by recognizing ways in which analogous results would be generated if we modified our language."[31] This is connected with extending mathematical language by generalization. A well-known example is Cantor's generalization of finite arithmetic. Generalizations may explain by showing how a generalized language and theory are obtained within which results analogous to those we have already accepted are forthcoming. From the point of view of the generalization we can see the old theory as its special case, and at the same time the generalization may improve our understanding of the old theory.[32]

The second type of explanation is connected with a clarification of language and of techniques of reasoning, and it is called 'rigorization' by Kitcher. Obviously one of the most dominant features in the historical developments of mathematical practice is that it has become more and more rigorous, and, hence, examples can be easily found, for instance, in the development of analysis. Rigorization can be explanatory in that it removes ". . . previous inability to recognize the fine structure of connections." The third type mentioned by Kitcher is associated with what he calls 'systematization', by which he means such activities as axiomatization and conceptualization, where the latter term refers to a modification of mathematical language ". . . so as to reveal the similarities among results previously viewed as diverse or to show the common character of certain methods of reasoning."[33] Systematization is explanatory in that it yields unification.

Explanations of the mentioned kinds at least satisfy the most important requirement that all explanations have to satisfy, i.e., they increase our understanding. On the other hand, Kitcher does not make it very clear what the explanation-seeking questions would be like that these explanations are assumed to answer. Each of the types is so general that we have to look at their specific applications in order to see the corresponding questions, but it is obvious, however, that instances of generalization, rigorization, and unification mean mathematical progress and, hence, may provide answers to why-questions of some kinds. Similar patterns of explanation may also occur in scientific practice, but in so far as the explanation of mathematical *theories* is concerned, the notion of reduction – whose special case generalization obviously is – seems to be the most central tool. On the other hand, our notion of correspondence includes the most important notions of reduction used for various metamathematical purposes.

In order to see whether actual cases of reduction in mathematics really have an explanatory import and how reductive explanations in mathematics would differ from those in science, we should work out detailed and comparative case studies. There exist developments in mathematics – and they seem to be less controversial than many of the much discussed changes in science – which can be considered progressive and where reduction seems to play a similar explanatory role as in science. On the other hand, however, this role is not always very obvious. It is not quite clear, for example, in what sense a reduction of arithmetic to set theory really explains arithmetics even if it may increase our understanding. As usually acknowledged, this reduction, among many other reductions in mathematics, is ontologically and methodologicaly important. According to Bonevac (1982), its ontological import is, e.g., due to the fact that sets are epistemologically at least as accessible as numbers – since numbers are at least as abstract as the sets with which they are 'identified', i.e., to which they are reduced. Bonevac's notion of epistemic accessibility seems to be more or less empiricist, however – in the sense that ". . . our ability to have knowledge concerning the objects assumed to exist must itself be capable of being a subject for empirical, and preferably physiological, investigation"[34] – and, hence, could be criticized in the same way as empirists' views concerning the cognitive role of theoretical entities in science have been criticized during the last forty years or so.[35]

Whether or not ontological reduction is considered of epistemological weight in the sense advocated by Bonevac, its methodological value is undeniable at least in cases where reductions are part of systematization in Kitcher's sense. It is this methodological sense, rather than ontological or epistemological, in which, for example, the reduction of arithmetic to set theory may yield an explanation of the former theory and may advance our understanding of the role of numbers.

We can see now that even though the aims of reduction in mathematics are, at least in part, similar to those in science, its explanatory roles in these two fields are in many ways different. So far as mathematical progress is concerned, there hardly exist in (pure) mathematics any important cases of counterfactual correspondence and explanation, since in all important reductions in mathematics the reduced theories are not considered false – so it follows, in particular, that the extensional notion of deductive-nomological explanation is more readily applicable in mathematics, as can be expected. The role of pragmatic and intensional features of explanations in mathematical reductions is as minimal as it can be. Pragmatic features seem to occur more prominently in the kinds of explanation in which Kitcher is concerned. Limiting case correspondence is, on the other hand, the most important kind of reduction that is associated with actual theories in science.

Bonevac argues that in mathematics there exist no counterfactual reductions since in mathematical contexts the notion of counterfactual does not make much sense. It does not make sense, says Bonevac, since mathematical statements are necessarily true if they are true at all. This is a bit hasty conclusion, however, since it is easy to find mathematical statements that are not necessary nor false in any direct sense. Thus, for instance, the statement (formulated in an appropriate language) claiming that there exists a number that is greater than all natural numbers is counterfactual as far as standard models of arithmetic are concerned but true in its nonstandard models – and there are, of course, indefinitely many similar examples. Whether or not a given statement is necessary is, even in mathematics, more or less relative to context and to the logic within which it is considered, and therefore the reason why there exist no important counterfactual reductions in pure mathematics is not that the notion of counterfactual does not make sense.

In the philosophy of science much attention has been paid to the concept of explanation, but no agreement upon its form or import has resulted from this interest. In the philosophy of the humanities and in aesthetics understanding and interpretation have been central notions, and they have proved to be even more controversial than explanation. In the philosophy of mathematics, on the other hand, more explicit attention should perhaps be paid to these and related notions, so that we would be in a better position to understand the cognitive and symbolic nature of mathematical change. We have seen, for example, that it is far from clear what the cognitive import of reduction basically is in mathematics – in particular, that of ontological reduction – over and above the more or less empiricist views that have been popular in recent philosophy of mathematics. More importantly, the question concerning the nature of mathematical progress is to a great extent a matter of interpretation.

6.9. CORRESPONDENCE AND THE EXPLANATION OF LOGICS

There is another kind of theoretical change that is often considered more relevant to mathematics than to empirical sciences, and which is similar to, but in many ways even more fundamental than, paradigm changes in science. It consists of changing logical foundations of theories. The role of explanation is here even more problematic, since it is not always obvious how far (for instance, within the metatheory) such a change, especially if it is a radical one, would eventually go or should go.

Furthermore, philosophical discussions about logical change have been ambiguous in the sense that no clear distinction has been made between the logic of a given theory – i.e., the logic in which the theory is reconstructed, if in any, or whose principles the inferences of the theory and its interpretations are assumed to follow – and the logic of its metatheory.[36] It is difficult to see which role logic has been assumed to play here. The latter role has usually been even more obscure than the former. There is a considerable disagreement concerning the questions of what the logics are that underlie mathematical reasoning and natural and scientific languages – if there are any definite logics – and what they possibly should be. In what follows, I shall nevertheless assume, to be able to use explicit terms, that to each theory some logic is assigned in which the theory can be formalized, but which may change, and, further, that its metatheory, though not formalized, is dependent on some recognizable logical principles, which may also change.

So, instead of thinking about changes of theories in the first place, let us now emphasize logical change. This emphasis is more or less hypothetical since, for evident reasons, there exist no obvious examples concerning revisions of logics underlying actual empirical theories. There are usually no definite, explicitly characterized logics for actual theories, not before they are logically reconstructed – and more importantly, scientists themselves are not usually interested in such questions. On the other hand, philosophers of science have suggested revisions, as for instance, in the case of quantum logic where the suggestion is a consequence of a new interpretation of 'empirical reality'. Purely logical reasons for revisions have been proposed as well, which, however, have so far concerned mathematical theories rather than empirical ones and thus resulted in proposals for adjusting mathematical principles. Intuitionistic logic and relevance logic provide well-known examples.

Keeping to the aims of the present essay, I shall only consider the kind of change where an earlier logic is in a correspondence relation to a new one in the sense defined in Section 5.4, above. Evidently our general notion of correspondence is the weakest notion of reduction that covers different variants of reduction (or, rather, interpretation[37]) and yet makes it possible to say that the reduced logic is explained by the reducing one. Its explanatory import depends, again, on context and pragmatic constraints the relevant mappings satisfy.

If a logic L is in a correspondence to L', important properties transfer from L' to L, such as Compactness and Löwenheim. Since the translations of the valid sentences of L are logical consequences in L' of the sentences defining the class of models K' (i.e., the domain of the structural correlation),[38] L' may explain L (on appropriate pragmatic conditions) analogously to the kind of explanation discussed above. Furthermore, any theory T, mathematical or scientific, formulated in L, can be reformulated in L' in an obvious way, that is, if T is axiomatizable in L, it can be transformed into a theory T' in L', and therefore there is a correspondence of T to T', relative to $\langle L, L' \rangle$.

It can be shown, for example, that classical predicated logic is in a correspondence to intuitionistic logic, and *vice versa,* but we have to ask, again, whether, or in what sense, these reductions are explanatory. Therefore, they should be studied more closely from a pragmatic point of view. It is obvious, anyway, that when classical logic is reduced to intuitionistic logic, the character of a possible explanation depends on the philosophical role the Kripke semantics of intuitionistic logic is assumed to play. Thus, for example, if the semantics is given an epistemic construal, as often done, the reduction may explain the status of classical logic from that epistemic point of view which is represented by intuitionistic logic. In any case, the reduction increases our understanding of the relation of the two logics – whether they are considered from a formal or epistemic point of view. Furthermore, since there is a correspondence of a theory T, axiomatizable in classical logic, to its transformation in intuitionistic logic, T', it immediately follows that if the former logic is in some sense explained by the latter, or understood in terms of the latter, the corresponding fact also holds for the theories T and T'.

7

CASE STUDIES

In earlier chapters we have defined the general notion of correspondence relation and its special cases, the most important of which is counterfactual correspondence. This relation, in turn, was seen to be a global form of explanatory translation. In what follows, I investigate some cases of scientific change that are well known from the history of science, and point out that our notions are applicable to them – at least if appropriate pragmatic conditions are assumed to obtain. As we have seen earlier, whether an application of the correspondence relation, particularly of the counterfactual one, is of explanatory import is very much dependent on relevant pragmatic and hermeneutic conditions. I also examine the intricate – and perhaps not equally well known – problem whether between the phlogiston theory and modern chemistry one can define a reduction or correspondence sketch, and we shall see that the answer is not quite evident, since the alleged informal correspondence relation seems to split into local relations. On the other hand, I study a structuralist logical reconstruction of the former theory, which might yield a possibility to define an exact correspondence relation. The reconstruction may not be intuitively evident, however.

Furthermore, I sketch a limiting case correspondence between symbolic and connectionist representations in cognitive science, and study the problem of how propositional knowledge may arise from non-propositional knowledge. I consider the question concerning what consequences this enterprise might have when we try to understand some intricate and much-discussed relations between the connectionist and symbolic paradigms and, on the other hand, relations within the connectionist paradigm. As will be seen, the investigations also have implications concerning epistemic logic.

Finally, the well-known relation between propositional modal logic and first order predicate logic is studied in order to see how the present framework of correspondence applies there.

Detailed logical reconstructions of some of the relations studied here are only presented in Part Three, below, where most of the formal investigations are concentrated. In the present chapter, only the gist of some formal reconstructions is given. For the mathematical and logical notions used here, the reader should consult Section 8 in Part Three, whenever needed.

7.1. CLASSICAL PARTICLE MECHANICS AND RELATIVISTIC PARTICLE MECHANICS

Let CPM and RPM be the axiomatic theories of classical particle mechanics and relativistic particle mechanics as formalized in Chapter 9, below.[1] Their classes of models are essentially composed of all models of appropriate many-sorted types satisfying Newton's second law and Minkowski's force law, respectively, and relevant mathematical conditions concerning the functions, individuals, and other entities needed. The models are of the form $\mathfrak{M} = \langle \mathfrak{B}; \ldots \rangle$, where $\mathfrak{B}$ is a standard or nonstandard model of analysis, that is, a superstructure (or Zermelo structure), or an appropriate elementary extension of one. Somewhat more completely, the models of CPM (of type τ, say) are of the form $\mathfrak{M} = \langle \mathfrak{B}; P, T; s, m, f \rangle$, where P and T represent a set of particles and a time interval, respectively, s and f are functions representing the position and force of the particles at each time instant in T, and m is a function representing their masses. Similarly, the models of RPM (of type τ') are of the form $\mathfrak{M} = \langle \mathfrak{B}; P, T; s, m, f, c \rangle$, where c represents the velocity of light. $\mathfrak{M}$ itself is called standard or nonstandard according to whether $\mathfrak{B}$ is standard or nonstandard.

CPM and RPM can be axiomatized, thus their classes of models defined, in a suitable infinitary logic $L_{\kappa\omega}$.[2] Infinitary formulas are needed to define certain standard sets in nonstandard models and for the notion of finiteness.[3] Clearly these theories are local in the sense that they only represent small parts of what is generally included in classical and relativistic particle mechanics. However, the forthcoming correspondence relation is global in the sense presupposed earlier.

Now consider the model-theoretic definitions of correspondence and limiting case correspondence as presented in Sections 5.3 and 5.5, above, and the notation used there. Then the model-theoretic construction (which is formally presented in Section 9.1, below) of a limiting case correspondence $\langle F, I \rangle$ of CPM to RPM (relative to $L_{K\omega}$) can be briefly described as follows:

(i) The class K is (essentially) the class of all standard models of CPM where the velocities and accelerations of the particles are bounded and accelerations are continuous;[4] (ii) K' is a class of nonstandard models of RPM which are 'near-standard' in the sense that they have appropriate finiteness properties so that they have standard approximations in K. The elements of K' (which in this setting can be thought of as the result of applying the respective minimizing transformation[5]) represent limiting cases in the sense that the velocities and accelerations of their particles are very small compared to the velocity of light, that is, the ratio is infinitesimal; (iii) K and K' can be defined in $L_{K\omega}$; (iv) the correlation mapping in this correspondence, $F \colon K' \to K$, is such that $F(\mathfrak{M})$ is obtained from $\mathfrak{M}$ by forming the standard approximation of the latter;[6] and (v) F is associated with the translation I, which (essentially) describes syntactically the process of forming the standard approximation of a model. In particular, the translation of Newton's second law says that the law 'almost' holds, which means in the language of nonstandard analysis that it holds with an infinitesimal accuracy. The translation of the law can be derived from Minkowski's force law and sentences describing the conditions in (ii), above, and some mathematical principles.

It will be of some interest to see that the semantic transformation (which is the combination of the minimizing transformation and correlation F, as described in (ii) and (iii), above) takes ultimately certain standard models of RPM (here regarded as actual models in the sense discussed earlier, and intended models in the sense of philosophy of science) into standard models of CPM *via* nonstandard models of RPM. The first transformation makes the ratio mentioned in (ii) infinitesimal and the second removes the nonstandard elements of models of RPM when transforming these models into models of CPM. The assumption that the ratio is infinitesimal means that if in a model in K' the velocity of light is finite, then the absolute values of the velocities and accelerations of the particles in the model are infinitesimally small (possibly zero). This is intuitively a trivial case since it means that the particles are at rest or almost at rest. In the less trivial, but highly counterfactual case the velocity of light is infinite, and this is the case we are interested in.

The conditions in (i), that the velocities and accelerations of the classical particles in question be restricted to bounded ones, are stated for technical reasons, as can be seen in Chapter 9. But they are not technical in any *ad hoc* sense. First, the restrictions reflect the fact that we cannot have a reduction here – that is, a correspondence relation whose correlation function would be *onto* the class of all models of CPM – but only a correspondence in a weaker sense. It is conceivable that this is typical of relations between theories across revolutionary scientific changes. It is usually said that in a correspondence relation the new theory should explain the success of the old one. Now classical particle mechanics has been successful with respect to relatively slowly moving and slowly accelerating particles. That velocities and accelerations are supposed to be bounded in our reconstruction does not mean that they are small in any physicists' sense of the word 'small', but the small ones can be found among the bounded ones. On the other hand, one cannot expect, for instance, that RPM explains everything that is explained by CPM, since the latter is heavily wrong from a point of view of RPM about the systems of particles where the velocities are high. The third condition in (i), saying that the accelerations are continuous, is likewise natural, as far as I can see. Thus we may say that the class K, defined by (i), is as close to what could be called the intended models of CPM as can possibly be defined by extensional, formal methods. On the other hand, the class K', the domain of the correlation function F, is defined in such a way that it will be as much as possible like K.

It can also be proved that Curie's Principle for Correspondence holds here when the symmetries in question are Galilean and Lorenz transformations (appropriately reconstructed).[7] Let G and G' be the classes of all Galilean and Lorentz transformations, respectively, such that they are restricted to K and K' and that in the latter the velocities of the respective coordinate systems are infinitesimally small compared to the velocity of light. Then $\mathbf{K} = \langle K, G \rangle$ and $\mathbf{K}' = \langle K', G' \rangle$ are categories whose objects are in K and K' and morphisms in G and G', respectively. Then, as the principle requires, F can be extended to a functor from $\mathbf{K}'$ to $\mathbf{K}$. The exact treatment of this result can again be found in Chapter 9.

Since there is a limiting case (thus counterfactual) correspondence of CPM to RPM, the notion of counterfactual explanation, defined in Chapter 6, above, provides an explanation of the former theory if the pragmatic constraints presented in 6.4.6 (Section 6.4) hold. In particular, the conditional of the form (6.4.1) is in the present case of something like the following form (informally), if the less interesting alternative that the particle velocities are infinitesimal is ignored:

If the velocity of light were infinite, then Newton's second law would almost hold.

As we have seen, pragmatic constraints are in cases of this kind rather involved and dependent on paradigmatic and even subjective elements. Particularly interesting is here the cotenability condition (d) in (6.4.6), saying that the premises — one of which is Minkowski's force law — from which the translation of Newton's second law is inferred must be cotenable, at a given world, with the counterfactual assumption that the velocity of light is infinite.[8] A world that is relevant in this case is obviously a standard model of RPM satisfying appropriate conditions, since such a world can be thought of as being an actual world with respect to RPM. Now Minkowski's law is a law of nature (or at least a lawlike sentence), and as argued by Lewis (1973), laws of nature tend to be cotenable with counterfactual suppositions since they are important to us. More generally, whether the premises are cotenable is relative to the notions of possibility at the world and similarity of possible worlds. The answer to the question of cotenability depends on how these notions are construed and how the world is chosen in this context. In Chapter 6, an example was presented that is closely related to the case study at hand, and there it looked obvious that the premises in question are cotenable with the condition that the velocity of light is infinite.[9]

This is a clear-cut example of the prominent role played by the principles of interpretation and by the interaction of the syntactic and semantic transformations. The above clauses (ii), (iv), and (v) of the description of the correspondence of CPM to RPM give us a hint of the fine-grained nature of this interaction, but its real character can only be seen by studying its formal version forthcoming in Section 9.1. The complexity of the relation of the theories is increased by Curie's Principle, as we shall see, but, on the other hand, it is evident that working out the principle will also add to our understanding of the relationship between the theories.

One may try to define another correspondence relation between the two theories in a more direct way, as, e.g., by defining a translation I that indicates how classical notions of mass and force are related to the relativistic ones, respectively. I shall skip exact formulations here, but roughly speaking it would mean the following. If **m** is an appropriate formal symbol for the classical mass,[10] $\mathbf{m_0}$ for the rest mass, **f** for the classical force,[11] **v** for the velocity of a particle,[12] and **c** for the velocity of light, then

$$I(\mathbf{m}) = \mathbf{m_0}(1-\mathbf{v}^2/\mathbf{c}^2)^{-1/2}$$

and

$$I(\mathbf{f}) = (1-\mathbf{v}^2/\mathbf{c}^2)^{1/2}\mathbf{f}.$$

If, furthermore, atomic formulas involving **m** and **f** are translated accordingly, and other terms and atomic formulas in a relevant way, then, e.g., Newton's second law will be translated into Minkowski's force law. What would be a more adequate and illuminating translation than this one? No nonstandard analysis is needed in defining the translation, which complies with what we know about how the meanings of mass and force change when we move from classical to relativistic mechanics.

However, the translation may not be feasible for the following reason. It seems that there is no way to define a correlation F that would harmonize with I in such a manner that $\langle F, I \rangle$ would yield a satisfactory explanation of Newton's law by means of the relativity theory. There seems to be no natural semantic transformation of the models of RPM into those of CPM that would reflect the translation in question, and thus no way (that is not *ad hoc*) to define the range and domain of F (that is, classes K and K'). This means that the above syntactic transformation does not lead to a global explanatory translation in the sense of correspondence.

7.2. KEPLER'S LAWS AND NEWTONIAN GRAVITATION

Next I try to briefly describe the relation between (certain formulations of) Newton's theory of gravitation and Kepler's theory of planetary motion, as presented in Pearce and Rantala (1984b).[13] It is well known that Kepler ignored the influence of the other planets upon a given one, and on this fact the treatment of the relation is based. The present case is analogous, and in many respects even similar, to the preceding case. There is one logical difference, but it is innocent from a physical point of view. We have to consider a trivial extension of Newton's theory (trivial in the sense of just expanding the type without new axioms) since Kepler's theory contains terms (representing Kepler's constant and the sun) that cannot be related to expressions of the Newtonian language.

Assume that KTP is Kepler's theory of planetary motion and NTG Newton's theory of gravitation as formulated by Scheibe (1973). Their central axioms are, of course, Kepler's laws of planetary motion and, respectively, Newton's law of gravitation. The trivial extension of NTG is called NTG', and it is obtained from NTG by expanding its language but inserting no new axioms, i.e., by just adding two individual constants that are assumed to refer to Kepler's constant and to the sun. The models for the infinitary logic needed to axiomatize KTP and NTG' are of the form $\mathfrak{M} = \langle \mathfrak{B}; \ldots \rangle$,[14] where $\mathfrak{B}$ is as in the previous case, which respectively satisfy (formalized) Kepler's laws and Newton's gravitational law, and some other conditions as specified in Pearce and Rantala (1984b). Now the steps – similar to those in the preceding section – of the model-theoretic construction of the limiting case correspondence of KTP to NTG', relative to the logic in question, are as follows:

(i) The class K is, essentially, the class of all standard models of KTP satisfying the condition that the distance between any two bodies has a nonzero lower limit; (ii) K' is a class of nonstandard models of NTG' that are near-standard in the same sense as in the preceding case study, that is, their standard approximations belong to K. However, this time that the models in question are near-standard means, in the first place, that the masses of their planets[15] are infinitesimal, which are then neglected in the approximations; (iii) K and K' are again definable in $L_{K\omega}$; (iv) $F(\mathfrak{M})$ is the standard approximation of $\mathfrak{M}$; (v) as in the preceding case, the translation I is induced by F so that they fit together as the definition of correspondence requires. Furthermore, Curie's Principle is valid here, but this time both categories are composed of Galilean transformations of a certain kind.

It is easy to see that the description of the correspondence relation in question at this intuitive level is similar to the preceding one, but when in the preceding case the correlation transforms the infinitesimal ratios of the particle velocities and the velocity of light into zero, in the present case the infinitesimal masses of planets are similarly transformed. In both cases the assumptions that the respective quantities are infinitesimal are counterfactual in the same degree, so that even in this sense the two correspondence relations are analogous. This implies that intertheoretic explanations in both cases are of the same complexity and dependent on similar logical, paradigmatic, and subjective presuppositions. For instance, on such presuppositions one is led to the following conditional:

> If the masses of the planets were infinitesimal, then Kepler's laws would almost hold.[16]

However, our notion of counterfactual explanation may play slightly different epistemic roles in the two cases, in the following sense. Relativistic particle mechanics and the respective paradigm, which establish the former explanation, are supposed to be correct according to the current scientific knowledge, whereas Newton's theory of gravitation and the Newtonian paradigm, establishing the latter explanation, are not, at least if compared to relativity theory. On the other hand, since the explanations are counterfactual in any case, the distinction may not be as significant as it would be if instead of counterfactual explanation we could here speak of deductive-nomological explanation, in which the explaining theory or law is usually supposed to be true.

7.3. CLASSICAL MECHANICS AND QUANTUM MECHANICS

One way to consider the distinction between classical mechanics and quantum mechanics is to say that the former does not correctly apply to small systems. In the former theory it is supposed that observations do not disturb systems, or if they do, the resulting disturbances can be corrected in principle. According to the latter, however, it is assumed that this is not the case when small systems are observed, but if a system is large enough, then the resulting disturbances can be neglected in practice. In the limit of sufficiently large systems, classical mechanics can be correctly applied, at least for all practical purposes. This is the essential, but rough, meaning of the original Correspondence Principle, discussed in Chapter 1, above. One criterion of the size of a system is its quantum number, which means that the system is large if its quantum number is large. In the limit this number would be (counterfactually) infinite (or approach infinity). Another way to characterize the meaning of 'size' is to say that for large systems Planck's constant can be neglected, and in the limit it would be (counterfactually again) infinitesimal (or approach zero).

One consequence inferred by physicists from the assumption that disturbances are not avoidable in the case of small systems was to include operations of observations explicitly in the theory.[17] Thus, for example, such classical variables as momentum and position are replaced by respective quantum-mechanical operators of momentum and position. For a model-theoretic reconstruction of the part of quantum mechanics under consideration, one must therefore introduce a similarity type, say υ', that contains symbols for the operators involved and for the state (wave) functions to which the operators are thought to be applied.

As usual in quantum-mechanical representations of motion, let us consider just a single system (particle), which therefore need not be denoted by any symbol in υ'. In our above representation of classical mechanics and relativistic mechanics, a similar assumption means that the domain P and the variables referring to elements of P are omitted, so that in the case of classical particle mechanics the models, of type υ, say, are now of the form $\mathfrak{M} = \langle \mathfrak{B}; T; s, m, f \rangle$ — and the functions involved are simplified correspondingly. Thus Newton's second law, as (nonformally) represented in $\mathfrak{M}$, is now of the following form: For all $t \in T$,

(7.3.1) $m d^2 s(t)/dt^2 = f(t)$.

Now let **s** be the (formal) symbol in υ for the position function, interpreted in the above model $\mathfrak{M}$ as s, and let **f** be the symbol for the force (here f); and, furthermore, let CPM_υ be the formalized theory of this slightly simplified classical particle mechanics.

On the other hand, in order to conform to our notation above and later in Chapters 8-9, we deviate from the notation of the position operator that is common in quantum mechanics by using, as above, the symbol **s** for the representation of the operator in υ' and 's' for its interpretation in a (given) model. Thus a model of a theory of quantum mechanics QM, involving a state function Ψ,[18] a position operator, and a potential function V, some other operators and individual constants (distinguished elements),[19] and the time-dependent expectation function <o> for each operator o, is of the form

$$\mathfrak{M}' = \langle \mathfrak{B}; T; s, \ldots, \Psi, <s>, m, f, <f(s)>, \ldots, V \rangle.$$

Hence the type υ' must contain symbols for Ψ, $<s>$, and $<f(s)>$. Let the symbols for the last two functions be **<s>** and **<f(s)>**. As usual, the force is defined in terms of the potential, as the negative of the derivative of the potential with respect to the position:

(7.3.2) $f(s) = -dV(s)/ds.$

In CPM_υ the position is a function of time, but it will be assumed here that the position operator of QM is time-independent, so that we refer to the so-called Schrödinger picture rather than the Heisenberg picture. However, each <o> is time-dependent, whence it is a function on T; and it is defined (for all $t \in T$) by

$$(7.3.3) \quad <o>(t) = \int_{-\infty}^{+\infty} \Psi^*(s,t)\,o\,\Psi(s,t)\,ds$$

(whenever the integral exists), where Ψ^* is the complex conjugate of Ψ. In addition, we shall suppose that the models of QM satisfy the Schrödinger equation and the usual equations concerning all the other operators involved.

As pointed out by some physicists, there are several relations between (parts of) classical mechanics and quantum mechanics that can be said to exemplify limiting case correspondence. The following treatment, which is very rough omitting many fine details, indicates that Newton's second law is related to a certain equation that holds in quantum mechanics between *expectation values* in such a way that one seems to be entitled to speak about two different correspondence relations in the sense of this book. I shall not try to present any exact proofs for these conjectures, which would be very tedious,[20] but only try to make the conjectures plausible. The treatment is based on Gasiorowicz's (1996) mathematical exposition of the relation.[21] He derives from (7.3.3) the following equation, for all $t \in T$:

(7.3.4) $md^2<s>(t)/dt^2 = <f(s)>(t)$,

where $f(s)$ is as in (7.3.2), and s is a time-independent quantum-mechanical operator as explained above.

The first correspondence relation is a limiting case correspondence. Gasiorowicz points out that for large systems, for which the uncertainty of the position is small, $<f(s)>(t)$ is close to $f(<s>(t))$, which means that the left-hand side of (7.3.4) is close to $f(<s>(t))$. If the uncertainty is infinitesimal, it is infinitesimally close to it:

(7.3.5) $md^2<s>(t)/dt^2 \approx f(<s>(t))$.

This suggests, analogously to Sections 7.1 and 9.1,[22] that under appropriate conditions on some nonstandard models of QM and standard models of CPM_υ the standard approximations of the models in the former class, say K', belong to the latter class, say K. If $\mathfrak{M}'$, as above, is such a model of QM, its standard approximation (within the present framework) is of the form

$$^0\mathfrak{M}' = \langle \mathcal{Q}; {}^0T; {}^0<s>, {}^0m, {}^0f<s> \rangle,$$

where $\mathcal{Q}$ is the standard model of analysis, ${}^0T$ is the standard interior of T, $f<s>$ is the composition of f and $<s>$, and ${}^0m$, ${}^0<s>$, and ${}^0f<s>$ are the standard parts of m, $<s>$, and $f<s>$, respectively. The meanings of these notions are explained Section 8.6, below. It follows from (7.3.5) and certain considerations in Chapter 9.1[23] that ${}^0\mathfrak{M}'$ satisfies Newton's second law. Hence this operation of standard approximation establishes the required correlation mapping from K' onto K.

Now the character of the required translation is obvious. It is otherwise analogous to the translation in the correspondence of CPM to RPM, but it first translates **s** into **<s>**. That the characteristic condition (5.3.1) holds is here as obvious as in that case. But now the meaning changes involved in the correspondence are more readily understood since the translation replaces the classical concepts of position and force of CPM_υ by the expectation values of the measurements of the quantum-mechanical notions of position and force. The obvious counterfactual conditional, analogous to the earlier ones, that is derivable under appropriate pragmatic and logical conditions is now of the following kind:

If the uncertainty of the position were infinitesimal, then Newton's second law would almost hold.

The second kind of correspondence is not counterfactual, but rather a reduction. Since a model $\mathfrak{M}'$ of QM, as described above, satisfying appropriate conditions concerning, e.g., the derivatives, integrals, and the other mathematical operations involved also satisfies (7.3.4), the model $\mathfrak{M} = \langle \mathfrak{B}; T, s_1, m, f_1 \rangle$, obtained from $\mathfrak{M}'$ by defining $s_1(t) = <s>(t)$ and $f_1(t) = <f(s)>(t)$, is a model of CPM. This defines an appropriate correlation mapping, and the respective translation translates **s** into **<s>** and **f** into **<f(s)>**. Whether the two mappings really amount to a correspondence relation is not evident without a proof for condition (5.3.1). If the condition holds, it means in the present case that the expectation functions of the position and the force play the same role within QM as the classical variables (functions) of position and force play within CPM_U.

One weakness of these alleged results is that they do not involve the principal equation of motion in the Schrödinger picture of quantum mechanics – namely the Schrödinger equation of motion – as closely as our earlier correspondence between CPM and RPM involves Minkowski's force law of relativistic particle mechanics. However, it is pointed out, for example, by Matthews (1968) that one may start from the Heisenberg equation of motion, where operators are time dependent, and end up with the result that the time-dependent operators of position and momentum satisfy a direct generalization of Newton's second law and with the above result (7.3.4).[24] I shall not study the relevance of the former result for correspondence.

7.4. THE PHLOGISTON THEORY AND MODERN CHEMISTRY

A crucial presupposition for the possibility to define an explanatorily relevant correspondence sketch of the phlogiston theory to (informal) modern chemistry is that situations (worlds) where combustion consists of uniting with oxygen can be in an intuitive and smooth manner transformed into (imaginary) situations where it consists of releasing phlogiston. In other words, if it is natural for us to imagine such a transformation, then we have a translation of some central part of the language of the former theory into the language of the latter such that it fits together with the semantic transformation and is of some explanatory import – at least it would increase our understanding of the respective conceptual change. Whether this is the case is not evident, however, since these two phenomena are so much in opposition to each other, and, furthermore, there are other problems that are due to the global nature of the theories and the ambiguity of the term 'phlogiston'. We may be able to establish a set of local explanatory translations rather than a correspondence relation sketch in the global sense of correspodence.

I shall start by briefly considering the possibility of a correspondence sketch by studying Kitcher's (1978) proposal for a translation of the phlogiston theory and Kuhn's (1983) criticism of the proposal. According to Kitcher, many expressions of the phlogiston theory can be defined in terms of more modern chemistry in the sense that the referents of the former can be identified by means of the latter. Thus, some derivatives of the term 'phlogiston' can be understood as having the same referents as certain derivatives of the term 'oxygen' in twentieth-century chemistry. For example, 'dephlogistigated air' may refer to air enriched by oxygen or to oxygen itself, and, accordingly, 'phlogistigated air' to what is left when oxygen is removed from air, and 'a is richer in phlogiston than b' is coreferential with 'a has a greater affinity for oxygen than b', and so on. As can be expected, the term 'phlogiston' itself does not usually refer at all, as in the expression 'phlogiston is emitted during combustion'; but sometimes it refers to hydrogen. According to Kitcher, this process of identifying referents amounts to a translation and shows that there is no incommensurability in this case of scientific change.

According to Kuhn, the process is no translation, but it indicates how the old language can be learned by historians of science, and it helps them to understand the success of the old theory. It is a hermeneutic process, not a translation.[25] Kuhn does not accept the partiality of Kitcher's translation, which is due to nonreferring occurrences of 'phlogiston' and for which Kitcher does not provide a solution. In the phlogiston theory there are other terms, such as 'element' and 'principle' for which there are no equivalents in the modern theory, in the sense of having the same referents or meanings. Therefore, they are not replaceable by modern expressions. Kuhn also considers the possibility that such terms, whether or not referring, could be translated (in the sense of identifying referents) contextually, in the same way as the above compounds 'dephlogistigated air', etc., were given refer-

ents by Kitcher in modern terms: "'Phlogiston' would then sometimes be rendered as 'substance released from burning bodies', sometimes as 'metallizing principle', and sometimes by still other locu-
tions." However, this is not acceptable for Kuhn, for it would make
the resulting text incoherent. Unrelated or differently related expressions would replace expressions that are taken to be related (or identical) in the original text, and therefore it would suppress the beliefs
that the author of the original text tried to communicate by means of
that text. A reader would not understand why sentences that look unrelated were juxtaposed in the text.

Let us now relate the above discussion about contextual translation
– which, as we have seen, is not a translation according to Kuhn's
(1983) standards – to our notion of explanatory translation. That
there is a clear-cut connection is suggested by Kuhn's remark that
when Kitcher explains such juxtapositions by referring to beliefs of
the author of the original text and to modern theory, these explanations are glosses and do not belong to the translation itself. Part of the
idea that they are glosses seems to be close, in principle, to what our
notion of semantic transformation is supposed to accomplish (or perhaps they would be commentaries), though no exact connections can
be given here; another part, on the other hand, belongs to the hermeneutic environment of the translation, that is, to what we have been
calling pragmatic constraints.[26]

Therefore, it is obvious that Kitcher's contextual translations (of
expressions containing terms like 'dephlogistigated air', etc.) are explanatory translations in our earlier *local* sense of the word, at least if
the semantic transformation from situations involving oxygen into situations involving phlogiston (discussed at the beginning of this section) is intuitively acceptable. On the other hand, the phrases 'dephlo-
gisticated air', 'phlogisticated air', '*a* is richer in phlogiston than *b*',
and some others, all refer to a single role phlogiston plays in the old
theory, and their connections to certain phrases talking about oxygen
are intuitively analogous, that is, they are different sides of the same
question. If this is the case, we obviously have a more global relation
between the two theories, that is, from the phlogiston theory we might
separate some subtheory that is related to the oxygen theory by way
of a correspondence sketch. I shall not try to work out such a relation
here in any detail, since it should be rather evident on the basis of

Kitcher's article and of what we have said so far. Furthermore, it is not clear whether it would be of much explanatory import over and above that of the local translations.

Let us next consider Kamlah's (1984) reduction of the phlogiston theory to the oxygen theory, which he works out by first formally reconstructing the theories. I shall only present some remarks on it so that we can see why the conclusions concerning the reduction of the phlogiston theory to the oxygen theory are here so different from those in the above nonformal treatment. It is likely that one reason for the difference is that Kamlah's formal reconstruction of the phlogiston theory is not a direct formalization of the intuitions behind the nonformal theory as hinted above, which pertain to the ontological nature of phlogiston and are therefore close to the original, eighteenth-century ideas of the theory. Therefore some of his assumptions are artificial, as he admits himself, and even *ad hoc*. We can see in the next chapter, on the other hand, that the reconstructions of the theories of particle mechanics, for example, rather directly correspond to physical intuitions behind the respective force laws. One reason is, of course, the fact that the original theories are mathematical.

Kamlah's idea is to reconstruct both theories so as to present them as theories that are similar to each other as mathematical entities; and this is accomplished by representing them as theories of chemistry. Kamlah formulates the theories in a set-theoretic framework – this kind of formulation is advanced by what is often called the structuralist philosophy of science.[27] According to him, this similarity in the logical structures of the two theories does not justify Kuhn's argument that the scientific change in question is revolutionary. The theories differ only in numerical values of certain functions that are used in the reconstruction, and therefore there is a translation between the theories; and since there exists a translation, the existence of phlogiston and oxygen may not be characteristic assumptions of these theories. Existence was an important question for Kuhn and Kitcher, as we have seen.

It seems to me that the questions of existence in the metaphysical sense discussed by Kuhn can be largely got rid of when intertheoretic relations are subjected to formal treatments, since they can be replaced by questions of how certain mathematical entities can be defined. This feature, among other things, in fact shows how different theories and their relations may look in the two kinds of approach, formal and intuitive.

The (real-valued) functions Kamlah uses in his formalization of the theories describe mass density distributions of substances, mass density distributions of elements, molecular weights, atomic weights, and the numbers of atoms in molecules. Both theories are axiomatized by using these functions, indices (natural numbers) for substances, and a notion of space-time region. Then some of the functions of the phlogiston theory are defined in terms of functions of the oxygen theory by means of linear transformations, and the rest of the functions are given directly their values as real numbers. The feature that gives 'phlogiston' (i.e., the index for phlogiston) a special role here is that its 'atomic weight' is defined as being a negative number (that is, -16).[28] By means of these definitions a reduction of the phlogiston theory to the oxygen theory can be constructed, which is done by Kamlah by using Adams' (1959) structuralist notion of reduction.

Pearce (1987) points out that on certain technical conventions and plausible assumptions the structuralist reduction relation, as defined by Adams and here employed by Kamlah, induces a correspondence relation in our sense,[29] whence it follows that under those conventions and assumptions Kamlah's reduction of the phlogiston theory to the oxygen theory induces a correspondence relation (if an appropriate logic is introduced first) of the former theory to the latter. The correspondence relation would be a reduction in the traditional sense rather than, for instance, a counterfactual correspondence. But, as we have noted, this reduction does not respect the intuitive ideas about the historical relation of the two theories, unlike the formal correspondence between the two particle mechanics. The latter is faithful to the historical idea of a nonformal limiting case correspondence between the theories, in the sense advocated by many physicists.

7.5. CORRESPONDENCE IN COGNITIVE SCIENCE

7.5.1. A Turing Representation and Network Representation

Smolensky (1988) presents the idea that in cognitive science the symbolic representation can be seen as idealizing the subsymbolic representation. Both kinds of representation are important since they are complementary ways to understand and explain cognition. Therefore, we should not try to eliminate one of them in favor of the other, but rather think of their relationship as providing a cognitive correspondence principle, a principle analogous to the respective principle discussed in the philosophy of physics. The relation holding between the symbolic and subsymbolic theories is much like the relation between classical mechanics and quantum mechanics. If certain limit conditions are in effect, a subsymbolic theory approaches a symbolic one. The former theory describes a system satisfying parallelly and simultaneously 'soft constraints', and as soon as appropriate limit assumptions are made, it yields a macrotheoretic description of a system that is hard and rule-governed. Hence it follows that the subsymbolic paradigm brings about meaning changes of cognitive concepts.

If this is the case, then the cognitive correspondence principle is an instance or a special case of the more general principle discussed in Chapter 1, above. However, Smolensky's proposals concerning 'limitivism' are ambiguous, and therefore in Rantala and Vadén (1994) an attempt was made to give some of them a more definite form. The following study in this section and the formal investigation in Chapter 9 are based on that work and on Vadén (1995) and (1996).

Let us start by summarizing some basic features of the *symbolic*, which is often called classical, and *subsymbolic* models of cognition. The two accounts are even thought of as different paradigms of cognition, in something like Kuhn's sense of the word. A well-known characterization of the former is put forward by Fodor and Pylyshyn (1988), which, however, can be subjected to strict criticism since its picture of human cognition seems to be too narrow.[30] But it is simple and covers Turing computation, which is one element in the case study to follow. They maintain that the classical, symbolic model of cognitive architecture has the following characteristics and argue that the characteristics make a crucial difference between the classical and connectionist accounts.

The first characteristic is that mental representations, being symbolic, have combinatorial syntax and compositional semantics. By these properties they mean that, first, representations are structurally atomic or molecular, second, molecular representations have either atomic or molecular constituents, and, third, the semantic content of a representation is a function of the semantic contents of its syntactic parts, together with its constituent structure. This characterization implies that mental representations are symbol systems.

Another characteristic, which is dependent on the assumption that representations are structured and which is a cornerstone of symbol manipulation, is that mental processes are structure-sensitive. Mental operations, transforming representations into other representations, apply to representations by reference to their form. Briefly stated, the symbols of a symbol system have a semantics in the sense that they refer to and mean something else, and they are syntactic entities capable of being manipulated according to rules.[31]

Consider next the subsymbolic model of cognition as described by Smolensky (1988).[32] Subsymbolic representations and processes are taken care of by *subsymbols* rather than symbols, and they are realized by connectionist systems (networks), whose behavior is said to be rule-described — not rule-governed like symbolic representations and processes. While the so-called conscious rule application can be exactly described by means of the symbolic (classical) model, the subsymbolic one can be used to exactly describe intuitive cognitive processing. Rules in a system of the former kind are sometimes called hard constraints, whereas the constraints in a connectionist system are soft and they are provided by the connections between its units. The terms are used in the sense that a hard constraint is deterministic and does not leave room for violations, whereas the import of a soft constraint is more limited since it is just one among many.

Symbols are manipulated by syntactic operations, whereas subsymbols are not dealt with by means of symbol manipulations but processed by numerical computations, and they are context-sensitive microfeatures of symbols rather than context-free symbols. They need not have a semantic content individually in the same sense as symbols have. Generally, only a set or vector of subsymbols, which is distributed among units of a connectionist system, is collectively interpreted in the same way as a single symbol. Subsymbolic cognitive explanations are explanations in terms of subsymbols.

There is another distinction about a connectionist system made by Smolensky that will be relevant later on in this chapter. It is the distinction between the *conceptual* and *subconceptual levels* of analysis, and it is semantic, that is, it concerns the notion of semantic content. The conceptual level is the semantic level of descriptions in the ordinary sense of semantics, as, for instance, in the sense in which we speak about the semantics of natural language. The subconceptual level is a semantic level at which the activities of individual units in a connectionist system (and possibly units in subsymbolic systems of other kinds) are given a semantic content. Here the semantic content of a unit is a microfeature of something; it is a subconcept rather than concept. Only vectors (or ensembles) of units (or, rather, those of activation values of units), that is, representations distributed over a number of units, can generally speaking represent concepts and thus have meanings in the ordinary sense.

Now a connectionist system can be described at both semantic levels, of which the subconceptual level is fundamental. The conceptual level description of such a system is, according to Smolensky, an *approximation* of its accurate subconceptual level description.

There are aspects of cognition that can be described in an accurate way only at the subconceptual level by using notions pertaining to individual units (of a network), activation values (of units), connections (of units), weights (of connections), and the like. On the other hand, the conceptual level semantics refers to notions and elements of the task domain, and only this level is relevant (or at least preferred) to the purposes of a symbol system. The principal idea Smolensky advocates here is that in the description of cognition both the symbolic model and subsymbolic model are needed, and within the latter both the conceptual and subconceptual level.

As remarked above, Smolensky's claims about the meaning of the correspondence in cognitive science are ambiguous. They are so since it is not always very clear whether by limitivism in cognitive science he means the view that there is a limiting case correspondence between (i) the symbolic and subsymbolic paradigms themselves, or, rather, between two theories or models describing at the conceptual level two systems belonging to these paradigms; or between (ii) conceptual and subconceptual level descriptions *within* the subsymbolic paradigm, that is, descriptions of the levels as they occur in a connectionist system; or even between (iii) a subconceptual level description of a connectionist system and a conceptual level description of a symbol system. Be this as it may, I shall in the rest of the present section explore (i), and its possible consequences in the next section, and make some tentative remarks on (ii) and its possible implications in

Section 7.5.3.

Let us start with what Smolensky suggests about the relation between conceptual level descriptions. Here he is quite explicit:

> (7.5.1.1) If a symbol system (von Neumann computer) and a connectionist system are described at the conceptual level – not at the subconceptual or another 'lower' level – the descriptions are good approximations to each other.[33]

The following claim is relevant for (i), above, and is more or less based on an interpretation of what Smolensky might mean.[34] The symbolic account of cognition is a limiting case of the subsymbolic, which, in view of (7.5.1.1), obviously presupposes again the conceptual level:

> (7.5.1.2) The relation (at the conceptual level) between symbolic and subsymbolic theories (models) is a limiting case correspondence. Therefore, when certain idealizing assumptions are made, the latter approaches the former.[35]

Such assumptions could be, for example, that the number of units, or, alternatively, some connection weights, in a network approach infinity or are infinite.

Now in Rantala and Vadén (1994) a symbolic representation and a subsymbolic representation are constructed such that the former is a limiting case of the latter. However, it does not follow that the symbolic model in general is a limiting case of the subsymbolic model. Smolensky's proposal is too general to have an exact meaning, and therefore it cannot be verified in such an extensive form. Though the constructions involve special designs and thus the import of the case study may be limited as such, there are reasons to assume that they are typical and can be generalized, and therefore can help one to understand and make sense part of Smolensky's argument. In the construction, a Turing machine is used rather than von Neumann computer.

In what follows, I give a brief account of this case study, where it is shown that:

> (7.5.1.3) A conceptual level description (theory) of a Turing machine representation is in a limiting case correspondence to a conceptual level description of a connectionist re-

presentation.[36]

The former description is called the *theory of Turing representation*, TR in brief, and the latter the *theory of network representation*, NR.

Now the Turing machine, let it be *T*, described by TR represents expressions of a combinatorial language *L* in the sense that it is able to recognize whether or not a sequence of symbols, of a given finite length *n*, is an expression (e.g., a well-formed formula) of *L*, that is, belongs to the set of its expressions, Exp*L*. In a positive case, if an input of *T* represents an element of Exp*L*, *T* prints '1' on a square as its output, and in a negative case, '0'. As we can see in Chapter 9, TR can be given both a mathematical and model-theoretic formulation. Let θ denote here (either the mathematical or formal version of) the sentence describing the above input-output relation in the Turing machine *T* (i.e., describing how it decides whether or not a sequence belongs to Exp*L*). Informally, then, θ is of something like the following form: For all relevant inputs *i*,

(7.5.1.4) If *i* is in *E*, then the output is 1;

If *i* is not in *E*, then the output is 0,

where *E* represents (in the framework of *T*) the elements of Exp*L*.[37]

The connectionist system *N* representing expressions of *L* and described by NR is as follows. It is a network that is already trained to recognize expressions belonging to Exp*L* and to recognize whenever a sequence (of length *n*) of symbols is not in Exp*L*. Hence the description below belongs to the test mode of the network, rather than the training mode. In practice, *N* can actually be trained to recognize only a finite number of expressions in Exp*L*, but if it learns to generalize, it might learn an infinite set in principle. This might presuppose an infinite net, however. *N* is assumed to have only one output unit, whose activation values are in the open interval (0,1). Some version of the so-called logistic activation function can be used here. The assumption that there is only one output unit does not effect the main conclusions to follow, but makes the formalism a little simpler. The recognition of expressions belonging to Exp*L* now means that the numerical value of the output of the network (the activation value of the single output unit) is close to 1 when an input (a pattern of activation values) represents an expression belonging to Exp*L*, and close to 0 in the other case.

Let θ' denote the following (informal) sentence describing this input-output relation in N: For all relevant inputs i,

(7.5.1.5) If i is in E, then the output is $a^+(i, m)$;

If i is not in E, then the output is $a^-(i, m)$,

where m is a natural number (indicating the number of the units in N, or, alternatively, in a relevant layer of units), a^+ and a^- are real valued functions such that $a^+(i, m)$ is close to 1 and $a^-(i, m)$ to 0.[38] It is assumed, furthermore, that the former approaches 1 and the latter 0 (for all i) when m increases, and that they are, respectively, infinitesimally close to 1 or 0 when m is infinite. As we shall see below, the role of the parameter m is to ensure the existence of a limiting case correspondence, but this role is not entirely *ad hoc*, however, since it seems natural to expect (and assumed by Smolensky) that an increase of the number of units has an influence to the output of the kind assumed here. Clearly, when m changes, N changes to another network, and also i may change. Since N has been trained to recognize the relevant expressions, we can suppose that it is self-organizing, and therefore it is enough to have the respective inputs i as arguments of the functions a^+ and a^-, that is, explicit information about connection weights can be dispensed with at the level of description displayed in θ'.[39] If more information is needed, these functions can be replaced by functions where there is an additional argument representing the weight matrix of the network. Then we can have a description similar to θ' referring to the training mode, where the respective output functions, like a^+ and a^-, may get values farther off from the desired values 1 and 0, respectively; and similarly for cases where the network has to *generalize* from what it has learned. In what follows, I shall occasionally refer to such possibilities.

It will be assumed in the more formal version of this case study, in Section 9.3, that N is a deterministic system, but it would not change the net results if it were supposed to be stochastic, in the sense that its activation rules were stochastic, in so far as appropriate conditions (e.g., concerning the degree of variability) were satisfied. In such a case, the outputs, that is, the functions a^+ and a^-, are interpreted as probabilities. Then the relation between its inputs and outputs is probabilistic, and it may happen that different trials with the same input (and the same connection weights) result in slightly different outputs.

Now we may perhaps say that, at least if compared with the above Turing machine T, the connectionist system N recognizes the expressions in Exp$L$ only approximately since the output values are 1 or 0 only approximately. If the system can be realized, the recognition will possibly be approximate in another sense, too. If it is stipulated that a recognition of the fact that an input belongs to E takes place (sufficiently well) if the respective output exceeds a given value, close to 1, it may happen that the trained network occasionally gives an output value that is smaller than what was stipulated. This may take place if it has to generalize, that is, if it is actually trained to recognize a proper subset of Exp$L$, which means that E represents a proper subset of Exp$L$ and θ' describes the result of such a partial training. There are some experiments which indicate that such things happen in practice, as, for instance, Bechtel and Abrahamsen's (1991) simulation model to solve some logical problems, where trained networks have to generalize. However, when I in the present section consider the relation between TR and NR, I assume that in this purely theoretical model E encodes the whole Exp$L$, whence possible large errors need not be considered. On the other hand, if N were an actual or theoretical network of this kind but exhibiting small variability when tested with the same input (which means that a^+ and a^- and possibly other activations are probabilistic), it would not influence the consequences of the following overall behavior.

It is easy to see that θ' is a conceptual level description of the representation in question. The subconceptual level of N is of course important when one considers the transformation of an input into the respective output, and, furthermore, when one studies the problem of how the input-output relation of systems of the above kind would change if the number of units (that is, m) were increasing, and reasons for the change in the first place. Outputs depend on inputs, connection weights, and the like, but, as we shall see in Section 9.3, the structure of N is designed in such a way that keeping these parameters implicit does not influence our results of the present section.

Since it is assumed that the architecture of N is such that when m increases,[40] $a^+(i, m)$ gets closer to 1 and $a^-(i, m)$ closer to 0, the theories TR and NR are good approximations to each other, in the sense proposed in (7.5.1.1). The greater m is, the closer the respective net-

work is (in the functional sense) to the Turing machine T, and if m
goes to infinity, it goes to T, as can be seen from θ' and θ. Thus we
may say that the theory TR is a limiting case of NR in a mathematical
sense, but in order to study whether there exists a limiting case correspondence in the model-theoretic sense, as defined in Chapter 5, a-
bove, the theories must be formalized. Both the mathematical expositions of the theories and their model-theoretic versions (in the deterministic case) will be accomplished in Section 9.3.

The construction has a feature that may appear somewhat speculative. It does not address the question of whether the network N is
able to learn to recognize the elements of $\mathrm{Exp}\mathcal{L}$ as NR requires. It is
simply supposed that theoretically a training of the above kind can be
done. However, as will be seen in Section 9.3, only the last two layers of N are there considered (the ouput layer and the one preceding
it) in addition to the input one, and the hidden layers may be as many
as needed and have whatever connectionist architecture. Therefore it
is obvious that the recognition hypothesis is even realizable as soon
as $\mathrm{Exp}\mathcal{L}$ and the other relevant sets are finite and small. But if the sets
are recursively infinite, the system is purely theoretical in the same
way as Turing machines are purely theoretical. In this case the recognition hypothesis is, of course, more hypothetical than the similar
hypothesis concerning the Turing machine T, since, presumably, it is
not supported by anything that would be analogous to Turing's (or
Church's) thesis.

Let us briefly consider the question whether the notion of counterfactual explanation makes sense in this case, which in certain methodological and ontological senses is different from our earlier examples
of counterfactual correspondence. As pointed out earlier, in Chapter
6, if appropriate logical and pragmatic (and paradigmatic) conditions
are satisfied, then, the existence of the limiting case correspondence
of TR to NR yields a counterfactual explanation of TR, and, in particular, the conditional of the form (6.4.1) is true (at an appropriate
world),[41] where σ is now the sentence 'm is infinite' (or, rather, its
formal counterpart), θ is as above, and its translation $I(\theta)$ says (if informally construed) something like that a symbolic Turing representation 'almost' is the case (that is, with an infinitesimal accuracy).
Thus the conditional might informally be read as follows:

If a network (of the above kind) had infinitely many units, then a symbolic Turing representation (of the set Exp$\mathcal{L}$) would almost be the case.

Hence, if the conditions are satisfied, TR can be considered as an idealization and limiting case of NR in something like the sense suggested by Smolensky, as quoted in (7.5.1.2). However, this case may look basically different from the above examples from physics, since it is not perhaps appropriate to say that θ and θ' express scientific laws or are lawlike sentences in the same way as the respective sentences in the physical examples, as, e.g., the force laws of classical and relativistic particle mechanics. Therefore, even if we think that natural laws are cotenable with counterfactual suppositions, the cotenability of θ' with the condition that the number of units is infinite would not follow. This might mean that at the level of counterfactual explanation there are here even more speculative and subjective elements than in the examples from physics.

Furthermore, only if considered from an appropriate perspective, one could say, for instance, that TR is supplanted by NR; this much might say a connectionist who believes that the latter provides a better account of (human) cognitive representation so that the version that is provided by the former is even more idealized. Only from such a point of view the above informal paraphrase of the formal counterfactual conditional makes sense. It is clear that it is not meaningful if looked at from a purely metamathematical or logical point of view, since from that perspective the role of the notion of Turing machine, and, more generally, the role of symbolic representation and symbol manipulation, is fundamental, not that of subsymbolic. These pragmatic problems and other consequences of the limiting case correspondence between TR and NR will be discussed in the next section in greater detail.

7.5.2. *Beyond Functionalism and Reductionism in Cognitive Science*

Many conflicts in the philosophy of cognitive science, such as those between the classical and connectionist paradigms and between functionalism and reductionism, have been largely discussed in the literature in terms of their logical features. However, the arguments presented have often been speculative in the sense that they do not rest on, and they have not given rise to, relevant case studies. This seems to be true, in particular, about the arguments for and against reducibility of one cognitive model to another. One reason might be the fact that such case studies would often be logically extremely complicated if perfomed in detail; another the fact that they could not be worked out by means of classical notions of logic and reduction, but would rather require new and nonstandard insights. However, the classical, Nagelian notion of reduction has been referred to in cognitive science almost exclusively.

As we have seen, the most-discussed intertheoretic relations in science seem to be such that their reconstruction and understanding require that logical principles in a very general sense are used, and, in addition, pragmatic and hermeneutic explorations, including the principles of interpretation. I have tried to show in this essay that faculties belonging to these different philosophical traditions − logic, pragmatics, and hermeneutics − that are often considered as being in conflict with each other, can merge into approaches to intertheoretic relations which are applicable to actual cases of theory change, and, in particular, to problems of cognitive science. Such a fusion seems to be necessary for a realistic approach; but then, somewhat unexpectedly perhaps, many philosophical conflicts about intertheoretic relations evaporate. I shall in what follows investigate briefly one such conflict in cognitive science, the controversy between functionalism and reductionism.

In the most general sense of the word, a view or discipline is *functionalist* if it says that something can be identified or characterized by means of its functional or causal role, rather than by means of its substance or stuff, or hidden essence, or the like. Therefore, the latter is not theoretically important and can be ignored. Currently the term is most commonly used in cognitive science, but certain views within other disciplines can be taken to be functionalist in this broad sense. For example − if we consider the philosophy of art − Goodman (1968) maintains the idea that something can be identified as a work of art only if it plays an appropriate symbolic role. This opposes the view that a work is identified with, or reducible to, its material embodiment. In psychology, behaviorism has functionalist undertones; and in the philosophy of science, perhaps, instrumentalism.

Within cognitive science, functionalism has a variety of definitions and versions, and a variety of different applications. For example, according to Bechtel (1988), "[f]unctionalism maintains that mental events are classified in terms of their causal roles," and ". . . mental events can be recognized and classified independently of their physical constitution."[42] According to Bechtel and Abrahamsen (1991), the functionalist theory of mind says that ". . . mental states are characterized and to be understood in terms of their interactions with other mental states, not in terms of their physical embodiment . . .."[43] Churchland (1992) defines it as a discipline whose core idea ". . . is the thesis that mental states are defined in terms of their abstract causal roles within the wider information-processing system. A given men-tal state is characterized in terms of its abstract causal relations to environmental input, to other internal states, and to output"; and furthermore, ". . . functional kinds are specified by reference to their roles or relational profiles, not by reference to the material structure in which they are instantiated."[44] Thus it seems that functionalism is a methodological rather than ontological discipline.

Since all these versions hold that mental states are not identifiable in terms of their embodiment, it follows that they can be embodied in many ways and in many stuffs. Churchland and others see this feature as being the reason why functionalist have typically rejected reductionism. According to her construal, functionalists maintain that a reductive strategy could only succeed if a type of mental states were identical to a type of physical states, or rather, if there were a one-one relation between these two kinds of type. But this is not the case because there can be many embodiments. It is observed by several authors, however, that this is no reason for not having a reduction between descriptions of the two kinds available; on the contrary, multiple realizability would in fact be very much in agreement with certain views of the general notion of reduction.

Reasons for maintaining reductionist ideas are, in any case, often opposite to those motivating functionalism. In the philosophy of mind and in cognitive science, a reductionist may assume, e.g., that the physical reality is fundamental, whence descriptions of the mental should or could be reduced to physics, and cognitive and psychological theories to neuroscience theories.[45] Thus the former theories and the talk about the mental are less important, and can be ignored, even eliminated. As we saw, on the other hand, functionalism asserts that

mental events can be characterized in functional terms, whence neuroscience is not that important for this purpose. Therefore, reductionism
and functionalism usually represent conflicting views in cognitive science. This controversy can be roughly condensed as follows, if we
speak about it in terms of two theories, say, T and T'.[46] Let us assume in what follows that reductionism argues that T' is more basic
in cognitive science, whence T is to be reducible to T'. Then in some
ontological or methodological sense the reducible theory T is dispensable. Let functionalism maintain, on the other hand, that the alleged
more basic theory T' is dispensable in the sense that cognition can be
functionally characterized and understood by means of T alone.

This distinction between theories like T and T' can be made a little
more fine-grained if we say that the theory T gives a functional and T'
a structural characterization of the mind, or, rather, that T is *functional relative to* T' and T' *structural relative to* T.[47] Thus, for example, we may assume that classical cognitive theories (symbolic theories in various forms) are functional relative to connectionist (subsymbolic) theories and to theories of neuroscience; connectionist ones
in turn are functional relative to theories of neuroscience; and conversely for structural. According to some views, the latter theories
provide more structural descriptions of mental events than do the former. The actual controversy between functionalism and reductionism
usually refers to the relationship between the above groups of theories. Roughly speaking, of these pairs of theories the reductionist
may prefer the more structural one over the more functional one, and
thus would like to see the latter as reduced to the former; the functionalist prefers the more functional one, and would like to dispense with
the more structural one as far as cognition is to be characterized.

Let us study the controversy in the light of the limiting case correspondence between the symbolic and subsymbolic theories TR and
NR, surveyed in the previous section – which, to some extent at least,
seems to confirm Smolensky's suggestions as presented in (7.5.1.1)
and (7.5.1.2) – and also in the light of some proposals concerning
the relation between connectionist and neural theories. What has been
said of the latter relation in the literature so far is even more tentative
in the sense that there does not seem to exist a theoretical construction
which would display the relation and would be analogous to the relation between TR and NR. But what Smolensky (1988) argues about it

is instructive here. He proposes that the general principles of computation at the subconceptual level of description must apply to computation in the brain:

> (7.5.2.1) If a connectionist system and the brain are described at the subconceptual level,[48] the descriptions are good approximations to each other.

As far as I know, he does not propose that there might be a limiting case correspondence between connectionist and neural theories of cognition.[49] The reason is clearly that too little is known of the latter. It is known, and also mentioned by Smolensky, that connectionist models lack important features (e.g., spatial location) present in neural models, and hence contain idealizations when compared with neural models. This feature and (7.5.2.1) might suggest — and they even make it sound plausible — that the following will turn out to be true in the near future:

> (7.5.2.2) A counterfactual (possibly a limiting case) correspondence of some subsymbolic theory to some neural theory can be constructed (at the subconceptual level).

If (7.5.2.2) held, there would possibly be a counterfactual explanation of a subsymbolic theory by means of a neural one.

Given the above relationships, let us draw some tentative conclusions concerning the debate between functionalists and reductionists. What are, for instance, implications of there being a limiting case correspondence of a symbolic representation, like TR, to a subsymbolic one, like NR? Analogous conclusions can be drawn if a similar relation between a subsymbolic and neural theory can be found, at least in so far as the former itself is not considered a neural theory.

As we have learned in this and earlier chapters, the above kind of relation between the symbolic and subsymbolic representations TR and NR is not that of reduction in the classical sense, since, first, one of the special assumptions (that the number of units is infinite) needed is counterfactual, and, second, not the symbolic representation itself can be inferred from the subsymbolic one plus special assumptions, but rather its translation. It follows that if this kind of relationship can be generalized and if psychological laws are of a computational kind in the classical sense, those laws are not reducible, in the standard sense, to subsymbolic theories nor explainable in the classical deductive-nomological sense.[50] Then psychological laws enjoy some degree of autonomy, and the hard-core reductionist is shown to be wrong, and the functionalist, at least in part, right. It is pointed out by Richardson (1979) that there is no inconsistency between computational models and classical reductionism. He shows, e.g., that the plurality of physical realizations is not inconsistent with reductionism in the classical sense.[51] But if what we have done can be generalized, it follows that no important (or nontrivial), classical reductions between (classical) symbolic and connectionist models in fact exist, and this may also make it implausible that there would be any important reductions between classical and neurological models either. Even though the counterfactual correspondence between TR and NR is not an ordinary reduction, it connects symbolic and subsymbolic representations that are of typical kinds. This typicality seems to make the idea that there exist no such classical reductions plausible.

On the other hand, if there is only a counterfactual correspondence, and hence only a counterfactual explanation, available, then whether or not computational psychology is to be considered completely autonomous seems to depend on pragmatic and paradigmatic elements, not merely on any objective or logical features of the relationship between the two kinds of theory. The classicist – maintaining that the symbolic representation in question (when generalized) yields correct psychological characterizations – need not worry about there being a limiting case correspondence, since it is not a reduction in the classical sense and does not yield elimination. At least it does not yield any compelling logical reasons to give up the belief in autonomy. But the connectionist, if believing in the subsymbolic nature of mental events, can as well argue that the classicist is wrong since the symbolic representation is not reducible. He could also claim (by using the notion of counterfactual explanation) that subsymbolic theories are good in the sense that they show under what conditions the classicist would be right (or 'almost' right) and why the classicist is in fact wrong.

On the other hand, the above results, if true, make the cognitive import of functionalism questionable. If a (more) functional theory can be understood, even in a roundabout way, and counterfactually explained by means of a (more) structural theory, it is kind of conceptually dependent on the latter. Thus it seems that Smolensky is correct when he argues that the different theories (he is talking here about classical and connectionist models) should intimately collaborate to provide an understanding of cognition. The subsymbolic paradigm in cognitive science does not just implement classical accounts. He does not deny the worth of classical accounts either, because they provide explanations of an enormous range of classical phenomena for which lower-level accounts are infeasible and because they have historically guided the discovery of the lower-level principles. More generally, in view of our results − more or less hypothetical, though − it does not seem to make much sense to one-sidedly emphasize functionalist or reductionist attitudes.

7.5.3. *Propositional vs. Nonpropositional Knowledge*

It is often said, and this has been my assumption in the preceding two sections, that the symbolic representation of knowledge is external to connectionist systems.[52] It does not belong to their architecture, but they can learn to represent symbols and concepts by means of non-symbolic processes. If this holds and if connectionism models human cognition, the same can be said about human symbolic representations too. Whether it more or less correctly models human cognition is of course one of the most controversial problems in cognitive science, there being disagreement not only between the connectionist and symbolic paradigms of cognition but among researchers working on artificial neural networks. There are views according to which connectionist networks can model neural structures and processes in some approximate and abstract sense, and, in fact, human cognition in general. Smolensky's above conjecture (7.5.2.1) is one expression of such a view.

According to many connectionists, the processes belonging to human cognitive architecture that give rise to beliefs should not be thought of as logical operations of propositions nor mental states in terms of propositions. Instead, the formation of a new belief is analogous to what takes place in a connectionist network when it is presented with an object or pattern which causes that its units are activated. Knowing is analogous to the fact that the network recognizes this object or pattern, which means that the weights of its connections receive appropriate values.[53] Symbolic knowledge representations −

as described in the classical model of cognition and in the connectionist model at the conceptual level – concern propositional knowledge,
describable by phrases of the form 'x knows that . . .', whereas connectionist systems basically (i.e., at the subconceptual level) model
nonpropositional knowledge. The latter can sometimes be expressed
by means of phrases of the form 'x knows how . . .', but it is more
general in nature. It is sometimes called experiential knowledge and
sometimes skill, and, perhaps, tacit knowledge.

Let me assume from now on that one purpose of the subsymbolic
and symbolic paradigms is to model human cognition, no matter how
controversial the assumption might be.[54] Taking that for granted, I
next consider some consequences that the relation between two cognitive theories – such as TR and NR, discussed above in Sections 7.5-
6 – may have for the understanding of connections between propositional and nonpropositional knowledge and between individual and
cultural (or public) knowledge.[55] It seems that it is not unwarranted
to regard the model of symbolic representations in the classical sense
of the word, that is, as related to the symbolic paradigm, as a normative and idealized theory, whereas subsymbolic representations would
be seen as providing a more realistic picture of human knowledge and
learning, and their logic.

In Section 7.5.1 a limiting case (conceptual level) correspondence
between the symbolic and subsymbolic theories TR and NR was ab-
stracted, and its implications were discussed in the preceding section.
Let us now try to see if anything can be said of the relation between
the conceptual and subconceptual levels within a connectionist system
N of the kind constructed above, and of the possible consequences of
the relation. If it is assumed, as I tentatively did above, that the system models human cognition, then there are consequences in that direction, that is, one may learn something about how propositional
knowledge is connected to nonpropositional knowledge. If not, the
following investigation only concerns artificial networks.

The representation described by NR, by the sentence θ' in the first
place, is conceptual, as we have argued. At the subconceptual level,
the most important description would concern the transformation of
an input (which represents an element of Exp$L$ or else a sequence not
in Exp$L$) into the respective output, and, moreover, the question of
why this input-output relation changes as it does if the number of

units increases and goes to infinity.

It should be obvious that the relation between the conceptual and subconceptual level descriptions of N is not a limiting case correspondence, so the relation is not analogous to that between classical mechanics and quantum mechanics. There seems to be no evidence for such a relation. Occasionally Smolensky seems to suggest this analogy,[56] as I remarked in Section 7.5.1, though it is likely that he means something else, which I shall study a little later in this section.[57] The relation seems to be more like a reduction in the classical Nagelian sense. As soon as an input is given and the network is trained to 'know' how to adjust connection weights to correctly respond to this input, a certain output follows, at least with an appropriate probability. This means that the internal structure and activation rules of N determine, possibly in a statistical sense, the output when the input is given. Hence, if the rules are deterministic and they are known for all units, that is, if there is a subconceptual level theory S describing the architecture of N that is deterministic and can be specified by the explainer, he can explain the output from the input in the D-N sense of explanation; and if they are stochastic, in a statistical sense. Since θ' describes this input-output relation, it seems to follow in the former case, at least roughly speaking (that is, up to a logical reconstruction), that something like θ' – that is, (7.5.1.5) – is a logical consequence of S.[58] S could be a theory describing the process in which the weight matrices of N connect inputs and outputs.

To lay this consequence relation open, one should specify the exact logical structure of S. Furthermore, S should be a sufficiently complete subconceptual description of N. As can be seen from Sections 7.5.1 and 9.3, the architecture of N is not completely defined by NR since there we were only interested in the correspondence relation between TR and NR at the conceptual level. In general, if a network is deterministic, the respective mathematical description, like S, would be available if enough were known about the network. However, this much knowledge may not be available if there are hidden layers in the network. As connectionists, like Bechtel and Abrahamsen, sometimes argue, typically there is no simple description for the regularities of hidden units.[59] According to Smolensky, on the other hand, conceptual-level descriptions can be derived from subconceptual ones, but they are incomplete, imprecise, or informal in relevant senses of the words. If a network is stochastic, one must be satisfied with probabilities and with a statistical or probabilistic inference of the input-output relation from the possible subconceptual description of the network.

Despite this uncertainty concerning the relation between the subconceptual and conceptual level descriptions of networks, such as N, let us nevertheless try to investigate the connection between nonpropositional and propositional knowledge. Let us consider N for a while, and assume tentatively that a sufficiently complete subconceptual description S of N exists. Then it follows from what we have said above that S describes the mechanism of how an input is transformed into an output (e.g., how connection weights are adjusted), that is, it tells in a general way how nonpropositional knowledge works in N. It does not express the content of that knowledge itself, since it is nonpropositional – assumed to consist in dynamic processes of N and to be related to activation values of individual units and adjustments of connection weights.

Now it is sometimes said (metaphorically) in such a case that S describes N's capability of *knowing how* to transform a given input into the correct output, i.e., it describes nonpropositional knowledge of N. The piece of propositional knowledge of N resulting from an input may be said to consist in its ability to recognize in every test case whether or not the respective string of symbols belongs to the set Exp$L$. This amounts to saying (metaphorically) that if it manages to recognize it by giving an appropriate output value in every test case, it in fact *knows that* the string is or is not in the set, as the case may be. Some connectionists make a distinction of this kind between propositional and nonpropositional knowledge of connectionist (even artificial) networks.[60]

7.5.4. *Individual vs. Public Knowledge*

Let us now assume that the network N is typical, whence we can generalize and say that 'N' refers to any network of a similar kind. If what I have said about N models human cognition more or less correctly, as I have tentatively assumed, then knowledge of the above kind might be called, after Smolensky, *individual knowledge* – in the sense that "[i]t is not publicly accessible or completely reliable, and it is completely dependent on ample experience."[61] Now, if a theory S of the above kind is available, describing one's individual nonpropositional knowledge in a general way, it follows that it explains, either in the classical deductive or statistical sense, the piece of one's individual propositional knowledge that is brought about by an input, as, e.g., a retinal impression. Let us suppose that the content of this propositional knowledge is formulated by a sentence φ (of an appropriate mathematical or some other coded language) and the respective input and possibly other conditions needed by a sentence ψ. Then the following holds (if it is assumed that S is a sentence or a set of sentences):

(7.5.4.1) S, ψ / φ,

where '/' means inference in a deductive or statistical sense.[62]

Assume that the above 'more or less correctly' does not only mean what it directly says but it means what is proposed in (7.5.2.1)-(7.5.2.2), above. If S, as above, is a subconceptual level theory of N of the kind referred to in (7.5.2.1), it approximates an appropriate neural theory, say S', i.e., a subconceptual description of the brain. If (7.5.2.2) holds, there obtains a counterfactual, possibly even a limiting case, correspondence of S to S'. It follows from (6.2.2)-(6.2.3), in Section 6.2, that if these descriptions are exact enough to admit a representation in model theory (whence 7.5.4.1 amounts to a logical consequence in some logic) and if ψ is logically weak enough, then for appropriate sentences $\sigma_0, \ldots, \sigma_n$,

(7.5.4.2) S', $\sigma_0, \ldots, \sigma_n$ / $I(\varphi)$,

where I is a translation and some of the auxiliary conditions among $\sigma_0, \ldots, \sigma_n$ are counterfactual (with respect to what is considered holding in actual worlds). On the other hand, if a model-theoretic formulation is not possible, it may still be possible to have a correspondence sketch (in the sense of Section 5.1), whence (7.5.4.1)-(7.5.4.2) would mean more informal consequences. In any case, if pragmatic conditions of the kind discussed in Section 6.4 hold, the neural theory S' provides a counterfactual explanation (via the translation) of the piece of propositional knowledge represented by φ. As a special case, S' would provide a counterfactual explanation of the theory S itself:

(7.5.4.3) S', $\sigma_0, \ldots, \sigma_n / I(S)$.

Now one could conclude that in view of (7.5.1.1) conditions of the form (7.5.4.1) approximately describe, in terms of connectionist networks, the relation obtaining between propositional and nonpropositional knowledge of the humans, when the latter is (approximately) described at the subconceptual level, by means of S. But if hypothesis (7.5.2.2) holds in addition, then conditions of the form (7.5.4.2) show how a certain kind of propositional knowledge (i.e., representable in the form $I(\varphi)$, where φ belongs to the connectionist language) is precisely connected to nonpropositional knowledge, which is presented by a subconceptual neural theory.

This may be too good to be true. Massive connectionist networks with hidden units may not be, or usually are not, describable by such theories as S. More importantly, conjectures (7.5.4.1) and (7.5.4.2) are based on insufficient evidence. Even though they turned out to be correct at some point, it may not be possible to have a neural theory like S' that would be exact enough to admit a comparison with S in the sense of (7.5.4.2) and (7.5.4.3), or else expressible in such a language that a comparison of this kind would be practicable even in a sketchy way. Furthermore, the above results might yield too much to reductionism. But if anything of the kind suggested above is feasible, we might be a little closer to the understanding of the relation between the two kinds of knowledge of the individual level.

Let us now study the distinction between individual knowledge and *public knowledge* (or cultural knowledge), in Smolensky's sense, by exploring the infinite limit of N.[63] Let such a limit network of N be N_∞. It is described by the same condition, θ', as N, but now m is said to be an infinite (nonstandard) natural number, so (according to the conditions stated in Section 7.5.1 for a^+ and a^-) $a^+(i, m)$ is infinitesimally close to 1 and $a^-(i, m)$ to 0.[64] What Smolensky means by public knowledge is presumably propositional, and, moreover, it has such properties as being publicly accessible and expressible, reliable (in the sense that it admits of public checking), and universal. Public knowledge and its rules are analyzable at the conceptual level. It admits of procedures that different people can reliably execute by using linguistic step-by-step instructions, and in this respect it is related to Turing machines and von Neumann computers.

It is obvious that this notion of public knowledge is more restricted than what is commonly understood by the word, and therefore its relation to the above notion of individual knowledge seems to conform well to the distinctions we have made when studying the limiting case correspondence of TR to NR. Assuming again that T and N are typical representatives of Turing machines and connectionist networks, respectively, and (despite its possible infeasibility) that N (and its descriptions) can be used to approximately represent human individual knowledge, it is quite straightforward to say that the representation of public knowledge in Smolensky's sense is in a limiting case correspondence to the representation of individual propositional knowledge. Then one may, on relevant pragmatic and logical conditions, be able to provide a counterfactual explanation of the former in terms of the latter, and ultimately, in view of (7.5.4.1), in terms of the above representation of individual nonpropositional knowledge – and even in terms of the subconceptual neural theory, if (7.5.4.2) were true.

If this is the case, it implies, on one hand, that public knowledge enjoys some degree of autonomy, that is, independence with respect to individual knowledge. On the other hand, since there is a limiting case correspondence of the representation of the former to that of the latter (in the propositional sense), there is a certain conceptual relationship (inspite of meaning variance) between them, which is of the sort explained in earlier chapters. Therefore, the former is conceptually dependent on the latter, and if there also is a relation of the kind explained in the above between individual propositional and nonpropositional knowledge, the same holds for cultural knowledge and nonpropositional knowledge.

What this all really means remains to be investigated. But it seems to follow, assuming that communication is propositional, that something of communication breakdowns – in so far as they reflect subjective or other local differences in experience or in subsymbolic processes – can be explained by means of there being relations of the above kinds between different types of knowledge representation. Such local differences may cause conceptual differences at the cultural level, so to speak. Not everything of such breakdowns is so explainable, however, if cultural knowledge enjoys a degree of autonomy.

7.5.5. *Connectionist Epistemic Logic*

The problem of how individual and public knowledge are related to each other is relevant to certain much-discussed questions concerning epistemic logic, or, more generally, the logic of propositional attitudes. In what follows, I indicate some possible consequences of the relation for that logic. Even though some of the above ideas about the different cognitive models and their relations are hypothetical to some extent, they nevertheless may open new perspectives to logical problems – in so far as epistemic logic can be thought of as a descriptive theory of human epistemic processes. If we are not entitled to assume so, then the logic that I shall ouline below is only relevant to artificial networks. In any case, from the above kind of limitivist relation holding between symbolic and subsymbolic representations – realized by means of an appropriate Turing machine and connectionist network, respectively, such as T and N, above – it would follow, on some plausible assumptions, that a normative epistemic logic is a limiting case of a more realistic, descriptive epistemic logic. If we assume this distinction between normative and descriptive epistemic logics, then the limiting case correspondence, on one hand, and the above relation between conceptual and subconceptual descriptions, on the other, may explain why individuals are not logically omniscient and why they make mistakes even in relatively simple tasks.[65]

What I here call normative refers to a logic based on classical logic or something similar. It is, of course, controversial and probably unwarranted to say that something like classical logic yields an ideal that human reasoning should pursue, but I do so for historical reasons. Thus, for example, those working on epistemic logic commonly think so, and evidence to the contrary has raised arguments against the relevance of epistemic logic.[66] More importantly, however, if there is a distinction between public and individual knowledge, there must be a similar distinction between the logics they possibly obey, and saying that one of them is classical can also be thought of as a methodological device, which can be changed if needed.[67]

If there is a logic that can be associated with subconceptual (connectionist) knowledge, it is a kind of *internal* logic (of a connectionist system). In the same spirit, let us call *external* the logic (of the system) that conceptual individual knowledge possibly obeys. The latter

results when the system is taught to recognize some logical principles. If a classical symbolic representation (in a Turing machine) of
public knowledge is in a limiting case correspondence to a subsymbolic representation (in terms of a connectionist system) of individual
knowledge, as proposed above, we may expect that classical (propositional) logic (also representable by means of a Turing machine) is
a limiting case of the external logic of the connectionist system in
question. This is one way to ground the above proposal concerning
the limiting case correspondence of an epistemic logic in the normative sense to an epistemic logic in the descriptive sense. In what follows, I shall only sketch the external logic and an epistemic (propositional) logic of connectionist systems, not the limit procedures.
They are logics suggested to us by the kind of process by means of
which a system learns to recognize patterns.

Consider once again a connectionist network, say M, which is, if
needed, a variant of the network N discussed earlier – i.e., if needed,
we may suppose that it has two or possibly more output units and a
slightly different structure of connections. To simplify matters, we
shall assume, however, that M has only one output unit, whose activation values are determined by two activation functions, analogously
to what was assumed about N.[68]

Now think of the language L as a formal language of classical
propositional logic. In view of what was said about how N learns to
recognize whether or not a string of symbols is an expression of L, it
is evident that a network like M can equally well be trained to recognize whether a given atomic sentence of L is considered true or false
by the trainer and how the truth-value of a more complex sentence (a
well-formed formula of the language) is determined, that is, how the
truth-functional nature of classical connectives is determined. In practice, only a finite number of sentences can be handled, but if we are
willing to idealize, we may consider all atomic sentences. But it is
more important to notice that the network can learn to apply truth-tables to many cases of complex sentences and this way learn to recognize some logical truths, that is, classical tautologies in this case.
It is evident that something about substitution can also be taught to it,
and this skill it needs when generalizing.

If an input pattern of M or part thereof encodes a sentence φ of L, I
suppose for simplicity that the respective activation value of the out-

put unit indicates whether or not the truth-value of φ is recognized by
M. Analogously to what was said about the structure of the earlier
network N, we stipulate that M recognizes that φ is true if it yields an
output that is sufficiently close to 1, and false if the respective output
is sufficiently close to 0.[69] If M does not identify the truth-value of φ,
the output is somewhere in between.

Let w_0 be the 'actual world' to which truth and falsity are related. It
is thought of as a classical, i.e., 'normal' world, obeying the classical
truth-tables. Let us next consider a *state* of M, call it w. By a state I
mean, intuitively speaking, the ability of M to adjust (after some epochs of training or testing) its connection weight values so as to be
capable of propagating activations in an appropriate way (either in the
training mode or test mode). Since this notion of state would not be
very accurate, let us conform to the standard usage and say that a state
is a pattern (vector) of activation values of all units of M. A given
state is therefore determined (up to some probability, perhaps) by a
weight matrix of M and a pattern of input values.[70]

The state of M will change when a new input pattern is presented to
it and it will also change when a given pattern is presented repeatedly
if weights change in the process of learning. If M has learned to recognize an input, it is self-organizing, which means that it is capable of
adjusting the weight matrix so as to have a sufficiently good output.
In such a case, something like conditions (7.5.1.5) can be used to describe the input-output behaviour of M.[71]

In the context of epistemic logic, let us call a state a ('nonnormal')
possible world, that is, a world that is epistemically possible.[72] If a
sentence φ is presented to M, the respective output value (the activation value of the output unit corresponding to φ in the sense discussed
above) will be related to the valuation of φ at w, and it will be denoted
by $v(\varphi, w)$, or simply by $v(\varphi)$ if no confusion is going to arise. As
we have supposed, it belongs to the open interval (0, 1), whereas the
desired ouput is either 0 or 1, depending on whether φ is considered
false or true by the trainer, that is, false or true at w_0.

Now I shall say that

(7.5.5.1) φ is true at w iff $v(\varphi)$ exceeds a certain value, say r, that
              is greater than or equal to 0.5.

It is natural to stipulate that r is identical with the output value (determined by the trainer) indicating that M recognizes that φ is true. As usual, there are two alternatives to name a case that φ is not true at w, and they correspond to different intuitions concerning nontruth. The first one is to say that

(7.5.5.2) φ is false at w iff it is not true at w, that is, iff $v(\varphi)$ is at most r.

According to the other, more natural, option,

(7.5.5.3) φ is false at w iff $v(\varphi)$ is close to 0, which means that it is less than $1-r$, and undetermined iff it is neither true nor false, that is, iff $v(\varphi)$ is between r and $1-r$ (these values included).

The latter construal is more natural since we have stipulated that the network recognizes that φ is false if the output is close to 0. Therefore, I shall in what follows choose (7.5.5.3) rather than (7.5.5.2). However, it is important to notice that the kind of epistemic logic, to be defined below, to which the connectionist notion of learning leads, is independent of which alternative is chosen.

As far as the truth-functional role of classical propositional connectives is concerned, that is, their role at w_0, the aim of the trainer is to teach M their truth-conditions (ordinary truth-tables), and this implies that if M learns (relative to the given degree of the output value r or $1-r$, as the case may be) something of the role of the connectives in complex sentences, it also learns to recognize some classical tautologies. One way to teach M the connectives, given that the ideal, classical truth-values correspond to 1 and 0, is to teach it the following rules (familiar from some well-known many-valued logics), where φ and ψ denote sentences of L:

(7.5.5.4) $\quad v(\neg\varphi) = 1-v(\varphi);$

$\qquad v(\varphi \vee \psi) = \max\{v(\varphi), v(\psi)\};$

$\qquad v(\varphi \wedge \psi) = \min\{v(\varphi), v(\psi)\};$

$\qquad v(\varphi \to \psi) = v(\neg\varphi \vee \psi).$

$\qquad v(\varphi \leftrightarrow \psi) = v((\varphi \to \psi) \wedge (\psi \to \varphi)).$

From a point of view of the trainer, this many-valued semantics in terms of output activities is what the internal logic of M should obey, since in the cases of recognition—that is, when truth-values are true or false in the senses of (7.5.5.1) and (7.5.5.3) — these truth-conditions for connectives are consistent with the classical truth-tables. There are some extra problems here, however. Thus it is not clear, for example, whether (7.5.5.4) allows teaching $v(\varphi)$ independently of $v(\varphi \lor \psi)$, as it perhaps should be taught, and similarly for the rest of the dyadic connectives. On the other hand, even though from a logical point of view conditions (7.5.5.4) would express optimal ideas of what the internal logic should look like, they are not necessarily needed. A network can be trained (and more easily) to recognize something of the truth-functional role (truth-tables) of the connectives $\lor, \land, \rightarrow$, and $\leftrightarrow$ without such exact quantitative constraints at the level of output activities as those in (7.5.5.4). The more coarsegrained semantic level of the connectives (that is, the semantics with only the truth-values true, false, and undetermined, as defined in 7.5.5.1 and 7.5.5.3, above) does not entirely depend on such exact numerical details as those in (7.5.5.4). It is sufficient to teach the network, e.g., that $\varphi \lor \psi$ is true if one of the values $v(\varphi)$ and $v(\psi)$ exceeds r without teaching it that $v(\varphi \lor \psi)$ is to be identical with the bigger one of the values $v(\varphi)$ and $v(\psi)$, and similarly for the other dyadic connectives. In a sense, either semantics approaches classical semantics when the values of v approach 1 or 0.

It is evident that the combination of the syntax of L and the three-valued semantics defined by (7.5.5.1) and (7.5.5.3) can be construed as *a logic* in the sense of logic to be discussed in Chapter 8, below, even though it is not truth-functional; and similarly for the many-valued case as exhibited in (7.5.5.4). Clearly, then, the former can be considered as what was earlier called an external (individual) logic of M and the latter as an internal one.

It is easy to see that not all classical tautologies are true at a given w in cases where M has not yet learned to recognize them, as, for instance, if w belongs to the training mode or if it belongs to the test mode but M has to generalize from what it already knows. Whenever φ is a sentence whose truth-value has already been taught to M, we may suppose that $v(\varphi, w) > r$ if φ is true at w_0 and $v(\varphi, w) < 1-r$ if φ is false at w_0, for all relevant worlds w belonging to the test mode. That is, even if the network is stochastic, we may assume that once it has learned to recognize the truth-value of a sentence, all actual tests yield a correct output for that sentence and all possible tests would do so. Only in cases where it has to generalize, tests may occasionally yield wrong results. Thus, analogously to what we said in Section 7.5.3 about the propositional knowledge of the system N – about what it means to say that N knows that a string of symbols is an expression of a given language – we may now say that M *knows* at the actual world w_0 *that* φ if it recognizes its truth at every state w' of the test mode, that is, if $v(\varphi, w') > r$ for all such w'. Likewise, M knows at a world (state) w belonging to the test mode that φ if it recognizes its truth at every state w' following w. As will be seen below, however, for logical purposes it is necessary to abstract from actual tests and states of M, so as to admit possible but unrealizable states too, in some relevant sense.

Consider now a Kripke model $\mathfrak{M} = \langle W, R, w_0, V \rangle$, where W consists of w_0 and of nonnormal worlds (that is, network states in the above sense) of M, and R is an appropriate alternativeness relation in W, for example, such that wRw' when w is w_0 and w' belongs to the test mode, or when both w and w' belong to the test mode in such a way that w' is later than w. Let us assume, furthermore, that R is reflexive, since this assumption is natural in view of what has been said above about learning. There may exist other natural ways to define R, but it is not clear yet what they might be. V is a valuation complying with v for nonnormal worlds and having values 1 and 0 for the classical world w_0. Now it is a direct consequence of the above assumptions that it is natural to extend the notion of truth for the language that is obtained if an operator, say K, which corresponds to the frase N *knows that*, is added to L by stating, as usual:

(7.5.5.5) Kφ is true at w iff φ is true at every w' such that wRw'.

Furthermore, we define:

(7.5.5.6) A sentence φ is true in $\mathfrak{M}$ iff φ is true at w_0.

The reader may have noticed that we have not defined truth values at nonnormal worlds for complex sentences that are combinations of the connectives and the operator K, as, for example, for sentences of the form K$\varphi \vee$ Kψ, $\neg$Kφ, and the like, since conditions (7.5.5.1) and (7.5.5.3) only concern the language L. However, this (somewhat defective) kind of logic seems to result if we conform as closely as possible (by only abstracting to the extent to which it is necessary to have any logic at all) to connectionist principles of learning classical propositional logic. Now, if we do not want to go further with truth conditions, we have two options. The first one is to assume that L will be expanded by adding K in the usual way by accepting all admissible combinations of K and the symbols of L. Then we must accept the fact that there are sentences which either do not have truth values at all at a normal world or the fact that their truth values are arbitrarily chosen. Both positions would cause some additional logical problems and necessitate further conventions. The second option is to define a restricted notion of formula, in such a way that the inconvenient combinations are excluded, that is, within a scope of K the connectives are not applied to formulas in which K occurs. It seems that both alternatives provide realistic bases for epistemic logics in the present context, that is, for realistic, descriptive theories of connectionist knowing – and if that approximates human knowledge processes, of human knowing. The two options just correspond to two slightly different intuitions about the logic of knowing.

Let us consider the second option more closely. Think of all Kripke models of the above form and define the validity of a sentence in the usual way. This means that we have to abstract and generalize from networks proper. If Ω is the (nonempty) set of sentences of L from among which any (abstract) network has learned to recognize the respective true ones and the false ones,[73] then the Rule of Necessitation will be semantically valid in the following restricted form:

(RNΩ)   For all sentences φ in Ω, if φ is valid, then the sentence Kφ is valid.

If R is also assumed to be transitive, it is evident that all the axioms of S4, when restricted to the language in question, are valid, and *modus ponens*, as well, whence the present semantics affords a (sound) semantics for the subsystem of S4 in which the notion of formula and the Rule of Necessitation are restricted as indicated. If R is defined in such a way that it is only reflexive, we have a (sound) semantics for a subsystem of T that is a special case, say SΩ, of the system TΩ defined in my (1982a). There the notion of nonnormal ('impossible') world was to some extent *ad hoc*, but now we have seen that it can be given a natural construal by means of networks.[74] Whether the present semantics is complete and yet corresponds to connectionist learning in an abstract sense depends on what kinds of abstract networks are considered admissible. Thus, for instance, if we want to have a number of models available that together would play the role of something like a canonical model in the ordinary sense, we must admit abstract networks that learn the truth-values of infinitely many sentences of the language.

With such reservations, the logics corresponding to these subsystems seem to correspond in a natural way to what knowing means for connectionist networks. If network learning is at all analogous to human learning, as Bechtel and Abrahamsen (1991) suggests, then these logics can be considered as descriptive theories of human propositional knowing in the sense discussed above. They do not give rise to logical omniscience and they simulate important features of how one learns logic in the semantic sense. Even if human learning were different from connectionist learning, it is not too farfetched to imagine that it also contains epistemic states at which various sentences are recognized true or false. If this is the case, then the above results are *mutatis mutandis* applicable.

It should be obvious that the logic for the full system T (that is, a normative one) is in something like a limiting case correspondence to the logic for the present subsystem $S\Omega$ of $T\Omega$, and similarly for S4. One has to be careful here, however. Since the language of $S\Omega$ is restricted, it is evident that there is not an appropriate translation of the full language of T into the language of $S\Omega$, which means that one either has to consider the similarly restricted version of T, that is, the version of T whose language is restricted in the same way as that of $S\Omega$, or one must be satisfied with a partial translation of the language of T in the sense mentioned in Section 5.3, above.

7.5.6. *Propositional Knowledge and Emergence*

Let us next investigate the above theories in the light of the notion of emergence, as studied in Chapter 5. It is often said by cognitive materialists that consciousness, mind, psychological properties, and the like, emerge from the processes of the brain. As we have seen, emergence is often thought of as being incompatible with reduction, but we have noticed that whether they are to be considered mutually in-

compatible depends on the notion of reduction and of course on the
intuitive understanding of what emergence is.

Assume first that we are working on a connectionist network N,
similar to the ones considered above, which can be trained to recognize a pattern, say a sentence φ, coded as an appropriate input pattern.
So we assume that the input is fixed in all test trials. Let the natural
numbers n belonging to the closed interval $[n_0, n_1]$ represent the consecutive test trials for φ. In each trial, the respective state (determined
by a weight matrix and input) can be slightly different and we may assume that it depends on n. Let $S(n)$ be a description telling how in the
special case investigated here the output of N depends on its input and
state, for each test trial n. Let now $a(n)$ be the output function. Obviously $S(n)$ can be derived, by means of the assumptions referring to
this special case, from a more general theory of networks, but this
simplified notation is sufficient to help us to understand the notion of
emergence in this context. If N is trained to recognize φ, then, according to what was said above, something like the following holds for all
n in the interval indicated:

(7.5.6.1) $S(n) \,/\, a(n) > r$,

where r is a fixed real number determined by the trainer of N. Now
one cannot right away say that propositional knowledge is reducible
to the theory $S(n)$, or to a more general theory of the network, or explainable or otherwise derivable from it. We saw above that we may
stipulate that N knows that φ if the condition $a(n) > r$ obtains for all
cases n of the test epoch. If we now make the agreement, let it be δ,
that

(7.5.6.2) N knows that φ iff for all n in the interval, $a(n) > r$,

then it follows that

(7.5.6.3) $\forall n \in [n_0, n_1] S(n), \delta \,/\, K\varphi$,

where, as above, K is shorthand for 'N knows that . . .'.

Therefore we may obviously say that $K\varphi$ is *reducible* in Nagel's
sense to the theory $\forall n \in [n_0, n_1] S(n)$, in so far as the linking assumption δ is acceptable. The assumption, connecting the two languages of
quite different kinds, is clearly independent of the theory both theoretically and empirically, yet it is intuitively justified, to some extent at

least, and grounded by what connectionists are doing, so that it is not completely arbitrary. Hence it is conceivable, in view of what we saw in Section 5.10, that Nagel would be willing to say that the property of knowing in the propositional sense of the word is in the present context *emergent* relative to this theory of the network.

These questions become much more problematic if we assume, as we tentatively did earlier in this chapter, that this kind of approach is analogous to, or represents approximately, human cognition. In such a case, the propositional knowledge in question was called individual knowledge; and a crucial question is then whether that knowledge is reducible to or whether it is emergent relative to a neural theory – for instance, to something like $\forall n \in [n_0, n_1]S(n)$. Let us focus on the problem of indepence of assumptions, such as δ. In the context of connectionist networks, they are almost entirely stipulative (nominal definitions), though not perhaps *ad hoc*. The crucial problem with respect to neural theories is whether in the future the brain research will be sufficiently advanced so as to be able to suggest, e.g., that a person knows something when some measurable 'output', whatever it might be, exceeds a certain value. Such a result must also be empirically justified, and, therefore, the respective linking assumption, analogous to δ, would be a real definition rather than a stipulation. There is no doubt that then one would be entitled to say that this cognitive feature is reducible to brain processes. The linking assumption would not be independent in the same sense as the above assumption δ concerning artificial networks and the earlier one linking temperature and kinetic energy.[75] Thus it is questionable whether we could anymore speak about emergence.

Back to networks. We must ask whether we are really entitled to say that a network knows that something is the case if it is successful with respect to some finite, even small, number of test trials. We may at most say that it knows *relative* to that test epoch. This agrees with saying, as in the previous section, that something is known at a certain world of a Kripke model. What would it mean then to say that it 'definitely' or 'absolutely' knows? One interpretation is that knowing takes place relative to all possible test epochs, i.e., the validity in an appropriate class of Kripke models. Another interpretation is that we counterfactually extend the test epoch to infinity and suppose that the network is successful in all trials in that sequence. Then knowledge would be a limiting case of that sequence of the relative kinds of knowledge. We may perhaps say, referring to Primas,[76] that knowledge emerges in the limit as a new property.

7.6. MODAL LOGIC AND CLASSICAL LOGIC

The reduction of propositional modal logic to classical predicate logic $L_{\omega\omega}$ is rather simple and well known, but it may of some interest to consider the question in terms of the present notion of correspondence relation. This will be here the only case study of correspondence relations of logics, since it suffices to illuminate the definition in Section 5.4.

Consider any modal logic L whose models are Kripke models of the form

(7.6.1) $\mathfrak{M} = \langle W, R, V \rangle$

and whose syntax consists of propositional variables $\mathbf{p}_0$, $\mathbf{p}_1$, . . ., ordinary connectives, say $\neg$ and $\wedge$, and the box $\square$. Then a predicate logic L' to which L is in a correspondence relation has in its type a dyadic predicate symbol $\mathbf{R}$ and countably many unary predicate symbols $\mathbf{P}_0$, $\mathbf{P}_1$, ... So its models are of the form

(7.6.2) $\mathfrak{M}' = \langle M; R', P_0, P_1, \ldots \rangle$,

where R', P_0, P_1, . . . are the interpretations of the predicate symbols in $\mathfrak{M}'$. Now if $\mathfrak{M}'$ is any model of that form, then it is readily transformable into a model of the form (7.6.1) by taking $W = M$ and $R = R'$, and defining $V(\mathbf{p}_i) = P_i$ for all $i = 0, 1, \ldots$. This defines the required correlation F, whose domain is the the class of all models of L' in which the relation R' satisfies the same condition as the alternativeness relation R characterizing the logic L. Conversely, if $\mathfrak{M}$ is any model like (7.6.1), it is clear how a model of L' is obtainable. So the range of F is the class of all models of L.

Now a translation I_0 of the L-formulas into L'-formulas is defined as follows:

(7.6.3) $I_0(\mathbf{p}_i) = \mathbf{P}_i(\mathbf{x})$;

$I_0(\neg\varphi) = \neg I_0(\varphi)$;

$I_0(\varphi \wedge \psi) = I_0(\varphi) \wedge I_0(\psi)$;

$$I_0(\Box\varphi) = \forall \mathbf{y}(\mathbf{R}(\mathbf{x},\mathbf{y}) \to I_0(\varphi)\mathbf{y}/\mathbf{x}),[77]$$

where $\mathbf{x}$ is a fixed variable of L' (e.g, the first one in the alphabetic order), $\mathbf{y}$ is any variable different from $\mathbf{x}$, and $I_0(\varphi)\mathbf{y}/\mathbf{x}$ means that $\mathbf{y}$ is substituted for $\mathbf{x}$ in $I_0(\varphi)$. Now it can be shown by induction that

(7.6.4) $\mathfrak{M}, a \models_L \varphi \Leftrightarrow \mathfrak{M}' \models_{L'} I_0(\varphi)[a],$

where $\mathfrak{M}$ and $\mathfrak{M}'$ are as above, $\mathfrak{M} = F(\mathfrak{M}')$, φ is any L-formula, and the left-hand side means that φ is true at a world $a \in W (= M)$ in $\mathfrak{M}$. It follows that

(7.6.5) $\mathfrak{M} \models_L \varphi \Leftrightarrow \mathfrak{M}' \models_{L'} I(\varphi),$

where

$$I(\varphi) = \forall \mathbf{x} I_0(\varphi),$$

so that condition (5.3.1) holds for $\langle F, I \rangle$. Therefore, L is not only in a correspondence to L' but even reducible in the sense of Chapter 5. That is, the domain and range of F are, respectively, a class of models of L' and the class of all models of L. The gist of this argument is, of course, that their respective models, like $\mathfrak{M}'$ and $\mathfrak{M} = F(\mathfrak{M}')$, are essentially identical.

It can be seen that what we have done above is extending a little bit what is usually called the *correspondence theory* in modal logic, in which conditions on the alternativeness relation R are translated into nonmodal predicate logic.[78] One should also notice here that similar arguments, but somewhat more complicated than those above, would show that a given modal predicate logic is reducible to a classical two-sorted predicate logic; and similarly, for example, for an intuitionistic predicate logic.[79]

—— PART THREE ——

THE FORMAL BASIS

OF

THE CORRESPONDENCE RELATION

THEORIES AND LOGICS

This chapter gives an overview of some formal principles and notions used earlier. They will also be needed in the exact treatment of some case studies in the next chapter. Logically educated readers will find most of the principles familiar from the literature, but the details are presented in the form that best suits my purposes. The basic idea is here that the logical tools chosen should not excessively constrain metascientific studies, but should provide sufficiently exact means whenever formal accuracy is needed. The former requirement means, first, that the logical framework employed has to have enough expressive power, and, second, that it must be adaptable so as to be able to combine with philosophical and pragmatic investigations. In addition to formal details, I consider, in the first section and at the end of the second, some questions of this kind, as related to the recent developments in logic that generalize the whole notion of logic (or, rather, of *a* logic). Furthermore, I make a brief historical digression, in Section 3, to explore Rudolf Carnap's anticipation, in the thirties, of what happened in logic and logical metascience somewhat later in the twentieth century. Carnap's enterprise seems to be of historical interest that is not generally recognized.

8.1. LOGIC AND APPLICATIONS: PROBLEMS

There has been some difference of opinion as to whether there is any point to suppose that philosophical research can or sometimes should be done by means of formal logic. In what follows, I shall discuss this intricate issue mainly in terms of the philosophy of science, but similar problems occur in other philosophical areas where formalizations are attempted, such as the philosophy of language and the analysis of propositional attitudes – and conceptual analysis in general. It seems, moreover, that there is now a general tendency among philosophers to avoid formal considerations as too restrictive, even wrong-headed, in so far as philosophical aims are concerned. Since part of the present book applies formal tools, this section investigates something of the controversy by evaluating drawbacks and advantages of using them.

Consider first the old question of why many philosophers of science used to prefer to employ first order (elementary) logic. Here are some obvious reasons. Because elementary logic is (both syntactically and semantically) well-behaved, as logicians often say, it seems to offer us a nice conceptual framework for metascientific studies. For example, the following arguments can be presented in its favor. When a theory is formalized in it, one knows precisely what is meant by such basic notions as 'primitive term', 'axiom', 'deduction', 'theorem', and so forth. Furthermore, since one now knows what deduction exactly means, one can try to explicate the meanings of such notions as deductive explanation and prediction. In the framework of elementary logic, such an old question as whether a theoretical term is definable by means of observational (nontheoretical) terms or whether it is eliminable in another sense receives an exact meaning.

On the other hand, any formalism associated with an empirical theory is of course thought of as being about those actual phenomena, even though by way of idealization, which the theory itself was created to describe. Here, then, the notion of model enters into the metatheory. It is known that the proof theory of elementary logic has several good properties, such as compactness and interpolation properties, which are now available in metascience, and it is also known that it has a very rich and workable model theory. It is complete, has various definability properties, the Löwenheim-Skolem property, and so on. Consequently, once a theory is thought of as being formalized in elementary logic, one does not only know what its models look like but one has much knowledge about their behavior in their relation to each other and to the theory.

Apart from its mathematically nice properties, the model theory of elementary logic can be efficiently used as a conceptual framework. It can be utilized if one wants to exactly define concepts that are felt to be metatheoretically important. Therefore, the model theory of elementary logic has been believed to do the same kind of useful work on one side of the coin as its proof theory on the other. All told, elementary logic seems to form a very natural formal framework for metascientific problems.

Now it would be possible to answer the question whether it is in some sense desirable to formalize an empirical theory in elementary logic (either directly or via set theory), rather than in some other logic. This could be done by studying which of its nice characteristics have been actually used in methodological investigations and which of them could be used, and, in particular, which of them are really relevant in connection with empirical sciences. But it is not my purpose to present any thorough study here – we know that many of the nice features of elementary logic have been employed, as, for instance, definability properties – since its having those nice properties has a negative side, which I next consider.

There has been some tension in methodology caused by the fact that what philosophers seem to have wanted from the syntax of first order theories, on one hand, and their semantics, on the other, are incompatible. This tension is similar to that in metamathematics, but philosophically more difficult to settle. It is connected with the 'weak' expressive power of elementary logic. If a theory is axiomatized in elementary logic in order to formally describe an empirical phenomenon or a mathematical structure, the resulting formal theory has models that intuitively seem to have too little in common with intended or standard models. This is due to those nice properties of first order model theory mentioned above. Results that express restrictions in the expressive power of elementary logic are central in its model theory. In metamathematics, the existence of nonstandard models has been put to work very efficiently, and this has led to fertile new theories and concepts.[1] As we have seen earlier, and shall see in a more exact way in later chapters, the same can also be attempted in metascience. An extra problem in metascience is the fact that the concept of nonintended or nonstandard model is not intuitively as clear there as in metamathematics, even though with respect to theories of mathematical physics, for example, the problem can evidently be handled more easily than with respect to many others.[2] That there are nonintended, even pathological, models is the prize that metascience has to pay for the nice properties of elementary logic.

Much energy has been consumed in efforts to get rid of such models, in order to obtain empirical interpretations of theories. Hence it seems that what philosophers often would like to do is to make use of good syntactic and proof-theoretic properties of elementary logic but not to accept the weak expressive power of its semantics. This holds even for logically simple theories that could be directly formalized in elementary logic in a perspicuous and suggestive way, but such empirical theories tend to be rather rare. Even if a theory can be subjected to a direct standard formalization, it is often the case (e.g., because of a strong mathematical basis needed) that the actual formalization would be far from being perspicuous, suggestive, simple, or whatever qualities are generally required from an axiom system. It is often the case that the weaker the undelying logic is, the more ingenuity is needed to construct an axiomatization, and the more difficult it is to intuitively understand the axioms. It is not the standard axiomatization, nor formalization at all, that makes one understand a theory. Rather, one must understand its laws and mathematical basis in some intuitive or mathematical way before one can understand it as a formal entity.

Moreover, formalizations in elementary logic may sometimes presuppose recourses to set theory. It seems, however, that besides their often being complicated in practice, it is not desirable to embed models of set theory into the models of a theory that is not intrinsically set-theoretic in character. In addition, a formalization via set theory always presupposes a decision concerning which set theory to use. This means that the theory as a formal entity — not only its metatheory — becomes tied to specific set-theoretic assumptions, and this is hardly the purpose of its axiomatization. The purpose is to find out what are the direct and minimal assumptions concerning the laws of the theory and its mathematical basis. On the other hand, one should not perhaps be too categorical in this matter, since whichever kind of formalization one chooses, one is anyway committed to those specific principles that belong to the underlying logic.

Although the expressive power of elementary logic is rather weak in the sense of allowing a large class of nonintended or nonstandard models, this feature can often be efficiently suppressed, as one can seen from Montague (1961). One may give a direct, standard formalization to rather complicated theories and obtain significant metatheoretical results by means of standard models (relevantly defined). Thus it seems that neither our above discussion about the complexity of actual formalizations nor the observation that Montague's work provides practically the only actual formalization – at least there are not many – of full-blooded scientific theories are entirely relevant arguments against preferring standard formalization. They are relevant only if one needs metatheoretical results concerning specific (or a specific kind of) theories. Then it is possible that one needs to know what the formal basis of a theory exactly is. But more often one is only concerned with more general kinds of metascientific questions, and has only to know whether a theory can be subjected to standard formalization. On this assumption, many general results have been obtained that are more or less dependent on the characteristics of elementary logic.

It is not clear whether such results that strictly depend on specific properties of elementary logic are relevant for the study of empirical sciences. It seems that much of the tendency to use elementary logic, perhaps together with its slight second order extensions, is due to its familiarity. If one thinks that formalizations are needed for some purposes, it could be worth investigating whether the merits of elementary logic are greater than those of some stronger, weaker, or deviant logic as far as the study of empirical sciences is concerned. If, for instance, the notion of deduction in elementary logic is considered a natural explication of what is meant by proof in mathematics, and if that explication is thought of as being important and relevant for the philosophy of empirical theories also, elementary logic can obviously be regarded as a natural tool. But if one wants to get closer to standard or intended models, or is interested in characterizing closer relations between models by syntactic means, and is ready to accept more general forms of deduction, one is committed to stronger logics. On the other hand, if a formalization of an axiom system is not the most important thing, it is even possible to regard a collection of classes of models as a starting point of an underlying logic and consider syntax as a secondary element.

These problems are analogous to certain well-known problems in pure logic. It has turned out that elementary logic – which for long was considered the paradigmatic logic – is incapable of capturing several central notions of mathematics and thus reflecting mathematical practice accurately enough. Therefore, logicians began to search for its extensions whose expressive power and applicability to foundational studies would be greater. Part of that research was directed at attempts to find extensions that, in addition, would have a workable model theory, for it was realized that properties of such logics as higher order ones, whose expressive power is strong enough, are not very well known, or those logics were not considered well-behaved. Somewhat later, logicians became more interested, for various purposes, in weaker and deviant logics and model theories, such as relevance logic, finitary model theory, and nonmonotonic logics. Such studies, those of extensions in the first place, created a new theory of logics, commonly called abstract logic or general model theory. In its framework it became possible not only to state a general definition of a logic, but investigate properties of logics and their interrelations.[3]

It may be illuminating to observe at this point that logicians have been uncertain about the relevance of the good properties of elementary logic outside pure logic. Thus Feferman (1974a) argues that the compactness theorem is out of place for ordinary reasoning and wonders if there is an area of reasoning where both the compactness theorem and the Löwenheim-Skolem theorem would be appropriate. He also refers to Kreisel's claim that the latter is not appropriate for languages intended for expressing mathematical reasoning.

Clearly mathematical physics, for instance, or any other empirical science using mathematics, is not such an appropriate area if mathematics itself is not. To be able to answer the question of which logic would be the most natural and workable extension of elementary logic for metascientific purposes, one should know what model-theoretic properties are relevant to metascience and what the proper mode of reasoning is there. As argued by many, something more is involved than mere mathematical reasoning. Any logic chosen is just a tool of analysis and more or less extensional. Thus one has to pay for attempts to keep an eye on both purely logical aspects and intensional aspects of an empirical theory. There should be a balance between the aims of a metascientific study and the logic used for the study. The rest of the present section considers this problem a little more closely, but refers to a more general range of philosophical applications, not just metascience.

When a formal logic is used for philosophical purposes, it has to fulfill two sets of criteria, somewhat opposite to each other, to be satisfactory. The first one pertains to its being logically adequate in the sense of satisfying certain minimum requirements as a logic, such as having well-defined syntactic and semantic rules, being consistent, possibly complete and sound, reasonably monotonic and extensional, and the like. These requirements are what logicians traditionally expect a logical system to satisfy.[4] Second, it must be intuitively adequate in the sense that its rules correspond, to a reasonable extent, to what is commonly thought of the nature of the subject matter in question. If, for example, the picture it gives of the subject matter is excessively idealizing or its semantics does not correspond to it well enough, it is no longer useful for explanatory and descriptive aims.

It seems evident that no complete correspondence between logical principles and our intuitive conceptions is possible, since the former are formal and exact and the latter are not. There is a tension between these two sets of criteria, as already noted. For instance, a formal logic obeying main principles of extensionality seems to violate intuitive adequacy, and if it is intuitively adequate, it violates extensionality.[5] Thus there appears to be a controversy between the two sets of criteria. For example, if we are to construct a formal modal logic that would adequately describe and explain epistemic notions, that is, that could be regarded as a descriptive and explanatory formal theory of epistemic notions, not too strongly idealized, we must give up some extensionality principles. Does it mean that we have to give up logical adequacy?[6]

There exist no definite conventions or intuitions concerning what conditions a logic should satisfy in order to be adequate as a logic. One may dispense with some criteria of logical adequacy without destroying logical adequacy in some minimal sense. One cannot dispense with the clarity of the basic syntactic and semantic definitions, nor with consistency and the like, but it seems that one can compromise with extensionality. Such compromising – which may increase applicability – usually results in a reduction of logical elegance and logical fertility, and may cause semantic troubles. Thus, for example, in epistemic logic yielding enough to intensionality possible worlds semantics in its familiar, original form has to be given up, and similarly for the other standard kinds of semantics. On the other hand, many of the solutions to this controversy presented during the last forty years or so seem to comply with at least some minimal standards of logical adequacy.

8.2. THEORIES AND THE ABSTRACT NOTION OF A LOGIC

Logics

For my purposes in this book, it is enough to characterize a *logic L* as something that is determined by (i) a nonempty set Typ_L of similarity *types* (*admitted* types of *L*); and for each type $\tau \in \mathrm{Typ}_L$: (ii) a collection of *formulas* $\mathrm{Form}_L(\tau)$, (iii) a collection of *models*, $\mathrm{Mod}_L(\tau)$, and (iv) a *satisfaction* relation $\models_L$.[7] A type is assumed to be a set consisting of sorts and (nonlogical) symbols. The collection of all *sentences* (i.e., elements of $\mathrm{Form}_L(\tau)$ with no free variables) of *L* will be denoted by $\mathrm{Sent}_L(\tau)$.[8] The models in $\mathrm{Mod}_L(\tau)$ are called *admitted* models of *L* of type τ, or simply *models for L*(τ). A formula belonging to $\mathrm{Form}_L(\tau)$ or $\mathrm{Sent}_L(\tau)$ may also be called an *L*(τ)-formula or *L*(τ)-sentence, respectively. Here and elsewhere in this book, variables and nonlogical symbols contained in types will be denoted by bold-face letters, and formulas by Greek letters.

The above is, roughly, Feferman's (1974b) notion. We assume, after Feferman, that a (many-sorted) model of type τ is a function assigning to each sort in τ a nonempty set (a domain) and to each symbol in τ a relation, function or individual of a specified rank and sort (to be specifed in each case) over appropriate domains. As usual, a model is often written as a structure $\mathfrak{M} = \langle M_1, \ldots ; R_1, \ldots ; a_1, \ldots \rangle$ where $M_1, \ldots$ are domains, $R_1, \ldots$ relations or functions, and $a_1, \ldots$ individuals. Furthermore, if **s** is a symbol in τ, then $\mathfrak{M}(\mathbf{s})$, which is often called the *interpretation* of **s** in $\mathfrak{M}$, may be written as '$\mathbf{s}^{\mathfrak{M}}$', sometimes simply as '*s*' with possible indices. If $\mathrm{Mod}(\tau)$ is the collection of all models of type τ, it is assumed that $\mathrm{Mod}_L(\tau)$ is a subclass (possibly proper) of $\mathrm{Mod}(\tau)$. If $\varphi \in \mathrm{Sent}_L(\tau)$, then $\mathrm{Mod}_L(\varphi)$ is the class of all models in $\mathrm{Mod}_L(\tau)$ in which φ is true; similarly for a set of sentences $\Sigma \subseteq \mathrm{Sent}_L(\tau)$. Such models are called *models of* φ or models of Σ, as the case may be.

There are other and earlier characterizations, but Feferman's gives rise, perhaps, to the most general notion.[9] Feferman presents a number of examples of logics and models of different kinds. Moreover, it admits, in principle, of logics that do not satisfy the isomorphism condition below (which is, however, usually assumed) and of logics that are not extensions of elementary logic. Furthermore, it applies in current attempts to define weak logics for the purposes of artificial intelligence. It admits of logics L and types τ such that in no sentence contained in $\text{Sent}_L(\tau)$ all the symbols of τ occur.[10] Hence, in addition to the logics extending elementary logic, like infinitary and second order ones, the definition admits of such deviant logics as modal logic and intuitionistic logic. For example, consider an ordinary Kripke model $\mathfrak{M} = \langle W, R, V \rangle$ for propositional modal logic. It can be construed as a function from the type $\tau = \{1\} \cup \{\mathbf{p}_0, \mathbf{p}_1, \ldots\} \cup \{\mathsf{R}\}$ such that $\mathfrak{M}(1) = W$, $\mathfrak{M}(\mathsf{R}) = R$, and $\mathfrak{M}(\mathbf{p}_k) \subseteq W$ for each $k = 1, 2, \ldots$.[11]

The following notions will be needed below. Firstly, if $\tau \subseteq \tau'$ and $\mathfrak{M} \in \text{Mod}(\tau')$, let $\mathfrak{M}|\tau$ be the restriction of $\mathfrak{M}$ to τ. Secondly, a (1-1 and onto) mapping $i\colon \tau \to \tau'$ $(\tau, \tau' \in \text{Typ}_L)$ is a *name changer* if it gives rise to a *renaming* of τ, that is, if it maps the sorts and symbols of τ to sorts and symbols of corresponding kinds in τ'.

Now conditions (8.2.1)-(8.2.3), below, are often stated for logics. Feferman (1974b) says that a logic (or a model-theoretic language, as he calls it) L is *regular* if it satisfies the conditions. The first condition is as follows:

(8.2.1)   *Renaming Condition.* For each name changer i, there is a mapping i transforming each sentence $\varphi \in \text{Sent}_L(\tau)$ into a sentence $i(\varphi) \in \text{Sent}_L(i(\tau))$ such that if $\mathfrak{M} \in \text{Mod}_L(\tau)$ and $\mathfrak{M}_i \in \text{Mod}(i(\tau))$, where $\mathfrak{M}_i(i(\mathbf{s})) = \mathfrak{M}(\mathbf{s})$ for each symbol $\mathbf{s} \in \tau$, then $\mathfrak{M}_i \in \text{Mod}_L(i(\tau))$ and $\mathfrak{M} \models_L \varphi \Leftrightarrow \mathfrak{M}_i \models_L i(\varphi)$.

This expresses a kind of invariance property; if it holds, then the truth relation is invariant with respect to renaming. Most of the familiar logics have this property, for instance, elementary logic and its infinitary extensions and normal modal logics with Kripke semantics. What this condition roughly says is that if the corresponding symbols occurring in the two sentences indicated have the same extension, then the sentences have the same extension (truth value). Hence, (8.2.1) can also be considered an extensionality principle.

The two conditions in (8.8.2), which Feferman states, require that a logic be in some sense *invariant* with respect to language (or, rather, to type).

(8.2.2)   *Expansion Conditions.* (i) If $\tau \subseteq \tau'$, then $\text{Sent}_L(\tau) \subseteq \text{Sent}_L(\tau')$; (ii) If $\tau \subseteq \tau'$ and $\mathfrak{M} \in \text{Mod}_L(\tau')$, then $\mathfrak{M}|\tau \in \text{Str}_L(\tau)$, and if $\varphi \in \text{Sent}_L(\tau)$, then $\mathfrak{M} \models_L \varphi \Leftrightarrow \mathfrak{M}|\tau \models_L \varphi$.

One may also say that what (8.2.2) requires is a kind of independence of (linguistic and other) contexts, which is again a feature of extensionality.

The last one is the familiar isomorphism condition:

(8.2.3) *Isomorphism Condition.* If $\mathfrak{M} \in \mathrm{Mod}_L(\tau)$, $\mathfrak{N} \in \mathrm{Mod}(\tau)$, and $\mathfrak{M} \cong \mathfrak{N}$, then $\mathfrak{N} \in \mathrm{Mod}_L(\tau)$ and $\mathfrak{M} \models_L \varphi \Leftrightarrow \mathfrak{N} \models_L \varphi$.

Hence, by means of an extensional language one cannot distinguish between different models if they are isomorphic. In a sense, therefore, from the point of view of an extensional logic there cannot exist intended or priviledged individuals. Such individuals can only be referred to intentionally; and the language in which the reference is expressed is intensional.

These are the conditions that Feferman states for regular logics. They are usually required of formal logic even in a very general sense of logic. There are other conditions, some of which overlap the preceding ones, and some of which are more general. I shall next consider some of the most discussed principles of extensionality. As I remarked earlier, there cannot be any definite meaning of extensionality; it is a vague notion that can be formally explicated in different ways. There are, however, some explications which are rather commonly accepted among philosophers and logicians – it is likely that (8.2.1)-(8.2.3.) belong to them.

Condition (8.2.1) can be regarded as a version of what is known as Frege's Principle (if the principle is relativized to a model):

(8.2.4) *Frege's Principle.* If a sentence ψ is obtained from φ by replacing an occurrence of any (nonlogical) symbol in φ, or any group of symbols, by a similar symbol, or respectively by group of symbols, having the same extension (in a model $\mathfrak{M}$), then φ is true (in $\mathfrak{M}$) if and only if ψ is true (in the model $\mathfrak{N}$ that is obtained from $\mathfrak{M}$ by changing the interpretations of the symbol or symbols in question in the obvious way).

If what reads inside the parentheses is taken into account, we have a model-theoretic version of Frege's Principle. There are various ways to formulate this version more precisely, but it is more important here to see that it is a generalization of (8.2.1) and to acknowledge other special cases and modifications of the principle. If it, e.g., holds for every model $\mathfrak{M}$ (admitted by L) of an appropriate type, we might replace in (8.2.4) the word 'extension' (*Bedeutung*) by the word 'intension' (*Sinn*) and 'true in $\mathfrak{M}$' by 'L-valid'. This implies, in particular, that the sentence below in (8.2.5) is L-valid, and hence an L-theorem if L is adequately axiomatizable:

(8.2.5)   *The Indiscernibility of Identicals*

$$\mathbf{c} = \mathbf{d} \rightarrow (\varphi(\ldots \mathbf{c} \ldots) \leftrightarrow \varphi(\ldots \mathbf{d} \ldots)),$$

where '$\ldots \mathbf{c} \ldots$' indicates an occurrence of an individual constant $\mathbf{c}$ in φ, and similarly for $\mathbf{d}$. Another special case is the validity of the following, discussed, e.g., by Carnap (1937):

(8.2.6) $(\sigma \leftrightarrow \vartheta) \rightarrow (\varphi(\ldots \sigma \ldots) \leftrightarrow \varphi(\ldots \vartheta \ldots)),$

and its more important consequence, a transformation rule, one form of which is important in modal logic:

(8.2.7)   *The Rule of Extensionality*

If $\sigma \leftrightarrow \vartheta$ is valid, so is $\varphi(\ldots \sigma \ldots) \leftrightarrow \varphi(\ldots \vartheta \ldots)$.

There are other criteria for extensionality, both syntactic and semantic ones. One is the truth-functionality of sentences (by Carnap and Wittgenstein), others are the invariance of deduction with respect to language, the rule of universal substitution, etc. Some of the above conditions are overlapping or consequences of others. Though all of them seem to characterize the extensionality of logics, it is somewhat difficult to tell which of these conditions a logic should satisfy in order that it could be called extensional rather than intensional. There hardly is any strict, intuitive borderline between the two notions. One way to informally characterize extensionality – in view of the conditions – is to say that an extensional logic emphasizes form at the expense of content. This is, of course, what is traditionally required of logic. It is known, however, that, e.g., modal logics do not satisfy all the conditions, most noticeably those which are not what Segerberg (1971) calls classical modal logics, and logics employing variants of possible worlds semantics that are not standard, in the sense that they rely on 'impossible', i.e., nonclassical kinds of possible worlds.[12] Furthermore, we have applied the logic of counterfactual conditionals in connection with counterfactual explanation, in Chapter 6, above, and in the formal treatment of that logic certain extensionality principles are also skipped. In the present essay, other deviant logics that dispense with some extensionality principles were discussed in Chapter 7.[13]

Theories

In the formal case studies of this book, forthcoming in the next chapter, my concern will mostly be expressive power rather than other model-theoretic properties or extensionality. In this sense, the present essay serves metascientific purposes with no purely logical aims, that is, I shall not much explore consequences or logical sophistication which model-theoretic properties (such as different versions of compactness, interpolation and definability properties) might provide for metascience.[14]

I shall not try to define the concept of *theory*, but only assume in what follows (as in earlier chapters) that all relevant principles and axioms of a theory can be expressed, if needed, as sentences of some logic, to be specified in each case. Several notions of theories, both syntactic and structural, are defined in the philosophy of science, but they are not needed in this essay. The most important elements of a formal theory will be, e.g., the similarity type in question, and hence the structure of the models of the theory, that is, models in which the axioms are true, and the nature of the logic employed. We may recall that a correspondence relation between two theories is determined by a correlation of models and a translation, and the elements mentioned are, therefore, involved in the correspondence.

In the formal case studies of Chapter 9 − where model-theoretic axiomatizations depend on the fact that theories can be originally axiomatized in a mathematical metalanguage − we only indicate the logics in which the theories can be formalized and the requisite classes of models defined; we shall not work out formalizations. The main methodological advantage of such an approach is that one need not be involved in laborious and complex formalizations, and yet one knows that formal languages are available when needed − as, for instance, in translations. However, one has to have enough knowledge about the logics used and their models in order to be able to work out relevant translation and correlation mappings.

8.3. CARNAP'S PROGRAM: A HISTORICAL DIGRESSION

In his *Logische Syntax der Sprache*, Rudolf Carnap introduces a *general syntax* to provide ". . . an exact syntactical method by means of which the results of logical analysis could be exactly formulated and the problems of the logic of science worked out."[15] It is of some interest to relate Carnap's idea to the more modern developments discussed earlier in this chapter, whose motivation and aims are evidently analogous to what Carnap had in mind. Carnap's enterprise must of course be pondered in the light of the philosophical goal of the Vienna Circle and bearing in mind the conception of logic then prevailing, but it contains elements whose historical import can only be understood if seen against the background of those recent developments in logic.

To supply evidence of how one may fail to appreciate Carnap's endeavor when that background is not available, I consider first what Sellars (1953) says of Carnap's account of inference. Carnap introduces what he calls extralogical rules or *P-rules* (physical rules) of transformation, contrasting them with logico-mathematical rules or *L-rules*:

> We may, however, also construct a language [i.e., a logic or calculus] with *extralogical rules of transformation*. The first thing which suggests itself is to include amongst the primitive sentences the so-called laws of nature, i.e. universal sentences of physics ('physics' is here to be understood in the widest sense).[16]

Thus he suggests that a law of nature can be made a primitive sentence, that is, a logical axiom, of a language (i.e., of a calculus). But he goes still farther and says that we can admit as primitive sentences empirical observation-sentences, even sentences that are "momentarily acknowledged."[17]

The methodological and philosophical role of P-rules seems to have remained somewhat unclear, however. For example, Sellars remarks that Carnap fails to explain the status or the specific contribution of P-rules; there is no mention in Carnap's book of whether P-rules authorize any linguistic activity that cannot be authorized by L-rules alone.[18] It is well known that P-rules can function as inference tickets which justify inferences from singular statements to other singular statements. For instance, let *S* be a language with English words used in their ordinary meaning[19] in which the general statement

(8.3.1)   All ravens are black

is made a primitive sentence. Then the following is a valid inference in S:

(8.3.2) a is a raven; hence a is black.

In Sellars' view, this kind of linguistic activity seems to be the only import of P-rules which is acknowledged in Carnap's book, and that activity can be authorized by L-rules as well.[20] Thus inference (8.3.2) can, of course, be replaced by the obvious inference in which (8.3.1) is the second premise; and that inference is valid according to ordinary predicate calculus.

Although Sellars may be correct in that Carnap fails to make it clear what is authorized by P-rules, it seems to me that it is philosophically more significant to ask whether Carnap makes it clear what authorizes them, that is, what entitles one to employ them as primitive sentences. Sellars fails to pose this question, which is in fact a question of crucial importance, since, as noted above, Carnap seems to anticipate more recent developments in logic. That Sellars does not consider the question is evidently due to the fact that those developments only began after Sellars' article appeared. In his book, Carnap says, for example:

> The range of possible language-forms and, consequently, of the various possible logical systems, is incomparably greater than the very narrow circle to which earlier investigations in modern logic have been limited.[21]

The fact that no attempts had been made to venture still further [than, e.g., to many-valued and intensional logics] from the classical forms is perhaps caused by the widely held opinion that any such deviations must be justified – that is, that a new language-form must be proved to be "correct" and to constitute a faithful rendering of "the true logic."[22] Furthermore, Carnap argues that

> . . . the view will be maintained [in this book] that we have in every respect complete liberty with regard to the forms of language; that both the forms of construction for sentences and the rules of transformation (the latter are usually designated as "postulates" and "rules of inference") may be chosen quite arbitrarily.[23]

These passages clearly show that Carnap's attitude in regard to logic (calculus) is in this book very liberal, and this is why he calls this standpoint the Principle of Tolerance. There is no true logic for him, whence P-rules are in some sense logically as acceptable as L-rules. Thus, although Carnap mostly uses the term 'language' (by which he means a calculus, a system of formation and transformation rules), it is a general definition of a logic (logical system) at which Carnap inevitably is aiming. Such a general notion entitles one to use P-rules as well as L-rules.

Carnap emphasizes syntax in his definition of a general logic, since he sees it as providing a sufficient degree of accuracy,[24] unlike semantic constructions, which are concerned with relations of meanings. When a syntactic treatment is combined with the Principle of Tolerance, it follows that a logic can even be defined by choosing postulates (in modern terms: logical axioms) and rules of inference arbitrarily. This makes the question of meanings secondary, since the choice of postulates and rules determine what meanings should be assigned to the logical symbols involved.[25]

In accordance with the common fashion of defining syntactical systems of logic, Carnap wants to distinguish between formation rules and transformation rules. From a methodological point of view, it is of some interest to notice his apparently economical way of defining the crucial concepts of any logic (or language or calculus, as he also calls it).[26] The most basic notion is 'direct consequence', and the rest are defined in terms of it. Carnap characterizes this notion and a logic in something like the following manner. Carnap calls it a definition, however:[27]

(8.3.3)   A *logic S* is a syntactical system determined by a relation of direct consequence. φ is a *direct consequence* of Σ in *S* if (i) φ and every expression in Σ have one of the following forms: . . .; and (ii) φ and Σ fulfill one of the following conditions: . . . .

Here φ is an expression of *S* and Σ a set of its expressions. Both φ and Σ can be finite or infinite. Now the formation rules of *S* are included in (i) and its transformation rules in (ii). The former characterize the sentences of *S*, and by means of the latter the notions of derivation (i.e., deduction), proof, primitive sentence (i.e., logical axiom), and other related notions are defined more or less in the usual way.

A logician who is accustomed to accurate definitions may consider the definition of direct consequence unsatisfactory. However, since Carnap made an early attempt to define a general logic, he can be seen as a predecessor of the logicians who from the early sixties on have been generalizing that notion and studied it in the framework of abstract logic, and somewhat later in connection with artificial intelligence and cognitive science. The loose way of defining direct consequence (or, rather, not defining it at all) is in agreement with Carnap's idea that the notion can in principle be chosen arbitrarily. For the sake of comparing Carnap's way of defining a general logic and his degree of accuracy in that enterprise with the corresponding aspects in those more recent developments, I shall next quote a well-known definition of a logic in the framework of abstract logic. This definition is somewhat simpler than, but not as general (as for the notion of model) as Feferman's one in the preceding section. The simpler version makes the comparison a bit more transparent. It will be easy to see that the current enterprise is similar in spirit to Carnap's attempts to define a general syntax. A main difference between the two definitions is that while Carnap's notion is syntactic, in abstract logic semantic considerations are dominant.

Barwise (1972) defines a logic, or, as he calls it, an abstract logic, as follows.[28] First are defined the notions of language (as a set of symbols) and model for a language (i.e., of a type) in the usual way. Then:

(8.3.4)   A *logic L* for a language L is a pair $\langle \Sigma, \models_L \rangle$, where Σ is a set (or proper class) and $\models_L$ is a relation between models for L and elements of Σ.

The elements of Σ are called the *sentences* of L and $\models_L$ the *satisfaction* relation for L. The isomorphism condition is always assumed to hold.

For our aims here it is important to notice the following things. Barwise's definition (8.3.4) is somewhat more precise and detailed than Carnap's definition (8.3.3), due to the fact that it is presented in set-theoretic terms. Carnap mentions the possibility of using arithmetized syntax but does not use this device in his general definition, nor a set-theoretic device. His definition does not say very much. It is conceptually loose, which is a consequence of the extreme generality he is aiming at. Both definitions leave the notion of sentence undefined, (8.3.3) the notion of deduction, and (8.3.4) that of satisfaction. They are both extremely general, which means that they cover all known logics whose notion of model is appropriate here, and both leave room for indefinitely many others, in addition. More precisely, Carnap's definition admits of all syntactically definable logics and Barwise's all semantically definable ones of the kind mentioned, and they are in conformity in cases where the syntactic and semantic features of logics can be defined so as to be equivalent, that is, when the logics in question are complete and sound. At that historical stage of progress in logic, Carnap himself could not, of course, consider semantic questions of this kind.

As a simple example of equivalence, let us again consider P-rules, that is, a case where a given syntactic system is strengthened, in accordance with Carnap's suggestions, by adding to its primitive sentences (logical axioms) a new one, as, for example, a formal sentence representing a scientific generalization or an "observation-sentence," or no matter what.[29] Let S be a logic (syntactic system) in the sense of (8.3.3) whose deduction relation is denoted by '$\vdash_S$', and let θ be a sentence that is added to the primitive sentences of S. Assume S^* is the new logic obtained, whose primitive sentences are those of S together with θ, and $\vdash_{S^*}$ its deduction relation. Then, if σ is a sentence of S, it is obvious that S and S^* are related in the following way:

(8.3.5) $\sigma \vdash_{S^*} \varphi \Leftrightarrow \theta, \sigma \vdash_S \varphi.$

Now compare this way of extending a given syntactic system with the following fashion of extending model-theoretically a given logic. It is a slight modification of an example in Barwise (1972). Suppose $L = \langle \Sigma, \models_L \rangle$ is a logic for a language L in the sense of (8.3.4) and K a class of models for L closed under isomorphism. Consider the new logic $L^* = \langle \Sigma, \models_{L^*} \rangle$ whose sentences are those of L, but for which a satisfaction relation $\models_{L^*}$ is defined as follows:

(8.3.6) $\mathfrak{M} \models_{L^*} \varphi \Leftrightarrow \mathfrak{M} \models_L \varphi$ and $\mathfrak{M} \in K,$

for all models $\mathfrak{M}$ for L and all sentences $\varphi \in \Sigma$. It is obvious now that if θ defines the class K in L, that is, $K = \mathrm{Mod}_L(\theta)$, and if S and L are equivalent in the sense that S provides a complete and sound axiomatization of L, then for all sentences σ, φ:

(8.3.7) $\sigma \models_{L^*} \varphi$ if and only if $\sigma \vdash_{S^*} \varphi$,

where '$\sigma \models_{L^*} \varphi$' means, as usual, that φ is a logical consequence of σ in L^*. Hence, under the conditions indicated, S^* and L^* are equivalent, that is to say, Carnap's and Barwise's ways to define an extension agree.

Trivial as this example is, it nevertheless indicates the historical importance of Carnap's general syntax as a predecessor of more recent developments. In particular, it may shed some new light on the role played by his notion of P-rule.

Carnap's struggle for extreme generality was clearly motivated by methodological and metascientific aims. He argues, for example, that defining syntactic concepts that are applicable to languages of any form is very important, and he presents various applications. As we saw above, the motivations for developing generalized logics in the framework of abstract logic were methodological as well; in a more restricted sense, however, since their applicability to metascientific purposes other than metamathematical ones was only seen later.[30] Then it was observed – and it will again be seen in the next chapter – that generalized logics and concepts from abstract logic provide powerful tools for handling intricate instances of intertheoretic relations, such as the relation of limiting case correspondence.[31]

There are also purely logical reasons for both developments. In Carnap's view, the degree of clarity and exactitude of logical expositions are often unsatisfactory, and therefore he emphasizes the importance of developing an exact method for syntactical studies.[32] It is equally important to notice his idea that general syntax provides a method of comparing different languages, that is, in the framework of general syntax it is possible ". . . to state exactly the characteristic differences between [a given language-form] and the other possible language-forms."[33] According to Barwise, on the other hand, abstract logic emerged from attempts ". . . to bring some kind of order to the whole study of generalizations of the lower predicate calculus."[34] It made it possible to state and study various properties and relations of logics in an exact way.

There is another analogy, which pertains to the purely logical side in abstract logic but more to the methodological side in Carnap. Since there is no true logic for Carnap, one generalization should be as acceptable as another. But then we can ask why it is necessary to distinguish between different kinds of logic, in particular, between different kinds of rule, L-rules and P-rules. An obvious explanation is that although for Carnap there is no true logic, one logic may still be better than another. Carnap observes, for instance, that it may be disadvantageous to adopt P-rules since they can frequently make us alter the language. It seems that Carnap's view here has to do (among other things) with truth, interpretation, and other factors which for the earlier Carnap are not purely logical. On the other hand, generalizations within abstract logic led logicians to ask whether new logics have a nice model theory and good properties, when compared with elementary logic. Thus, in a purely logical sense logics were not considered equally good, and this made logicians to search for logics which are more powerful than elementary logic but which still would have a workable model theory, as we noted above. Here, then, seems to be a further analogy which may increase the historical import of Carnap's account.

So it seems, furthermore, that our construal of Carnap's endeavor as a predecessor of current developments in logic makes his contribution philosophically and historically more important than does Sellar's construal which fails to acknowledge Carnap's aim to generalize, and therefore takes the main import of P-rules as being the reduction of the number of premises in inferences.

8.4. COUNTERFACTUAL CONDITIONALS

Next I summarize, both in formal and philosophical terms, some basic characteristics of Goodman's (1947) and Lewis' (1973) analyses of counterfactual conditionals. As I indicated in Chapter 6, above, I am concerned with the relation between such conditionals and valid arguments. Therefore, I shall mainly consider features having to do with that relation. It is pointed out by Nute (1984) and some others that Lewis' theory, which presupposes a kind of minimal deviation from the 'actual' world, is not good for all counterfactuals. However, it is obvious on the basis of what we have seen in Chapter 6 and elsewhere that it is appropriate for counterfactual explanation, since there we assume that the deviations caused by the counterfactual special conditions in question are minimal. The theory is in harmony with our principles of interpretation, studied in Chapter 2, above, especially with what we called the minimization principle.

According to Goodman's theory, and other theories that Lewis calls metalinguistic, a counterfactual conditional is true if there exist appropriate conditions from which in conjunction with the antecedent the consequent can be inferred. Hence, a main problem of counterfactuals is to define such conditions. The key notion is here 'cotenability': the conditions must be cotenable with the antecedent in order to be relevant. Cotenability can be thought of as compatibility, but in a strong sense of the word, or, alternatively, as necessity in a weak sense. As we saw in Chapter 6, this notion plays a central role in my approach to counterfactual explanation. The notion has slightly different meanings in Goodman's and Lewis' approaches. The latter approach will be considered here.

The problem is handled by Lewis in the following way. Cotenability is a relative notion: it is relative to how the notion of possibility and that of similarity of phenomena – or, rather, similarity of possible worlds – are defined. In order to discuss it and related notions in precise terms, let us consider a Kripke model of the form

$$\mathfrak{M} = \langle W, R, \leq, V \rangle,$$

where W is a set of 'possible worlds', R is an 'accessibility mapping' assigning to each $w \in W$ a subset of W, $\leq$ is a comparative (three-place) similarity relation of worlds, and V is a valuation function on an appropriate language.[35] If a triple of worlds, say $\langle w,v,u \rangle$, belongs to $\leq$, this is denoted by ' $v \leq_w u$ ' and reworded by saying that v is *at least as similar* to w as u is.[36] If a sentence φ of the language is *true at* a world w in $\mathfrak{M}$, we denote ' $\mathfrak{M},w \models \varphi$' or briefly ' $w \models \varphi$'. If $u \models \varphi$ for some $u \in R(w)$, then φ is *possible at* w, and if $u \models \varphi$ for all $u \in R(w)$, then φ is *necessary at* w. It is assumed here that (i) if $u \in R(w)$ and $v \leq_w u$, then $v \in R(w)$; (ii) $u \in R(u)$; and (iii) $w \leq_w u$ for all $u \in W$.[37]

A model of the above kind will be called a *Lewis model*.[38] In what follows, I shall assume that the requisite concepts are defined in relation to a Lewis model which is somehow fixed and that the sentences considered belong to a fixed language. If I want to emphasize that a concept is assumed to be relative to a model $\mathfrak{M}$, I shall add the expression 'relative to $\mathfrak{M}$'.

In this framework, the notion of cotenability is defined as follows. Assume first that a sentence φ is possible at a world w. Then a sentence σ is *cotenable with* φ *at* w if and only if for some $u \in R(w)$ such that $u \models \varphi$, the following holds: $v \models \sigma$ for every $v \in W$ such that $v \leq_w u$. If φ is not possible at w, that is, $u \not\models \varphi$ for all $u \in R(w)$, then σ is cotenable with φ at w if σ is necessary at w. Thus in the former case, which is the nontrivial alternative, the sentences φ and σ are jointly true at some world u accessible from the given world w, and σ is true at every world v that is more similar to the given one than u is. Thus such a world v is also accessible from the given world w, and σ is true at w.

Intuitively, then, the sentence σ is 'so strongly' possible at w that it is compatible with φ, no matter how 'implausible' the latter is. One may want to say, with Lewis, that, e.g., laws of nature tend to be in this sense cotenable with counterfactual suppositions. We can also see that the above definition (if related to the actual world) is stronger than Goodman's (counterfactual) description of cotenability, according to which the cotenability of σ with φ means that it is not the case that the former would not be true if the latter were. Lewis' notion entails that if the former were true, then the latter would be true, which in turn entails Goodman's condition.

Now consider a counterfactual whose antecedent is φ and consequent ψ:

(8.4.1) $\varphi \,\square\!\!\longrightarrow\, \psi$

and a valid argument

(8.4.2) $\sigma_1, \ldots, \sigma_k, \varphi \,/\, \psi$

where the stroke '/' signifies logical inference.[39] Then counterfactual (8.4.1) is said to be *backed* at a world w by argument (8.4.2) if each of the premises $\sigma_1, \ldots, \sigma_k$ is cotenable with the antecedent φ at w. A truth condition for counterfactuals can be defined by means of the notion of backing:

> Counterfactual (8.4.1) is *true at* a world w iff there exists a valid argument of the form (8.4.2) backing it at w.

This condition is a precise formulation of the idea that a counter-

factual is true if there are suitable additional premises from which together with the antecedent the consequent can be derived. There is
also a direct but equivalent truth condition, which, however, seems to
correspond to a different intuition. According to that intuition, (8.4.1)
is true at w if the consequent ψ is true at every world accessible from
w at which φ is true, but which otherwise resembles w 'as much as
possible'. More precisely:

> A counterfactual (8.4.1) is true at a world w iff the following con-
> dition holds: If there exists an accessible world u at which the ante-
> cedent φ is true, then for some such world u, the material implica-
> tion $\varphi \to \psi$ is true at every world that is at least as similar to w as u
> is.

If, in particular, the antecedent is not possible at w, (8.4.1) is true at
w in the trivial sense.

Instead of speaking of the similarity of worlds, one can use the
term 'closeness' − which may in some contexts express a different
intuition. It is more important to notice, however, that similarity and
closeness are relative notions, relative to a point of view. In this sense
they resemble the notion of relative identity.[40]

Counterfactuals were discussed above in terms of Lewis' theory,
and thus in terms of possible worlds. As I have already remarked,
'possible world' is a notion that in philosophical investigations must
be given a very general and abstract scope. It may refer to such things
as situations and events of various kinds, phenomena, physical systems, or model-theoretic models. In order to avoid confusions, the
notion should be understood broadly enough, so as to cover not just
complete and 'consistent' universes but such things as situations, fictional and belief worlds, *Lebenswelt*, and what not. When the notion
is used in literary connections or in belief contexts, worlds can be fictional, or they do not obey principles of classical logic; when used in
speech act contexts, in the sense of Chapter 2, it may refer to 'small
worlds', that is, situations, and so forth. In other words, possible
worlds are not necessarily such comprehensive entities as the term
may mistakingly suggest. Furthermore, the term may refer to things
that are − in some specified senses depending on context − actual or
only possible, standard or nonstandard, and which are always considered as providing a conceptual device that is somehow determined

and structured by our theories and conceptual frameworks. They are theory-relative rather than given.

8.5. SYMMETRIES AS CATEGORIES

It is characteristic of symmetry transformations related to actual scientific theories that they have clear-cut algebraic structures. They are groups or something similar. On the other hand, in our model-theoretic treatment the classes of models considered are large collections, usually even proper classes. As we saw in Section 5.6, it is therefore appropriate to characterize a symmetry as a category whose objects are models and morphisms (or arrows) are transformations of models. For example, the symmetry of a theory was defined as a category whose objects are the models of the theory and whose morphisms are transformations of the models in the class; and a subsymmetry of the theory was defined as its subcategory. To make this talk intelligible, and also for later purposes,[41] I give now a brief description of the most basic notions of category theory.[42]

A *category* is a pair $\mathbf{C} = \langle C, G \rangle$, where C is a collection of *objects* and G a collection of *morphisms*, such that:

(i)   If $f \in G$, there are $a, b \in C$ such that $f: a \to b$;[43]

(ii)   If $f: a \to b$ and $g: b \to c$, their *composite* $gf: a \to c$ exists;

(iii)   Given an $a \in C$, there is an *identity morphism* of a, $1_a: a \to a$, such that if $f: b \to a$ and $g: a \to c$, then $1_a f = f$ and $g1_a = g$;

(iv)   If $f: a \to b$, $g: b \to c$, and $h: c \to d$, then $h(gf) = (hg)f$.

A category $\mathbf{D} = \langle D, H \rangle$ is a *subcategory* of $\mathbf{C} = \langle C, G \rangle$ if $D \subseteq C$ and $H \subseteq G$. It is *full* if for any $f: a \to b$ such that $f \in G$ and $a, b \in D$, it holds that $f \in H$.

Let $\mathbf{C} = \langle C, G \rangle$ and $\mathbf{D} = \langle D, H \rangle$ be categories. Then a mapping F on $C \cup G$ is a (covariant) *functor*, denoted by $F: \mathbf{C} \to \mathbf{D}$, if the following conditions hold:

(v)   If $a \in C$, then $F(a) \in D$;

(vi)   If $f: a \to b$, then $F(f): F(a) \to F(b)$;

(vii)   If $f: a \to b$ and $g: b \to c$, then $F(gf) = F(g)F(f)$;

(viii) $F(1_a) = 1_{F(a)}$.

In addition to, or instead of, symmetry transformations, one may explore other relations between models of a theory.[44] In model-theoretic terms, such a relation can be thought of as being a collection of tuples of the form $\langle \mathfrak{M}_1, \ldots, \mathfrak{M}_n; h_1, \ldots, h_k \rangle$ where $\mathfrak{M}_1, \ldots, \mathfrak{M}_n$ are models and $h_1, \ldots, h_k$ are relations on the union of various domains of these models (relations holding between individuals of the models). A tuple of his kind can in turn be regarded as a (many-sorted) model. Since a symmetry can be expressed as a collection of tuples of the form $\langle \mathfrak{M}_1, \mathfrak{M}_2; h \rangle$, where h is a transformation $h: \mathfrak{M}_1 \mapsto \mathfrak{M}_2$, we have here a generalization of symmetry. In some specific cases, such more general relations can of course be construed as categories.[45]

8.6. NONSTANDARD ANALYSIS

Though there exist several approaches to nonstandard analysis, they all are based on Abraham Robinson's fundamental ideas.[46] I rely here on the framework and notation developed in Machover and Hirschfelt (1969) and Bell and Machover (1977). This approach is slightly different from the standard treatment,[47] but it is logically simpler and convenient for present purposes – although it may have some mathematical weaknesses.

Let U be a set. Define:[48]

(8.6.1) $A_0 = U$;

$\qquad A_{n+1} = \mathrm{P}(\bigcup_{i \le n} A_i) \ (n \in \omega)$;

$\qquad A = \bigcup_{n \in \omega} A_n$,

where 'P' denotes the power set operation. A is usually called the *superstructure* over U. U can be thought of as containing all the basic objects needed for a given theoretical (e.g., mathematical) study. The elements of U, whose set-theoretic structure is not relevant for the study, are often described as 'urelements' or 'points'. All the other elements of A are considered as sets within this framework. A contains, in the set-theoretic form, all entities that are usually needed in a mathematical study of the basic objects in U, i.e., it contains all the relations and subsets of U, relations of relations, etc. For example, if the set of all real numbers is included in U, A contains all objects needed in real and complex analysis.

Assume A is a superstructure and τ_0 a one-sorted type whose non-logical constants are a two-place predicate symbol $\in$ and an individual constant $\mathbf{a}$ for each $a \in A$ (and a different constant for different elements). We associate with A the model

$$\mathcal{U} = \langle A; \in, a \rangle_{a \in A}$$

of type τ_0, where $\in$ is the membership relation (that of the metatheory, restricted to $(A \backslash U)^2$) and each $a \in A$ is a distinguished element.[49] Thus for each individual constant $\mathbf{a}$ in τ_0, its interpretation (denotation) in $\mathcal{U}$ is a.

Let $\text{Th}(\mathcal{U})$ be the set of all $L_{\omega\omega}(\tau_0)$-sentences that are true in $\mathcal{U}$. It is known that $\mathcal{U}$ can be isomorphically (and uniquely) embedded in any model of $\text{Th}(\mathcal{U})$ and that every model of $\text{Th}(\mathcal{U})$ which is an extension of $\mathcal{U}$ is its elementary extension. Now from the standpoint of $\text{Th}(\mathcal{U})$, all its models that are isomorphic to $\mathcal{U}$ are *standard models*; all the other models of $\text{Th}(\mathcal{U})$ are *nonstandard*. In what follows, we shall mainly consider the relation between $\mathcal{U}$ and its nonstandard extensions. In view of the embedding result, these considerations are then applicable to the other standard models and their respective extensions. An arbitrary nonstandard extension of $\mathcal{U}$ will be denoted by

$$^*\mathcal{U} = \langle ^*A; ^*\in, a \rangle_{a \in A}$$

Thus A is a subset of $^*A$ and $\in$ is a subset of $^*\in$. The difference $^*\in \backslash \in$ need not be (part of) the membership relation.[50]

The following notation (which is related to $^*\in$) from Bell and Machover (1977) will be used. For any $a \in {}^*A$, we let

$$^\wedge a = \{ b \in {}^*A \mid b \; ^*\in a \}.$$

Let 'Q' denote any relation (function or operation) of elements of A which is $L_{\omega\omega}$-definable in $\mathcal{U}$, that is, definable by an $L_{\omega\omega}(\tau_0)$-formula φ.[51] Then '$^*Q$' denotes the relation between elements of $^*A$ which is definable by the very same formula, that is,

$$\bar{a} \in {}^*Q \Leftrightarrow {}^*\mathcal{U} \models \varphi[\bar{a}].^{52}$$

This means that 'Q' may be used to form a metalinguistic translation of φ in the standard sense (i.e., with respect to $\mathcal{U}$) and '$^*Q$' in a nonstandard sense (i.e., with respect to $*\mathcal{U}$). For instance, since the familiar set-theoretic relations and operations are $L_{\omega\omega}$-definable in $\mathcal{U}$, expressions like '$\subseteq$', '$*\subseteq$', '$\cup$', and '$*\cup$' will have definite meanings (given $\mathcal{U}$ and $*\mathcal{U}$).

Conversely, any metalinquistic (mathematical) statement containing only expressions for $L_{\omega\omega}$-definable relations (in $\mathcal{U}$ or $*\mathcal{U}$) can be translated into an $L_{\omega\omega}(\tau_0)$-sentence. Thus we may use 'semiformal' expressions, instead of $L_{\omega\omega}(\tau_0)$-formulas, whenever we known that a translation into the formal language is possible. This is a common practice in nonstandard analysis. It simplifies the discussion by yielding the so-called *transfer principle*:

(8.6.2.) Whenever statements α (concerning $\mathcal{U}$) and $*\alpha$ (concerning $*\mathcal{U}$) are metalinquistic statements translatable into the same $L_{\omega\omega}(\tau_0)$-sentence, then α if and only if $*\alpha$.

Furthermore, it will enable us to use ordinary mathematical metalanguage and familiar mathematical notation, instead of formal.

Let A be a superstructure as above and let $V \in A$. Let $\mathcal{T}$ be a topology on V such that $\langle V, \mathcal{T}\rangle$ is Hausdorff. Consider an arbitrary nonstandard model $*\mathcal{U}$ that is an extension of $\mathcal{U}$.[53] We shall generalize Robinson's notion of *standard part* (which is usually defined only for some elements of $^\wedge V$).[54] A natural way to do it is to consider the nonstandard extensions in $*A$ of such elements of A that can be constructed from those of V. So we need the superstructure over V:

$$B_0 = V;$$

$$B_{n+1} = \mathrm{P}(\textstyle\bigcup_{i \leq n} B_i);$$

$$B = \textstyle\bigcup_{n \in \omega} B_n$$

Each $B_n \in A$, and $B \subseteq A$; furthermore,

$$\mathcal{U} \models \forall \mathbf{x}(\mathbf{x} \in \mathbf{B_{n+1}} \leftrightarrow \forall \mathbf{y}(\mathbf{y} \in \mathbf{x} \rightarrow \mathbf{y} \in \mathbf{B_0} \vee \ldots \vee \mathbf{y} \in \mathbf{B_n})).$$

Since each B_n is in A, this sentence is an $L_{\omega\omega}(\tau_0)$-formula, whence it is true in $*\mathcal{U}$. Thus $a \in {}^\wedge B_{n+1}$ if and only if for every $b \in {}^\wedge a$, there is an $i \in \{1, \ldots, n\}$ such that $b \in {}^\wedge B_i$.

For each $a \in V$, the *monad* of a is

$$\mu(a) = \bigcap \{{}^{\wedge}Y \mid Y \text{ is a neighborhood of } a\}$$

(Robinson, 1961). Since $\langle V, \mathcal{T} \rangle$ is Hausdorff, the monads of any two distinct elements of V are disjoint. Any element $a \in {}^{\wedge}V$ such that $a \in \mu(b)$ for some $b \in V$ is *near-standard*. Let $X_0 \subseteq {}^{\wedge}V$ be the set of all such elements.

Now the generalized notion of standard part will be defined for appropriate elements of $\bigcup_{n\in\omega}{}^{\wedge}B_n$. Let a partial function st from $*A$ into A be defined inductively as follows:

(8.6.3) (0) If $a \in \mu(b)$ for some $b \in V$, let $\mathrm{st}(a) = b$;

　　　　　. . .

　　　　(n+1) Let X_n be the largest subset of $\bigcup_{i\leq n}{}^{\wedge}B_i$ for which st has been defined and let $a \in {}^{\wedge}B_{n+1}$. If ${}^{\wedge}a \subseteq X_n$, let $\mathrm{st}(a) = \{\mathrm{st}(b) \mid b \in {}^{\wedge}a\}$.

The function st will be called the *standard part function*. Its domain is $X = \bigcup_{n\in\omega}X_n$. Since for every $a \in X$, $\mathrm{st}(a) \subseteq B_n$ for some $n \in \omega$, it follows that $\mathrm{st}(a) \in B$; hence $\mathrm{st}(a) \in A$.

The definition implies, in particular, that st is obtainable in a natural way for any $*f(a_1, \ldots, a_n)$ where f is a familiar set-theoretic operation on A and $a_1, \ldots, a_n \in X$. Thus, for instance, $\mathrm{st}(*\langle a_1, \ldots, a_n \rangle) = *\langle \mathrm{st}(a_1), \ldots, \mathrm{st}(a_n) \rangle = \langle \mathrm{st}(a_1), \ldots, \mathrm{st}(a_n) \rangle$ and $\mathrm{st}(a *\cup b) = \mathrm{st}(a) *\cup \mathrm{st}(b) = \mathrm{st}(a) \cup \mathrm{st}(b)$.

In view of the embedding result, the above definition of st is immediately applicable to any nonstandard model.

8.7. STANDARD AND NONSTANDARD MODELS OF A THEORY

Let us assume that for every theory T to be considered, a unique superstructure, as in (8.6.1), and a unique Hausdorff topology $\langle V, \mathcal{T} \rangle$ on a set $V \in A$ are associated. For instance, if T is based on a mathematical theory, the superstructure associated with it is assumed to contain the standard objects of that mathematical theory (in the set-theoretic form). Furthermore, let τ be a many-sorted similarity type including τ_0 and let L be a logic in which T can be formalized such that L admits τ. A model of type τ is called *standard* (relative to A) if it is isomorphic to a model $\mathfrak{M}$ such that all domains of $\mathfrak{M}$ are included in A. It follows that if $\mathbf{S}$ is a predicate or function symbol in τ, its interpretation in $\mathfrak{M}$, $\mathbf{S}^{\mathfrak{M}}$, is a subset of A, $\mathbf{S}^{\mathfrak{M}} \subseteq A$; and if $\mathbf{c}$ is an individual constant, $\mathbf{c}^{\mathfrak{M}} \in A$. But if the domain and range of $\mathbf{S}^{\mathfrak{M}}$ are elements of A, then also $\mathbf{S}^{\mathfrak{M}}$ is in A. All the other models of this type are *nonstandard* (relative to A).

Whenever intertheoretic relations are studied here, the *same* superstructure and the same topology are associated with all theories that are compared with each other.

In the rest of this section, let a superstructure A and Hausdorff topology in A be fixed; likewise we assume that a nonstandard extension $*\mathcal{Q}$ of $\mathcal{Q}$, in the sense of the preceding section, is fixed. Thus I first consider standard and nonstandard models whose domains, relation, etc., are composed of elements of A and $*A$, respectively. In addition, all nonstandard notions we need are related to $*\mathcal{Q}$ and to the corresponding extension of the topology $\langle V, \mathfrak{T} \rangle$, in the manner explained in the preceding section.

Assume that τ and τ' are types such that $\tau \subseteq \tau'$.[55] Let $\mathfrak{M}$ be a standard model of type τ and let $\mathfrak{M}'$ be a nonstandard model of type τ'. Let the domains of $\mathfrak{M}$ and $\mathfrak{M}'$ be $M_i (i \in I)$ and M_i' $(i \in I'; I \subseteq I')$, respectively; so that for each $i \in I$, M_i and M_i' are domains of the same sort. Let us denote, for a moment: $\vec{a} = \langle a_1, \ldots, a_n \rangle$, $\vec{a}' = \langle a_1', \ldots, a_n' \rangle$ (for an arbitrary $n > 0$). We now say that $\mathfrak{M}$ is a *standard approximation* of $\mathfrak{M}'$ if for each $i \in I$ and for each predicate and function symbol $\mathbf{S}$ and individual constant $\mathbf{c}$ in τ,

(i) $\forall a \in M_i \; \exists a' \in M_i' : a = \mathrm{st}(a')$ or $a = a'$;[56]

(ii) $\forall \vec{a} \in \mathbf{S}^{\mathfrak{M}} \; \exists \vec{a}' \in \mathbf{S}^{\mathfrak{M}'} \; \forall i \in \{1, \ldots, n\}:$
           $a_i = \mathrm{st}(a_i')$ or $a_i = a_i'$;

(iii) $\mathbf{c}^{\mathfrak{M}} = \mathrm{st}(\mathbf{c}^{\mathfrak{M}'})$ or $\mathbf{c}^{\mathfrak{M}} = \mathbf{c}^{\mathfrak{M}'}$.

This notion is extended for classes of models in a natural way, as follows. If K and K' are classes of models of type τ and τ', respectively, then K is a *standard approximation* of K' if the following condition holds:

(iv)   For each $\mathfrak{M} \in K$, there is an $\mathfrak{M}' \in K'$ such that $\mathfrak{M}$ is a standard approximation of $\mathfrak{M}'$.

It follows from the remarks in the preceding section concerning isomorphism that these definitions are immediately applicable to any pair of standard and nonstandard models.

8.8. MODELS OF ANALYSIS

From now on it will be assumed that A (see 8.6.1) is the superstructure over the set of all real numbers $\mathbb{R}$, that is, $U = \mathbb{R}$.[57] Furthermore, we assume that $\mathcal{T}$ is the usual topology on $\mathbb{R}$, whence $\langle \mathbb{R}, \mathcal{T} \rangle$ is Hausdorff.

As stipulated in Section 8.6, the nonlogical constants of the formal object language, that is, in the type τ_0, are $\in$ and $\mathbf{a}$ for each $a \in A$. In particular, if an element of A has a commonly used name in the ordinary mathematical language, the corresponding name will be used in the object language. For instance, the relation $<$ is at the same time an element of A, so that $<$ is an individual constant in the object language. Thus the relation $<$ is $L_{\omega\omega}(\tau_0)$-definable in $\mathcal{U}$. It can be defined by an $L_{\omega\omega}(\tau_0)$-formula whose only nonlogical constants are $\in$ and $<$. Hence the metalinguistic expression '$*<$' has a definite meaning in a nonstandard model $*\mathcal{U}$, as explained in Section 8.6. More generally, one may in this and later sections use 'semiformal' expressions for formal ones, and indicate, whenever needed, the formal language in which their formal counterparts can be found. So these expressions become meaningful for $\mathcal{U}$ and $*\mathcal{U}$, and their appropriate expansions.

Any model of Th($\mathcal{U}$) will be called a *model of analysis*.[58] To avoid trivialities, we shall only consider nonstandard models of analysis in which relevant objects exist. As it is known to follow from the compactness of $L_{\omega\omega}$ and some embedding results, there exist nonstandard models of analysis that are extensions of $\mathcal{U}$, and in which the following condition, expressible as an $L_{\omega_1\omega}(\tau_0)$-formula, is true:

$$(8.8.1) \quad \exists x \left(\bigwedge_{a \in \mathbb{Z}} a *< x \right).$$

Here $\mathbb{Z} \subseteq \mathbb{R}$ is the set of all integers in $\mathbb{R}$.

In the rest of this section, let such a nonstandard extension $*\mathcal{U}$ of $\mathcal{U}$ be fixed, but arbitrary. Thus the subsequent notions and definitions will be related to $*\mathcal{U}$. Since $*\mathcal{U}$ is fixed, the following simplification will not cause any confusion. If $*Q$ is the extension (as explained in Section 8.6) in $*A$ of any familiar mathematical relation or operation of real numbers, we may omit the star. Thus we shall usually write, for example, '$|a|$' for '$*|a|$', '$a < b$' for '$a *< b$', etc., even if $a, b \in {}^\wedge\mathbb{R} \backslash \mathbb{R}$.

The following notation will be frequently used:

(8.8.2) $\mathbb{U} = \{a \in {}^\wedge\mathbb{R} \mid \forall b \in \mathbb{R}: b < a\};$

       $\mathbb{E} = \{a \in {}^\wedge\mathbb{R} \mid \exists b \in \mathbb{R}: |a| < b\}\ ;$

       $\mathbb{I} = \{a \in {}^\wedge\mathbb{R} \mid \forall b \in \mathbb{R}^+: |a| < b\};$

       $a \approx b \iff |a-b| \in \mathbb{I}.$

$\mathbb{U}$, $\mathbb{E}$, and $\mathbb{I}$ are the sets of all *infinite reals, finite reals,* and *infinitesimals,* respectively. Since (8.8.1) is true in $*\mathcal{U}$, it follows that there exist infinite reals, thus also infinitesimals. Furthermore, $\mathbb{I} = \mu(0)$; and $a \approx b$, i.e., a and b are *infinitely* or *infinitesimally close* to each other, whenever they belong to the same monad.

Next we consider briefly the question of definability of some important subsets of $*A$, introduced so far. For later purposes, we need to know in what logic (of type τ_0) they are characterizable. First, if a subset of $*A$ is of the form $^\wedge a$, where $a \in A$, then it is $L_{\omega\omega}$-definable in $*\mathcal{U}$.[59] However, many other important subsets that we need in what follows are not such natural extensions of standard entities, e.g., $\mathbb{E}$, $\mathbb{I}$, A, and $\mathbb{R}$.[60] Some of them, e.g., $\mathbb{E}$ and $\mathbb{I}$, are $L_{\omega_1\omega}$-definable in $*\mathcal{U}$, and hence also $\approx$.[61] Many of the subsets and elements of $*A$, such as A and $\mathbb{R}$, are not this way definable, but at least the logic $L_{\kappa\omega}$, where $\kappa = \beth_\omega$, is sufficient for their definitions.[62]

In what follows, we shall frequently refer to various entities of $*A$ and classes of models as $L_{\kappa\omega}$-definable (in terms of an appropriate type). This will be mostly justified by what we now know about the definability of the sets just mentioned. Here we have, then, a clear-cut example of how formal logic can be given a double role in metascience. For one thing, the so-called nice model-theoretic properties of one logic (here $L_{\omega\omega}$), resulting from the weakness of its expressive power, have been used to justify the existence of nonstandard entities, needed for metascientific purposes on a par with standard ones. For other metascientific aims, however, we shall need another logic, sufficiently strong to exclude entities the first logic justifies. As we shall see, this interplay provides us with a method of studying, for example, the relation between classical and relativistic particle mechanics in a proper model-theoretic sense, as described in Chapters 7 and 9.[63]

Let us consider generalizations for $*\mathcal{U}$ of some familiar notions of real analysis.[64] First, it should be clear what it means that something is an *interval* on $^\wedge\mathbb{R}$. In particular, if $T = \{r \in \mathbb{R} \mid a < r < b\}$ ($a, b \in \mathbb{R}$) is an interval on $\mathbb{R}$, its extension $^\wedge T$ is, by the transfer principle, the following interval on $^\wedge\mathbb{R}$:

$$^\wedge T = \{r \in {}^\wedge\mathbb{R} \mid a < r < b\};^{[65]}$$

and similarly for the other intervals on $\mathbb{R}$. Recall, that if f is a standard function $f: T \to \mathbb{R}$, then $^*f: {}^\wedge T \to {}^\wedge\mathbb{R}$.

Let T be an open interval on $^\wedge\mathbb{R}$, and let $a \in T$ and $b \in {}^\wedge\mathbb{R}$. Assume that a function f is defined on T, possibly a excepted, and that $f[T] \subseteq {}^\wedge\mathbb{R}$, where $f[T]$ is the range of f. Define:

$$(8.8.3) \quad \lim_{t \to a} f(t) = b \iff_{\mathrm{df}} \forall \varepsilon \in {}^\wedge(\mathbb{R}^+)\, \exists \delta \in {}^\wedge(\mathbb{R}^+)\, \forall t \in T{:}$$
$$0 < |t-a| < \delta \Rightarrow |f(t)-b| < \varepsilon.$$

The *definiens* of (8.8.3) can be formalized by an $L_{\omega\omega}(\tau)$-formula, for any expansion τ of τ_0 containing symbols for a, b, and f, and a new sort for T.[66] Thus it expresses the interpretation in $^*A$ of a first order formalization of the usual (standard) concept of *limit*.

Assuming that T and a are as in (8.8.3), f is defined on T, and the relevant limit exists, we obtain the corresponding first order generalization for $^*\mathcal{U}$ of the notion of *derivative* at a by stating:

$$(8.8.4) \quad \mathrm{D}f(a) = \lim_{t \to a} \frac{f(t)-f(a)}{t-a}$$

Accordingly we can define what it means in this generalized sense for f to be *differentiable* at an $a \in T$ or *on* T.[67] The meaning of the function $\mathrm{D}f$, that is, df/dt, is likewise obvious.

If f is a real valued *standard function* (i.e., belongs to A), defined on an interval T on $\mathbb{R}$, then to say that it is *differentiable* (at a, on T) means, of course, that it is differentiable (at a, on T) *in* $\mathcal{U}$, that is, differentiable in the ordinary, standard sense. Furthermore, notations '$\mathrm{D}f$' and 'df/dt' will be equally used for derivatives of standard and nonstandard functions. This unification of terminology and notation will not cause any confusion. A similar remark concerns the other first order notions below.

The next lemma is an immediate consequence of the transfer principle. It, and the others below, are needed in the next chapter, where we have to move between standard and nonstandard models.

LEMMA 8.8.1. Let $f: T \to \mathbb{R}$ be a standard function. Then f is differentiable on T iff $*f$ is differentiable on $^{\wedge}T$. If f is differentiable on T, then $D(*f) = *(Df)$.

Due to the ordinary first order form of the definition of derivative — since, in addition, the ordinary first order properties of $\mathbb{R}$ hold for $^{\wedge}\mathbb{R}$ — it is obvious that the *usual derivation rules* can be inferred in $*\mathcal{A}$, even for functions that are not natural extensions of standard functions.

Concepts of analysis can also be generalized for $*\mathcal{A}$ in a manner that is not first order expressible. Let T be an interval on $^{\wedge}\mathbb{R}$. Let a function f be defined on T and let $f[T] \subseteq {}^{\wedge}\mathbb{R}$. We say that f is *^{r}continuous* on T if

(8.8.5) $\forall a \in T \cap \mathbb{R} \ \forall t \in T: t \approx a \Rightarrow f(t) \approx f(a).$[68]

Furthermore, f is *bounded* on T when

(8.8.6) $\exists r \in {}^{\wedge}(\mathbb{R}^+) \ \forall t \in T: |f(t)| < r,$

and *^{r}bounded* on T provided that

(8.8.7) $\exists r \in \mathbb{R}^+ \ \forall t \in T: |f(t)| < r.$

It follows from what was said above that ^{r}continuity and ^{r}boundedness are $L_{\kappa\omega}$-definable in appropriate models of type $\tau \supseteq \tau_0$ containing symbols for T and f.

Now the following result is well known:

LEMMA 8.8.2 (Robinson, 1961). Let f be a standard real valued function defined on T. Then f is continuous on T iff $*f$ is ^{r}continuous on $^{\wedge}T$.

The next lemma is again a simple consequence of the transfer principle and definitions:

LEMMA 8.8.3. Assume f is a standard real valued function defined on T. Then f is bounded on T iff $*f$ is bounded on $^{\wedge}T$. Furthermore, if f is continuous on T, then f is bounded on T iff $*f$ is ^{r}bounded on $^{\wedge}T \cap \mathbb{E}$.

Let T be an interval on $^{\wedge}\mathbb{R}$. In what follows, its 'finite part' $T \cap \mathbb{E}$ is denoted by ^eT. If T is not infinitesimal and ^eT is not empty, let its *standard interior* be defined as follows:

$(8.8.8)$ $\quad {}^0T = \{t \in T \cap \mathbb{R} \mid \exists t_1, t_2 \in T \cap \mathbb{R}: t_1 < t < t_2\}.$

Let a function $f\colon T \to {}^\wedge\mathbb{R}$ be finite on ${}^0T$. Then its *standard approximation* ${}^0f\colon {}^0T \to \mathbb{R}$ is defined by

$(8.8.9)$ $\quad {}^0f(t) = \mathrm{st}(f(t)).$[69]

Naturally, ${}^0f \in A$. The following fact is also obvious; but we state it as a lemma, for later references:

LEMMA 8.8.4. *If* $f\colon T \to \mathbb{R}$ *is standard and* T *is open, then* ${}^0(*f) = f.$

If P is any nonempty set and f a function on $P \times T$, then we let for each $p \in P$ a function f_p on T be defined by

$(8.8.10)$ $\quad f_p(t) = f(p, t).$

That f is *differentiable* (at a, on T) means naturally that each f_p has the corresponding property. Similarly, the other notions defined above can be generalized for f in an obvious way. Hence, in what follows they can be applied to such two-place functions without any further trouble.

9

A FORMAL TREATMENT OF CASE STUDIES

In earlier chapters we have defined the notion of correspondence relation and explored its applications to some actual theories occurring in the history of science. Some of these applications are correspondence sketches, that is, the relevant syntactic and semantic entities are more or less vague. The rest of them can be exactly treated in model-theoretic terms, as proposed in Chapter 7, but only a few cases are studied here. The limiting case correspondence of classical particle mechanics to relativistic one and the respective instance of Curie's Principle are the only investigations that are presented here in full detail. The presentation is taken from Pearce and Rantala (1984a) – with some important corrections – where many-sorted infinitary logics and nonstandard analysis were engaged and certain well-known axiomatizations used as a starting point.[1]

As remarked earlier, such a presentation is methodologically useful since it enables one to rely on ordinary mathematical (meta)language and at the same time have a formal language and general model theory at one's disposal. It turns out, unfortunately, that reconstructions are formally complicated. Such complications are unavoidable, however, if one wants to exactly show, for example, that there is more to the intuitive idea of limiting case correspondence than what we saw in Chapter 1 Kuhn maintaining, namely, that it is more than a mere syntactic relation.

I also show in this chapter, following Rantala and Vadén (1994), how the limiting case correspondence of the theory of Turing representation to the theory of network representation – as sketched in Chapter 7 – can be mathematically and model-theoretically established in an exact way. A complete presentation is not given, since this and other model-theoretically definable examples of limiting case correspondence are very much analogous to the case of classical and relativistic mechanics, thoroughly worked out.

9.1. THE CORRESPONDENCE OF CLASSICAL TO RELATIVISTIC
PARTICLE MECHANICS

In elementary expositions of particle mechanics, one-dimensional systems (and transformations) are usually considered. This simplification is also assumed here. The full generalization is straightforward but formally somewhat tedious. I first present the axiomatizations of the theories and then the correspondence relation itself. As we shall see, the relation is constructed by using nonstandard analysis and infinitary logic. The correlation and translation mappings needed in the correspondence relation are defined in detail.

Classical Particle Mechanics (CPM)

In what follows, $\mathcal{C}$ is as in Section 8.8 (the specified standard model of analysis) and its type is τ. $\mathcal{B}$ will be a model of type τ_0 whose domain is B, and

$$\mathfrak{M} = \langle \mathcal{B}; P, T; s, m, f \rangle$$

is a three-sorted expansion of $\mathcal{B}$, having P and T as new domains such that $s, f: P \times T \to B$ and $m: P \to B$. Assume that τ is the type of $\mathfrak{M}$. For any $a \in B$, denote

$$a_{\mathcal{B}} = \{b \in B \mid b \in^{\mathcal{B}} a\}.$$

Thus $a_{\mathcal{B}}$ is of the form $^{\wedge}a$ if $\mathcal{B}$ is a proper elementary extension of $\mathcal{C}$, and a if $\mathcal{B}$ is $\mathcal{C}$. Let H_0 be the class of all models $\mathfrak{M}$, as above, of type τ such that[2]

(9.1.1) $\mathcal{B}$ is $\mathcal{C}$ or one of its elementary extensions satisfying (8.8.1);

(9.1.2) P is a finite nonempty set;

(9.1.3) T is an interval on $\mathbb{R}_{\mathcal{B}}$;

(9.1.4) $s, f: P \times T \to \mathbb{R}_{\mathcal{B}}$;

(9.1.5) $m: P \to (\mathbb{R}^+)_{\mathcal{B}}$;

(9.1.6) s is twice differentiable on T;

(9.1.7)   For all $p \in P$, $t \in T$, $m(p)D^2 s(p,t) = f(p,t)$.

Let H be the least class of models of type τ including H_0 and being closed under isomorphism. Any element of H is a (one-dimensional) *model of* CPM.[3] It will be called *standard or nonstandard* according to whether $\mathcal{B}$ is a standard or nonstandard model of analysis.

When defining concepts that are related to H and deriving results for them, we shall mainly refer to models in H_0. This is because in Section 8.6 we learned to consider analysis in terms of $\mathcal{A}$ and its extensions $*\mathcal{A}$, rather than in terms of their isomorphic counterparts. In principle, one could as well work with any standard model and its extensions in H. Anyway, our definitions and results are directly extensible for the latter models, and this possibility of generalizing will be assumed without any further notice; and similarly for the next theory. Later on, we consider formal languages in which one can define the classes of models that are here defined mathematically.

Relativistic Particle Mechanics (RPM)

Next we consider models of type $\tau = \tau \cup \{\mathbf{c}\}$, where $\mathbf{c}$ is a new individual constant:

$$\mathfrak{M} = \langle \mathcal{B}; P, T; s, m, f, c \rangle,$$

where $c \in B$ ($c = \mathbf{c}^{\mathcal{B}}$). Let H_0' be the class of all models as above satisfying (9.1.1)-(9.1.6) and the following conditions:

(9.1.8) $c \in (\mathbb{R}^+)_{\mathcal{B}}$;

(9.1.9) For all $p \in P$, $t \in T$, $|Ds(p, t)| < c$;

(9.1.10) For all $p \in P$, $t \in T$,

$$m(p)D\left(\frac{Ds(p, t)}{\sqrt{1-(Ds(p, t)/c)^2}}\right) = \sqrt{1-(Ds(p, t)/c)^2}f(p, t).[4]$$

Let H' be the least class including H_0' and being closed under isomorphism. Any element of H' is a (one-dimensional, standard or nonstandard) *model of* RPM. Again, we shall often consider H_0' instead of H' in defining concepts (see the similar remark, above, concerning H_0 and H).

Correlation

Next we define the correlation mapping needed for the correspondence relation in question. Let K_0 be the class of all standard models $\mathfrak{M} = \langle \mathfrak{B}; P, T; s, m, f \rangle$ in H_0 such that

(9.1.11) T is open;[5]

(9.1.12) Ds and $D^2 s$ are bounded on T;

(9.1.13) $D^2 s$ is continuous on T.

Assume K_0' is the class of all nonstandard models $\mathfrak{M} = \langle {}^*\mathfrak{A}; P, T; s, m, f, c \rangle$ in H_0' such that:

(9.1.14) T is a noninfinitesimal interval such that ${}^e T$ is non-empty;

(9.1.15) s and f are finite on ${}^e T$;

(9.1.16) m is finite and noninfinitesimal on P;

(9.1.17) Ds and $D^2 s$ are ^{r}bounded on ${}^e T$;

(9.1.18) s and $D^2 s$ are ^{r}continuous on ${}^e T$;

(9.1.19) Ds/c and $D^2 s/c$ are infinitesimal on T;

(9.1.20) ${}^0 s$ is twice differentiable and $D^2 ({}^0 s)$ continuous on ${}^0 T$, and $D^2 ({}^0 s) = {}^0 (D^2 s)$.[6]

Let now $\mathfrak{M} = \langle {}^*\mathfrak{A}; P, T; s, m, f, c \rangle$ be an arbitrary element of K_0'. By the definition of K_0' (and by the definition of the notion of standard approximation; see Section 8.6), there exists a unique model ${}^0 \mathfrak{M}$ of type τ such that

$$ {}^0 \mathfrak{M} = \langle \mathfrak{A}; P, {}^0 T; {}^0 s, {}^0 m, {}^0 f, {}^0 c \rangle. $$

This model is a standard approximation of $\mathfrak{M}$ in the sense of Section 8.6.

LEMMA 9.1.1. For all $\mathfrak{M} \in K_0'$, ${}^0 \mathfrak{M} \in K_0$.

Proof. Let $\mathfrak{M} = \langle {}^*\mathfrak{A}; P, T; s, m, f, c \rangle$ be in K_0'. It is obvious that conditions (9.1.2)-(9.1.6) and (9.1.11)-(9.1.13) hold for ${}^0 \mathfrak{M}$.

ad (9.1.7): It follows from (9.1.10), by (9.1.15) and (9.1.19), that for all $p \in P$, $t \in {}^e T$,

$$m(p)D^2s(p, t) \approx f(p, t).^7$$

Let $p \in P$, $t \in {}^0T$. Since the functions in question are finite,

$$\mathrm{st}(m(p))\mathrm{st}(D^2s(p, t)) = \mathrm{st}(f(p, t)),$$

that is,

$${}^0m(p){}^0(D^2s)(p, t)) = {}^0f(p, t)).$$

Thus by (9.1.20),

$${}^0m(p)D^2({}^0s(p, t)) = {}^0f(p, t)).$$

LEMMA 9.1.2. For all $\mathfrak{M} \in K_0$, there is an $\mathfrak{M}' \in K_0'$ such that ${}^0(\mathfrak{M}') = \mathfrak{M}$.

Proof. Let $\mathfrak{M} = \langle \mathcal{Q}; P, T; s, m, f \rangle$ be in K_0 and let $*\mathcal{Q}$ be an extension of $\mathcal{Q}$ as in (9.1.1). Let $*s$, $*f\colon P \times {}^\wedge T \to {}^\wedge \mathbb{R}$ be functions defined by

$$*s(p, t) = *(s_p)(t);$$

$$*f(p, t) = *(f_p)(t)$$

(cf. Section 8.6). Now it follows from the transfer principle that the model

$$*\mathfrak{M} = \langle *\mathcal{Q}; P, T; *s, m, *f \rangle$$

is a nonstandard model of CPM.

By (9.1.12), Lemma 8.8.1, and Lemma 8.8.3, $D(*s)$ and $D^2(*s)$ are bounded on ${}^\wedge T$. Thus there exists a $\mu \in \mathbb{U}$ such that for all $p \in P$, $t \in {}^\wedge T$,

$$|D(*s(p, t))| < \mu;$$

$$|D^2(*s(p, t))| < \mu.^8$$

Let $c = \mu^3$. Let $p \in P$, $t \in {}^\wedge T$. Then

$$k(p, t) = m(p)D\left(\frac{D(*s(p, t))}{\sqrt{1-(D(*s(p, t))/c)^2}}\right) : \sqrt{1-D(*s(p, t))/c)^2}$$

$$= m(p)(1-(D(*s(p, t)/\mu^3)^2)^{-2} D^2(*s(p, t)).$$

But $|D(*s(p, t)| < \mu$, and hence

$$1 \le (1-(D(*s(p, t))/\mu^3)^2)^{-1} < (1-1/\mu^4)^{-1} < 1+1/\mu^3;$$

thus

$$|m(p) D^2(*s(p, t))| \leq |k(p, t)| \leq |m(p)D^2(*s(p, t))| +$$
$$+ m(p)(2/\mu + 1/\mu^3)(|D^2(*s(p, t))|/\mu).$$

Since the latter addendum on the right is infinitesimal,

$$k(p, t) \approx m(p) D^2(*s(p, t)).$$

Since $*\mathfrak{M}$ is a model of CPM,

$$m(p) D^2(*s(p, t)) = *f(p, t).$$

It follows that

$$m(p)D\left(\frac{D(*s(p, t))}{\sqrt{1-(D(*s(p, t))/c)^2}}\right) \approx \sqrt{1-(D(*s(p, t))/c)^2}*f(p, t).$$

So there is a function $f'\colon P \times {}^\wedge T \to {}^\wedge \mathbb{R}$ such that for all $p \in P, t \in {}^\wedge T$,

(9.1.21) $f'(p, t) \approx *f(p, t)$

and

$$m(p)D\left(\frac{D(*s(p, t))}{\sqrt{1-(D(*s(p, t))/c)^2}}\right) = \sqrt{1-(D(*s(p, t))/c)^2}f'(p, t).$$

We claim that the model

$$\mathfrak{M}' = \langle *\mathfrak{A}; P, {}^\wedge T; *s, m, f', c\rangle$$

belongs to K_0'. It suffices to show that conditions (9.1.15), (9.1.17)-(9.1.18), and (9.1.20) hold for $\mathfrak{M}'$, since it is obvious that $\mathfrak{M}'$ is a nonstandard model of RPM and that (9.1.14), (9.1.16), and (9.1.19) hold.

   ad (9.1.15), (9.1.18): By (9.1.6), s is continuous on T; by (9.1.13) and (9.1.7), D^2s and f are continuous on T. Thus by Lemma 8.8.2 and Lemma 8.8.1, $*s$, $D^2(*s)$, and $*f$ are ʳcontinuous on ${}^\wedge T$. Since $*s(p, t) = *(s_p)(t) = s_p(t)$ and $*f(p, t) = *(f_p)(t) = f_p(t)$ for every $t \in T$, thus finite, it follows from the definition of ʳcontinuity that $*s$ and $*f$ are finite on ${}^e({}^\wedge T)$; hence (by 9.1.21) f' is likewise finite on ${}^e({}^\wedge T)$.

   ad (9.1.7): As shown above, $D^2(*s)$ is ʳcontinuous on ${}^\wedge T$, and similarly for $D(*s)$. Thus they are ʳbounded on ${}^e({}^\wedge T)$, by (9.1.12) and Lemma 8.8.3.

ad (9.1.20): This condition is a direct consequence of Lemmas 8.8.4 and 8.8.1.

Thus $\mathfrak{M}'$ belongs to K_0'. Finally, since T is open and each $m(p)$ is standard, it follows from (9.1.21) and Lemma 8.8.4 that

$$^0(\mathfrak{M}') = \mathfrak{M}.$$

Since K_0 is nonempty, it follows from this lemma that K_0' is nonempty.

Let K and K' be the least classes of models – of respective types τ and τ' – including K_0 and K_0', respectively, and being closed under isomorphism. In view of the isomorphism, the above notions and lemmas (concerning K_0 and K_0') can be generalized for the whole classes K and K'.

The correlation F will now be the mapping from K' that is defined by the condition

$$F(\mathfrak{M}) = \,^0\mathfrak{M}.$$

It follows from Lemma 9.1.2 and the isomorphism that F is a mapping *onto K*.

Translation

Let us associate with CPM a logic L in which H and K are definable. It is obvious, in view of Section 8.8, that we can let L be $L_{\kappa\omega}$, with an appropriate $\kappa \leq \beth_\omega$, whose type is three-sorted. The type τ gives rise to variables of three sorts. Let **x**, **p**, and **t**, with possible (non-bold) subscripts, be variables ranging over the domains B, P, T, respectively, and let **y** and **z** (with possible nonbold subscripts) be metavariables standing for variables of any sort. In addition to the nonlogical relation symbols and function symbols in τ, the logical language contains the identity symbol '=' that admits identities between terms of *any sorts*.[9] The very same logic L can be associated with RPM, since H' and K' are L-definable; but now the type is, of course, τ'.

Let $\theta(\mathbf{y}, \mathbf{z})$, $\vartheta(\mathbf{t})$, and $\sigma(\mathbf{x})$ be fixed $L(\tau_0)$-formulas formalizing the (semiformal) conditions indicated below. All the variables in the same formula are assumed to be distinct from each other, and naturally of the sorts indicated. '*x*', '*p*', and '*t*' are assumed to become **x**, **p**, and **t**, respectively, in the formalization.

$\theta(\mathbf{y}, \mathbf{z})$: $\exists x_1 \in \mathbb{E}\ \exists x_2 \in \mathbb{R}(x_1 = y \wedge x_2 = z \wedge \bigwedge_{a \in \mathbb{Z}^+} |x_1 - x_2| < 1/a$

$\qquad \vee\ \exists x_1 \in A\ \exists x_2 \in A\ (x_1 = y \wedge x_2 = z \wedge x_1 = x_2)$

$\qquad \vee\ \exists p_1 \exists p_2 (p_1 = y \wedge p_2 = z \wedge p_1 = p_2);$

$\vartheta(\mathbf{t})$: $\quad \exists x \in \mathbb{R}(x = t \wedge \exists x_1 \in \mathbb{R}\ \exists x_2 \in \mathbb{R}\ \exists t_1 \exists t_2 (x_1 = t_1$

$\qquad \wedge\ x_2 = t_2 \wedge x_1 < x < x_2));$

$\sigma(\mathbf{x})$: $\quad x \in A.$

Thus $\langle a,b \rangle$ satisfies θ in a nonstandard model of RPM if b is the standard part of a (and $a \in \mathbb{E}$, $b \in \mathbb{R}$) or $a = b$ (and $a, b \in A \cup P$); t satisfies ϑ if $t \in {}^0T$; and a satisfies σ if $a \in A$.

Let $\mathfrak{M} = \langle {}^*\mathcal{Q}; P, T; s, m, f, c \rangle$ be an arbitrary element of K_0'. Thus $F(\mathfrak{M}) = {}^0\mathfrak{M} = \langle \mathcal{Q}; P, {}^0T; {}^0s, {}^0m, {}^0f \rangle$ is in K_0 (Lemma 9.1.1). We expand the types τ and τ' by adding a new individual constant $\mathbf{q}$ of the sort of P for each $q \in P$ and a constant $\mathbf{r}$ of the sort of T for each $r \in {}^0T$.[10] Assume the resulting expanded types are τ_1 and τ_1', respectively, and let furthermore ${}^0\mathfrak{M}_1$ and $\mathfrak{M}_1$ be the expansions of ${}^0\mathfrak{M}$ and $\mathfrak{M}$ to the types τ_1 and τ_1' in which each $\mathbf{q}$, $\mathbf{r}$ is interpreted as q, r, respectively.[11]

Let $\mathbf{u}$, $\mathbf{u}_1$, $\mathbf{u}_2$, ... denote *terms* (of any sort) of the language of type τ_1. We first consider the following mapping I_1 of the $L(\tau_1)$-formulas into $L(\tau_1')$-formulas:

If $\mathbf{Q}(\mathbf{u}_1, \mathbf{u}_2)$ is an atomic formula,[12] then

$$I_1(\mathbf{Q}(\mathbf{u}_1, \mathbf{u}_2)) = \exists \mathbf{y}_1 \exists \mathbf{y}_2 \exists \mathbf{z}_1 \exists \mathbf{z}_2 (\mathbf{y}_1 = \mathbf{u}_1 \wedge \mathbf{y}_2 = \mathbf{u}_2$$
$$\wedge\ \theta(\mathbf{y}_1, \mathbf{z}_1) \wedge \theta(\mathbf{y}_2, \mathbf{z}_2) \wedge \mathbf{Q}(\mathbf{z}_1, \mathbf{z}_2)),$$

where $\mathbf{y}_i$ ($i = 1, 2$) is the ith variable (in the alphabetic order) of the same sort as $\mathbf{u}_i$ not occurring in $\mathbf{u}_1, \mathbf{u}_2$, or $\theta(\mathbf{y}, \mathbf{z})$; and $\mathbf{z}_i$ ($i = 1, 2$) is the $(i + 2)$nd variable of the same sort as $\mathbf{u}_i$ not occurring in $\mathbf{u}_1, \mathbf{u}_2$, or $\theta(\mathbf{y}, \mathbf{z})$;

$$I_1(\neg\varphi) = \neg I_1(\varphi);$$
$$I_1(\wedge\Phi) = \wedge \{I_1(\varphi) \mid \varphi \in \Phi\};$$
$$I_1(\forall \mathbf{t}\varphi) = \forall \mathbf{t}(\vartheta(\mathbf{t}) \rightarrow I_1(\varphi));$$
$$I_1(\forall \mathbf{x}\varphi) = \forall \mathbf{x}(\sigma(\mathbf{x}) \rightarrow I_1(\varphi));$$

$$I_1(\forall \mathbf{p}\, \varphi) = \forall \mathbf{p} I_1(\varphi)).$$

Correspondence

Next we show that F and the restriction of I_1 to I establish a correspondence relation of CPM to RPM in the sense defined in Chapter 5. The following result for $\mathfrak{M}_1$ (as defined above) is needed for that purpose:

LEMMA 9.1.3. For all $\psi \in \mathrm{Sent}_L(\tau_1)$, $^0\mathfrak{M}_1 \vDash \psi \Leftrightarrow \mathfrak{M}_1 \vDash I_1(\psi)$.[13]

Proof. The following notation will be used here: if $\mathbf{u}$ is any constant term of type τ_1, its interpretations in $\mathfrak{M}_1$ and $^0\mathfrak{M}_1$ will be denoted by 'u' and '^0u', respectively.[14] Notice, furthermore, that for every constant term $\mathbf{u}$ in the type τ_1, it holds by the construction of $^0\mathfrak{M}$ that $\langle u, {}^0u \rangle$ satisfies θ in $\mathfrak{M}_1$. Hence

(9.1.22) $\mathfrak{M}_1 \vDash \theta(\mathbf{u}, {}^0\mathbf{u})$.

Now the proof will proceed by induction on the complexity of the $L(\tau_1)$-sentences:

(1) Let ψ be an atomic sentence $\mathbf{Q}(\mathbf{u}_1, \mathbf{u}_2)$. Since $\langle \mathcal{Q}; P \rangle$ is a submodel of $\langle {}^*\mathcal{Q}; P \rangle$ and $T \subseteq {}^*A$,

(9.1.23) $\mathbf{Q}^{0\mathfrak{M}_1} = \mathbf{Q}^{\mathfrak{M}_1} \cap (A \cup P)^2$.

($\Rightarrow$) Assume that

(9.1.24) $^0\mathfrak{M}_1 \vDash \mathbf{Q}(\mathbf{u}_1, \mathbf{u}_2)$.

Then $\langle {}^0u_1, {}^0u_2 \rangle \in \mathbf{Q}^{0\mathfrak{M}_1}$, whence $\langle {}^0u_1, {}^0u_2 \rangle \in \mathbf{Q}^{\mathfrak{M}_1}$. This means that

$$^0\mathfrak{M}_1 \vDash \mathbf{Q}({}^0\mathbf{u}_1, {}^0\mathbf{u}_2).$$

On the other hand, by (9.1.22),

$$\mathfrak{M}_1 \vDash \theta(\mathbf{u}_1, {}^0\mathbf{u}_1) \wedge \theta(\mathbf{u}_2, {}^0\mathbf{u}_2).$$

Hence by existential generalization,

(9.1.25) $\mathfrak{M}_1 \vDash I_1(\mathbf{Q}(\mathbf{u}_1, \mathbf{u}_2))$.

($\Leftarrow$) Assume that (9.1.25) holds. Then $\langle a_1, a_2 \rangle \in \mathbf{Q}^{\mathfrak{M}_1}$ for some a_i ($i = 1, 2$) such that $a_i = \mathrm{st}(u_i)$ or $a_i = u_i$, and $a_i \in A \cup P$. According to the definition of $\mathbf{Q}^{\mathfrak{M}_1}$, $a_i = {}^0u_i$ in either case (and for the both values of i). Thus by (9.1.23),

$$\langle {}^0u_1, {}^0u_2 \rangle \in \mathbf{Q}^{0\mathfrak{M}_1},$$

wherefore (9.1.24) holds.

(2) The cases that ψ is $\neg\varphi$ or $\bigwedge\Phi$ are obvious.

(3) Let ψ be $\forall t\varphi$. By the induction hypothesis and the definitions of I_1 and ϑ, the following are equivalent:

$${}^0\mathfrak{M}_1 \models \forall t\varphi;$$

$${}^0\mathfrak{M}_1 \models \varphi\mathbf{r}/t \text{ for all } r \in {}^0T;$$

$$\mathfrak{M}_1 \models I_1(\varphi\mathbf{r}/t) \text{ for all } r \in {}^0T;$$

$$\mathfrak{M}_1 \models \forall t(\vartheta(t) \to I_1(\varphi));$$

$$\mathfrak{M}_1 \models I_1(\forall t\varphi).$$

(4) The cases that ψ is $\forall x\varphi$ or $\forall p\varphi$ are similar to (3).

Let I be the restriction of I_1 to the $L(\tau)$-sentences:

$$I = I_1 \!\upharpoonright\! \text{Sent}_L(\tau).$$

I is clearly a mapping into the $L(\tau')$-sentences. The following theorem, from which it follows that $\langle F, I \rangle$ is a limiting case correspondence of CPM to RPM, is a direct consequence of the previous lemma and of the fact that each model in K' is isomorphic to some model in K_0':

THEOREM 9.1.4. For all $\varphi \in \text{Sent}_L(\tau)$ and all $\mathfrak{M} \in K'$,

$$F(\mathfrak{M}) \models \varphi \Leftrightarrow \mathfrak{M} \models I(\varphi).$$

This case study of correspondence is the only one in this book that is formally worked out in full detail, but since it is a typical case, it would be unnecessary to repeat similar details in the other cases indicated in Chapter 7. By investigating the details of the above study, we may see why it does not run into such difficulties with counterfactual assumptions as the 'standard' approaches. The problem is that limit assumptions are often thought of as being counterfactual or in contradiction to the primary theory.[15] This problem has been avoided in the present study by using nonstandard analysis and employing some model-theoretic methods. Other methods to avoid it have been proposed by a number of authors, but they do not involve translation in our sense of the word.[16]

9.2. CURIE'S PRINCIPLE FOR CORRESPONDENCE
IN PARTICLE MECHANICS

The formal treatment below is a based on Pearce and Rantala (1984a), and its aim is to show that Curie's Principle, defined in Section 5.7, holds with respect to the correspondence relation in particle mechanics. We use notations that are familiar from the preceding section.

Assume that $\mathfrak{M} = \langle \mathfrak{B}; P, T; s, m, f \rangle$ is in H_0 and u is in $\mathbb{R}_\mathfrak{B}$. Consider the following Galilean transformation $t \mapsto \pi t,\ s \mapsto \pi s$. For each $t \in T, p \in P$,

$$(9.2.1) \qquad \pi t\ =\ t;$$

$$(\pi s)(p, \pi t)\ =\ s(p, t) - ut;[17]$$

$$(\pi f)(p, \pi t)\ =\ f(p, t);$$

This induces what we shall call a *Galilean transformation* of $\mathfrak{M}$, Π: $\mathfrak{M} \mapsto \pi \mathfrak{M}$ such that

$$(9.2.2) \quad \pi \mathfrak{M} = \langle \mathfrak{B}; P, \pi T; \pi s, m, \pi f \rangle;$$

$$\pi T\ =\ T;$$

$$\pi s,\ \pi f\colon P \times \pi T \to \mathbb{R}_\mathfrak{B}.$$

Then it holds that $\pi \mathfrak{M} \in H_0$.[18] A given value of the *parameter u* determines a transformation of $\mathfrak{M}$ uniquely. Now the following lemma is obvious:

LEMMA 9.2.1. K_0 is closed under the Galilean transformations.

Next consider a model $\mathfrak{M} = \langle \mathfrak{B}; P, T; s, m, f, c \rangle$ in H_0'. Suppose that $u \in \mathbb{R}_\mathfrak{B}$ is again a parameter, but now such that $|u| < c$, and denote: $\beta = u/c$. If $t \mapsto \rho t,\ s \mapsto \rho s$ is the Lorentz transformation corresponding to the parameter u, then for each $t \in T, p \in P$,

$$(9.2.3) \qquad \rho t\ =\ \frac{t - us(p, t)/c^2}{\sqrt{1 - \beta^2}};$$

$$(\rho s)(p, \rho t)\ =\ \frac{s(p, t) - ut}{\sqrt{1 - \beta^2}};$$

$$(\rho f)(p, \rho t)\ =\ \frac{1 - uDs(p, t)/c^2}{\sqrt{1 - \beta^2}} f(p, t).$$

Then P: $\mathfrak{M} \mapsto \rho\mathfrak{M}$ will be called the induced *Lorentz transformation* of $\mathfrak{M}$, where

(9.2.4) $\rho\mathfrak{M} = \langle \mathfrak{B}; P, \rho T; \rho s, m, \rho f, c \rangle$;

$\rho T = \{\rho t \mid t \in T\};$[19]

$\rho s, \rho f\colon P \times \rho T \to \mathbb{R}_{\mathfrak{B}}$.

Then $\rho\mathfrak{M} \in H_0'$. If now $\beta = u/c \approx 0$ and u is finite, we shall call P a *limit transformation*.

LEMMA 9.2.2. K_0' is closed under the limit transformations.

Proof. Let $\mathfrak{M} = \langle {}^*\mathfrak{A}; P, T; s, m, f, c \rangle$ be in K_0' and let P: $\mathfrak{M} \mapsto \rho\mathfrak{M}$, where $\rho\mathfrak{M} = \langle {}^*\mathfrak{A}; P, \rho T; \rho s, m, \rho f, c \rangle$ and ρT, ρs, and ρf are as indicated in (9.2.3)-(9.2.4), such that β is infinitesimal and u finite. In order to show that $\rho\mathfrak{M} \in K_0'$, it suffices to consider (9.1.15) and (9.1.17)-(9.1.20), since (9.1.14) and (9.1.16) are obvious.

ad (9.1.15), (9.1.17)-(9.1.18): We have to show that ρs and ρf are finite, $D(\rho s)$ and $D^2(\rho s)$ are ^{r}bounded, and ρs and $D^2(\rho s)$ are ^{r}continuous on ${}^e(\rho T)$. Let $p \in P$, $\rho t \in \rho T$; then:

$$(9.2.5)\ \ D(\rho s)(p, \rho t) = \frac{d(\rho s)(p, \rho t)}{dt} \cdot \frac{d(\rho t)}{dt} = \frac{Ds(p, t) - u}{1 - \beta Ds(p, t)/c};$$

$$D^2(\rho s)(p, \rho t) = \frac{d(D(\rho s)(p, \rho t))}{dt} \cdot \frac{d(\rho t)}{dt} = \frac{(1 - \beta^2)^{3/2} D^2 s(p, t)}{(1 - \beta Ds(p, t)/c)^3}.$$

If $\rho t \in {}^e(\rho T)$, then $t \in {}^eT$, because $t \approx \rho t$. Now, since $s(p, t)$, $Ds(p, t)$, $D^2 s(p, t)$, u, and t are finite and β, $Ds(p, t)/c$, and $D^2 s(p, t)/c$ infinitesimal, it is clear that

(9.2.6) $(\rho s)(p, \rho t) \approx s(p, t) - ut$;

$D(\rho s)(p, \rho t) \approx Ds(p, t) - u$;

$D^2(\rho s)(p, \rho t) \approx D^2 s(p, t)$;

$(\rho f)(p, t) \approx f(p, t)$.

Since s and f are finite, Ds and $D^2 s$ are ^{r}bounded, and s and $D^2 s$ are ^{r}continuous on ${}^eT$, the claims follow.

ad (9.1.19): That $D(\rho s)/c$ and $D^2(\rho s)/c$ are infinitesimal on ρT follows immediately from (9.2.5) and from the assumptions that Ds/c, $D^2 s/c$, and β are infinitesimal.

ad (9.1.20): Since P is a limit transformation, $^0(\rho T) = \,^0 T$. Let $\rho t \in \,^0(\rho T)$. It follows that $\rho t \in \,^0 T$ and $\text{st}(t) = \rho t$. Since, furthermore, s is ^{r}continuous on ${}^e T$, it follows that

$$\text{st}(\rho s)(p, \rho t) = \text{st}(s(p, t)) - \text{st}(u)\text{st}(t)$$
$$= \text{st}(s(p, \rho t)) - \text{st}(u)(\rho t).$$

Hence

$$^0(\rho s)(p, \rho t) = \,^0 s(p, \rho t) - \text{st}(u)(\rho t).$$

So, since $^0 s$ is twice differentiable on $^0 T$, it follows that $^0(\rho s)$ is twice differentiable on $^0(\rho T)$ and

$$D^2(^0(\rho s)(p, \rho t)) = D^2(^0 s(p, \rho t)).$$

Hence $D^2(^0(\rho s))$ is continuous on $^0(\rho T) = \,^0 T$. Furthermore, because $D^2 s$ is ^{r}continuous on ${}^e T$ and $D^2(^0 s) = \,^0(D^2 s)$ by assumption, it follows that

$$D^2(^0(\rho s)(p, \rho t)) = \,^0(D^2 s)(p, \rho t) = \text{st}(D^2 s(p, \rho t))$$
$$= \text{st}(D^2 s(p, t)) = \text{st}(D^2(\rho s)(p, \rho t))$$
$$= \,^0(D^2(\rho s)(p, \rho t)),$$

whence

$$D^2(^0(\rho s)) = \,^0(D^2(\rho s)).$$

Let $\mathfrak{M} \in K_0{}'$ and let P: $\mathfrak{M} \mapsto \rho \mathfrak{M}$ be any *limit* transformation, as in (9.2.3)-(9.2.4). Then by Lemma 9.2.2, $\rho \mathfrak{M} \in K_0{}'$. Consider the transformation $t \mapsto \text{o}t$, $^0 s \mapsto \text{o}(^0 s)$, $^0 f \mapsto \text{o}(^0 f)$, where for each $p \in P$, $t \in \,^0 T$,

(9.2.7) $\text{o}t = \text{st}(\rho t) = t;$

$\text{o}(^0 s)(p, \text{o}t) = \text{st}(\rho s)(p, \rho t) = \text{st}(s(p, t)) - \text{st}(u)t$
$$= \,^0 s(p, t) - \text{st}(u)t;$$

$\text{o}(^0 f)(p, \text{o}t) = \text{st}((\rho f)(p, \rho t)) = \text{st}(f(p, t)) = \,^0 f(p, t).$

(see 9.2.3-9.2.4). Since (9.2.7) is of the form (9.2.1), it induces a Galilean transformation of $^0 \mathfrak{M} \in K_0$ (where $\mathfrak{M} \in K_0{}'$ is as above):

$$^0\text{P}: {}^0\mathfrak{M} \mapsto \text{o}({}^0\mathfrak{M}).$$

LEMMA 9.2.3. If $\mathfrak{M} \in K_0'$ and P: $\mathfrak{M} \mapsto \rho\mathfrak{M}$ is a limit transformation, and $^0\text{P}: {}^0\mathfrak{M} \mapsto \text{o}({}^0\mathfrak{M})$, then $^0(\rho\mathfrak{M}) = \text{o}({}^0\mathfrak{M})$.

Proof. Let $\mathfrak{M} = \langle {}^*\mathfrak{A}; P, T; s, m, f, c\rangle$ and $\rho\mathfrak{M} = \langle {}^*\mathfrak{A}; P, \rho T; \rho s, m, \rho f, c\rangle$. Then

$$^0\mathfrak{M} = \langle \mathfrak{A}; P,\; {}^0T;\; {}^0s,\; {}^0m,\; {}^0f\rangle;$$

$$^0(\rho\mathfrak{M}) = \langle \mathfrak{A}; P,\; {}^0(\rho T);\; {}^0(\rho s),\; {}^0m,\; {}^0(\rho f)\rangle.$$

If now

$$\text{o}({}^0\mathfrak{M}) = \langle \mathfrak{A}; P, \text{o}({}^0T); \text{o}({}^0s),\; {}^0m, \text{o}({}^0f)\rangle,$$

we have to show that

(i) $^0(\rho T) = \text{o}({}^0T)$;

(ii) $^0(\rho s) = \text{o}({}^0s)$;

(iii) $^0(\rho f) = \text{o}({}^0f)$.

ad (i): $^0(\rho T) = {}^0T$, since P is a limit transformation. On the other hand, $^0T = \text{o}({}^0T)$, since ^0P is a Galilean transformation.

ad (ii): By (9.2.7), for all $\rho t \in {}^0T = {}^0(\rho T) = \text{o}({}^0T)$ it holds that $^0(\rho s)(p, \rho t) = \text{st}((\rho s)(p, \rho t)) = \text{o}({}^0s)(p, \text{o}t)$.

ad (iii): Similar.

Now we obtain by Lemmas (9.1.1)-(9.1.2) and (9.2.1)-(9.2.3) the following theorem:

THEOREM 9.2.4. Let F be the mapping on K' that is defined by the condition: $F(\mathfrak{M}) = {}^0\mathfrak{M}$. Then the range of F is K. Furthermore, for each $\mathfrak{M} \in K'$ and each limit transformation P of $\mathfrak{M}$, the following diagram commutes:

LEMMA 9.2.5. If G is the class of all Galilean transformations of the models in K and G' the class of all limit transfomations of the models in K', then $\mathbf{K} = \langle K, G \rangle$ and $\mathbf{K'} = \langle K', G' \rangle$ are categories.[20]

Proof. [21] If $P_1: \mathfrak{M}_1 \mapsto \mathfrak{M}_2$ and $P_2: \mathfrak{M}_2 \mapsto \mathfrak{M}_3$ are in G' and the respective values of the parameter are u_1 and u_2, then their composite $P_2P_1: \mathfrak{M}_1 \mapsto \mathfrak{M}_3$ is a limit transformation in G' with the parameter

$$u = \frac{u_1 + u_2}{1 + u_1 u_2 / c^2}.$$

Similarly, for the composite of two Galilean transformations in the class G with the parameter values u_1 and u_2 we have

$$u = u_1 + u_2. [22]$$

Furthermore, in both cases the composite is associative: If the values of the parameter for transformations O, Π, and P are u, v, and w, respectively, then for P(ΠO) and (PΠ)O they are

$$u_1 = \left(\frac{u+v}{1+uv/c^2} + w \right) : \left(1 + \frac{\frac{u+v}{1+uv/c^2}w}{c^2} \right);$$

$$u_2 = \left(\frac{v+w}{1+vw/c^2} + u \right) : \left(1 + \frac{\frac{v+w}{1+vw/c^2}u}{c^2} \right)$$

in the case of G' and

$$u_1 = u + (v + w);$$

$$u_2 = (u + v) + w$$

in the case of G.

Thus $u_1 = u_2$ in both cases. An identity morphism is a transformation in which the value of the parameter is 0.

THEOREM 9.2.6. Let G, G', $\mathbf{K}$, and $\mathbf{K'}$ be as in Lemma 9.2.5. If $\mathfrak{M} \in K'$ and $P \in G'$, then the mapping $F: K' \cup G' \to K \cup G$ such that $F(\mathfrak{M}) = {}^0\mathfrak{M}$ and $F(P) = {}^0P$ is a (covariant) functor $F: \mathbf{K'} \to \mathbf{K}$.[23]

Proof. Consider first Theorem 9.2.4. By it, if $P: \mathfrak{M} \mapsto \mathfrak{N}$, then

$F(P)$: $F(\mathfrak{M}) \mapsto F(\mathfrak{N})$. Furthermore, it is evident that $F(1_\mathfrak{M}) = 1_{F(\mathfrak{M})}$ (where $1_\mathfrak{M}$ and $1_{F(\mathfrak{M})}$ are identity morphisms). Thus we need to show only that for all transformations $P_1, P_2 \in G'$ it holds that if $P_1 \colon \mathfrak{M}_1 \mapsto \mathfrak{M}_2$ and $P_2 \colon \mathfrak{M}_2 \mapsto \mathfrak{M}_3$, then $F(P_2 P_1) = F(P_2)F(P_1)$, that is,

$$^0(P_2 P_1) = {}^0P_2 {}^0P_1.$$

Assume that the parameter values are u_1 and u_2, respectively. Since P_1 and P_2 are limit transformations, so is $P_2 P_1$, and $^0(P_2 P_1)$ is of the following kind (cf. 9.2.7):

$$0t = t;$$

$$0(^0s)(p, 0t) = {}^0s(p, t) - \text{st}\left(\frac{u_1 + u_2}{1 + u_1 u_2 / c^2}\right)t$$

$$= {}^0s(p, t) - (\text{st}(u_1) + \text{st}(u_2))t \qquad (u_1/c \approx u_2/c \approx 0)$$

$$= ({}^0s(p, t) - \text{st}(u_1)t) - \text{st}(u_2)t;$$

$$0(^0f)(p, 0f) = {}^0f(p, t),$$

and hence the claim follows.

This theorem says that the correlation F, when yielding standard approximations, preserves the categorial structure of $\mathbf{K}'$. This is (in a structural sense) what physicists' expression 'limiting case' in fact amounts to in particle mechanics when explicated in terms of nonstandard models.

9.3. THE CORRESPONDENCE OF THE TURING REPRESENTATION TO THE NETWORK REPRESENTATION

In this section the theory of Turing representation and the theory of network representation are presented in mathematical and model-theoretic forms and the (limiting case) relation between their mathematical representations indicated. I point out, furthermore, how the correspondence of the former to the latter can be established. The presentation of this section is based on Rantala and Vadén (1994).

The Turing Representation (TR)

Let us first recall the basic notions introduced in Section 7.5. We consider a 1-way infinite Turing machine,[24] such that each square of its tape is capable of having printed upon it any one of the primitive symbols of a combinatorial language L and symbols 0 and 1. Let n be a given natural number. Assume that the natural numbers $1, \ldots, n$ label the first n squares of the tape and 0 the $(n+1)$st square.[25] Let p be a function that determines how the $n+1$ first squares are printed. Thus p is a mapping from $\{0, 1, \ldots, n\}$ to $\mathrm{Prim}L \cup \{0, 1\}$, where $\mathrm{Prim}L$ is the set of the primitive symbols of L. Thus p determines a situation (i.e., condition) of the tape up to square 0.

Let $\mathrm{Exp}L$ be the set of all expressions of L of length n belonging to some recursively defined category. For instance, if L is a formal language, $\mathrm{Exp}L$ might be the set of its terms of lenght n or the set of its well-formed formulas of length n, etc. If L is an appropriate (recursive) fragment of natural laguage, it could be the set of all words of length n belonging to the fragment, an so on.

Since L is combinatorial, $\mathrm{Exp}L$ can be efficiently represented by a Turing machine of the above kind, schematically depicted in Figure 9.3.1, that satisfies the following conditions. It is is able to decide, for any printing on squares $1, \ldots, n$, that is, for any sequence of the form $s_1 \ldots s_n$,[26] considered as an *input*, whether or not that sequence belongs to $\mathrm{Exp}L$. Here each $s_i = p(i) \in \mathrm{Prim}L \cup \{0, 1\}$. In the positive case, let the machine print '1' on square 0 and then stop, that is, let the *output* be then $p(0) = 1$; in the negative case, let the output be $p(0) = 0$.[27]

1	$\cdot$ $\cdot$ $\cdot$	n	0	
s_1	$\cdot$ $\cdot$ $\cdot$	s_n	$p(0)$	$\cdot$ $\cdot$ $\cdot$

Figure 9.3.1

Thus it holds for all mappings p as above that

(9.3.1) If $p \!\upharpoonright \{1, \ldots, n\} \in \mathrm{Exp}L$, then $p(0) = 1$;

If $p \!\upharpoonright \{1, \ldots, n\} \notin \mathrm{Exp}L$, then $p(0) = 0$,

where $p \!\upharpoonright \{1, \ldots, n\}$ is the restriction of p to $\{1, \ldots, n\}$.

Since the machine is able to decide, for any printing (of the kind indicated) on the tape, whether or not it is an expression in $\mathrm{Exp}L$, one may say that $\mathrm{Exp}L$ is represented by the machine and − if one is willing to ignore philosophical difficulties involved − that the machine

is able to represent the concepts to which the expressions refer by
displaying the constituent features of the concept. Now the theory TR
is the above description – especially conditions (9.3.1) as its axioms
– of how the machine represents Exp$L$. It is only a theory that describes the surface structure of the representation, since many of the
most important and deepest details concerning general principles of
Turing machines are here dispensed with. It seems, however, that
this is the conceptual level at which representations in Turing machines (which are symbolic representations) and representations in
connectionist networks (subsymbolic representations) can be expected to approximate each other.

The Network Representation (NR)

Now we define a connectionist network of the kind preliminarily discussed in Section 7.5.1, above, which can be trained to recognize
whether or not a sequence of symbols from Prim$L$ belongs to Exp$L$.
Assume that the (possibly hidden) layer – hereafter called the λ-layer
– that is closest to the output layer is composed of two disjoint sets of
units, λ^+ and λ^-, both containing m units. λ^+ and λ^- are, in turn,
divided into n ensembles – which may overlap – such that the ith en-
semble as a whole represents a symbol s_i from Prim$L$, that is, the elements of such an ensemble represent microfeatures of the symbol. If
the λ-layer is hidden, one may not know the exact structure of the latter division.

   Assume, for simplicity, that there is only one output unit, call it
'0', which is trained to respond with a high activation value whenever
λ^+ is presented with a sequence belonging to Exp$L$ and with a low
activation value when λ^- is presented with a sequence (of length n of
symbols) not in Exp$L$. Because the λ-layer is divided into λ^+ and λ^-,
many problems of correlation will be avoided, since we assume, fur-
thermore, that the units in λ^+ are not feeded simultaneously with
those in λ^-; that is, whenever the former units are activated, the latter
ones are not, and conversely. Let the acivation values of the output
unit be in the open interval $(0, 1)$, whence we may use a logistic activation function for it. In Figure 9.3.2, below, the general structure
of the network is presented schematically. There the boxes stand for
units and straight lines for connections.

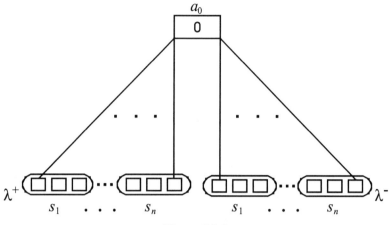

Figure 9.3.2

We take the logistic activation function in its simplest form:

$$a_0 = 1/(1+e^{-\text{Inp}}),$$

where Inp is the total input feeded by λ^+ or λ^- (as the case may be) into the ouput unit. In the literature, Inp is usually taken to be of the form $\text{Inp} = \Sigma_i w_i a_i$. If we follow this usage in the present case, w_i will be the weight of the connection of unit i to 0 and a_i the activation value (output) of i, assumed to be positive.[28] Since in the case of λ^+ the system is trained to maximize a_0, we can then also assume that after the training $w_i > 0$ for every $i \in \lambda^+$; and in the case of λ^-, it is trained to minimize a_0, whence $w_i < 0$ for every $i \in \lambda^-$.

However, if the λ-layer is a hidden layer, one may not exactly know how it feeds the output unit; whether, for instance, Inp really is the weighted sum as indicated above. Therefore, Rantala and Vadén (1994) made a somewhat weaker assumption that in the case of λ^+, $\text{Inp} = g(m)$, where g is some real-valued, monotonically increasing and positive function on the natural numbers such that $g(m) \to \infty$ when $m \to \infty$; and in the case of λ^-, $\text{Inp} = h(m)$, where h is a monotonically decreasing real-valued and negative function on the natural numbers such that $h(m) \to -\infty$ when $m \to \infty$. This much can be plausibly assumed even in case the structure and behavior of the λ-layer are not known in greater detail. It follows that when the units in λ^+ are activated after the training mode (but those in λ^- are not),

$$a_0 = 1/(1+e^{-g(m)}),$$

and when the units in λ^- are activated (but those in λ^+ are not),

$$a_0 = 1/(1+e^{-h(m)}).$$

Furthermore, if we let the number of units in the λ-layer increase, then in the case of λ^+, a_0 increases, and if $m \to \infty$, $a_0 \to 1$; but in the case of λ^-, a_0 decreases and $a_0 \to 0$.

Now let p be a mapping on $\{0, 1, \ldots, n\}$ such that $p(0) = a_0$ and $p(i) \in \mathrm{Prim}L$ for $i = 1, \ldots, n$. Here $p \upharpoonright \{1, \ldots, n\}$ is assumed to indicate with what symbols the system is presented. More precisely, if the sequence $p(1) \ldots p(n)$ belongs to $\mathrm{Exp}L$,[29] the units (the ensembles) in λ^+ are activated in the fashion described above, but if not, those in λ^- are activated. Thus it holds for all such mappings p:

(9.3.2)   If $p \upharpoonright \{1, \ldots, n\} \in \mathrm{Exp}L$, then $p(0) = 1/(1+e^{-g(m)})$;

   If $p \upharpoonright \{1, \ldots, n\} \notin \mathrm{Exp}L$, then $p(0) = 1/(1+e^{-h(m)})$.

NR is the theory axiomatized by (9.3.2).

Now it should be obvious in what sense TR and NR approximate each other. The more there are units in the λ-layer of the network, that is to say, the greater m is, the more closely (9.3.2) approximates (9.3.1), and if m approaches infinity, (9.3.2) goes to (9.3.1). Thus we may say that in a mathematical sense, and at the conceptual level as displayed in (9.3.1) and (9.3.2), TR is a limiting case of NR. But whether there obtains a limiting case correspondence between the two theories in the model-theoretic sense defined in Chapter 5 is not as obvious as the existence of the mathematical correspondence. If we look at the way in which axioms (9.3.1) and (9.3.2) were obtained, we may see that the meanings of the respective notions, particularly of 'p', occurring in the axioms are radically different, apart from their abstract mathematical meanings. What the terms mean in each case can only be seen by scrutinizing the two constructions, and they are of conceptually different kinds (perhaps even fundamentally different). One may then ask whether there is here a case of incommensurability in the sense of Kuhn (1962).

The Model-Theoretic Treatment[30]

To consider the question in the conceptual framework defined earlier, we next investigate briefly whether an appropriate limiting case correspondence of TR to NR can be defined and whether it results in an instance of counterfactual explanation in the sense described in Chapter 6. For that purpose, the above treatment is to be presented in model theory and nonstandard analysis. However, only an outline is given below.

Analogously to the way in which we proceeded in Section 9.1 with particle mechanics, we now present model-theoretic axiomatizations of TR and NR first by means of the standard model of analysis $\mathcal{U}$ and its elementary extensions, introduced in Section 8.8. They are of type τ_0. The following additional notation will be needed in the axiomatizations. If $\mathcal{B}$ is $\mathcal{U}$ or its elementary extension and n a standard or nonstandard natural number, then let $\mathbf{N}_n = \{1, \ldots, n\}$, that is,

$$\mathbf{N}_n = \{k \in^{\mathcal{B}} \mathbf{N} \mid 1 \leq_{\mathcal{B}} k \leq_{\mathcal{B}} n\},$$

where $\mathbf{N}$ is the set of natural numbers, $\leq_{\mathcal{B}}$ is the ordinary ordering relation $\leq$ if n is standard and its respective extension to the domain B of $\mathcal{B}$ if n is nonstandard. As in Section 9.1, denote $a_{\mathcal{B}} = \{b \in B \mid b \in^{\mathcal{B}} a\}$, where $\mathcal{B}$ is as above.

Now let $\mathcal{B}$ be a model of type τ_0. Consider a four-sorted expansion of $\mathcal{B}$, of type τ, say,

$$\mathfrak{M} = \langle \mathcal{B}; S, E, P; n \rangle,$$

where S, E, and P are new domains and n is a distinguished individual. Let H_0 be the class of all models $\mathfrak{M}$, as above, of type τ such that[31]

(9.3.3) $\mathcal{B}$ is $\mathcal{U}$ or one of its elementary extension satisfying (8.8.1);

(9.3.4) $n \in^{\mathcal{B}} \mathbf{N}$;

(9.3.5) S is a nonempty subset of A (i.e., of the domain of $\mathcal{U}$);

(9.3.6) E is a set of mappings from $\mathbf{N}_n$ to S;

(9.3.7) P is a set of mappings from $\mathbf{N}_n \cup \{0\}$ to $S \cup \{0, 1\}$;

(9.3.8)   For all $p \in P$,

If $p \upharpoonright \mathbf{N}_n \in E$, then $p(0) = 1$;

If $p \upharpoonright \mathbf{N}_n \notin E$, then $p(0) = 0$.

Condition (9.3.6) says that E is a set of sequences of length n of elements of S, and (9.3.4) that n is a standard natural number if $\mathcal{B}$ is $\mathcal{A}$ (i.e., standard); otherwise n is a standard or nonstandard (infinite) natural number.

Let H be the least class of models of type τ including H_0 and being closed under isomorphism. As we may see, the above conditions correspond (in a more abstract form) to the above, mathematical description of a Turing representation of expressions of L. Hence any element of H can be called a *model of* TR. In agreement with our earlier terminology, it is called standard or nonstandard according to whether $\mathcal{B}$ is a standard or nonstandard model of analysis. If L is a language as above, we may assume that its primitive symbols are sets,[32] even specified standard elements, in other words, elements of the superstructure A;[33] and therefore each expression of L is a sequence of elements of A.[34]

Conditions (9.3.3)-(9.3.8) can be formalized, and therefore H defined, in the four-sorted infinitary logic $L_{\kappa\omega}$, where κ is as before, that is, by means of $L_{\kappa\omega}(\tau)$-sentences. The kind of quantification in (9.3.8) notwithstanding (which is essentially a quantification over functions), the logic is first order (even though infinitary) in the sense that the functions in P are in fact considered as individuals of the respective sort (elements of the domain P). The class of standard models can likewise be defined in $L_{\kappa\omega}$.[35]

Now consider a model, of type τ', say, of the form

$$\mathfrak{M} = \langle \mathcal{B}; S, E, P, \mathbf{N}_{\mathcal{B}}, \mathbb{R}_{\mathcal{B}}; g, h, m, n \rangle,$$

where $\mathcal{B}$ is as above, S, E, P, $\mathbf{N}_{\mathcal{B}}$, and $\mathbb{R}_{\mathcal{B}}$ are domains, g and h are functions from $\mathbf{N}_{\mathcal{B}}$ to $\mathbb{R}_{\mathcal{B}}$, and m and n are distinguished individuals. Let H_0' be the class of all models $\mathfrak{M}$, as above, of type τ' such that (9.3.3)-(9.3.6) and the following hold:

(9.3.9) g (resp. h) is monotonically increasing (resp. decreasing) such that $g \upharpoonright \mathbf{N}$ ($h \upharpoonright \mathbf{N}$) is standard and unbounded;

(9.3.10) $m \in \mathbf{N}_{\mathcal{B}}$;

(9.3.11) P is a set of mappings from $\mathbf{N}_n \cup \{0\}$ to $S \cup \mathbb{R}_{\mathcal{B}}$;

(9.3.12) For all $p \in P$,

If $p \upharpoonright \mathbf{N}_n \in E$, then $p(0) = 1/(1+e^{-g(m)})$;

If $p \upharpoonright N_n \notin E$, then $p(0) = 1/(1+e^{-h(m)})$.

As can be seen, part of what (9.3.11) expresses is the condition (in an abstract form) that (for each $p \in P$) units $1, \ldots, m$ are grouped into n subsets so that each subset is a distributed representation of some symbol in S. Let H' be the least class of models of type τ' including H_0' and closed under isomorphism. It is definable in $L_{\kappa\omega}$, by means of $L_{\kappa\omega}(\tau')$-sentences. A model in H' will be called a *model of* NR, and whether it is standard or nonstandard can be described as above.

Next we outline a limiting case correspondence of TR to NR. Since the formal proofs are analogous to — but they are somewhat simpler than — those in Section 9.1, they will be omitted. To simplify the following treatment, we can, without any confusion (because of isomorphism), continue to use ordinary mathematical notation and terms (in the standard and nonstandard senses) even when working on arbitrary models of TR and NR.

Let K be the class of all standard models of TR in H and let K' be the class of all nonstandard models of NR in H' such that m is infinite (i.e., a nonstandard natural number) and n is a standard natural number (i.e., finite). Both classes are, again, definable in $L_{\kappa\omega}$. A mapping F is defined as follows. For each model $\mathfrak{M}$ in K' (of the form indicated above), let

$$F(\mathfrak{M}) = \langle \mathfrak{B}|\mathfrak{A}; S, E, {}^0P; n \rangle.$$

Here $\mathfrak{B}|\mathfrak{A}$ is the unique submodel of $\mathfrak{B}$ that is isomorphic to $\mathfrak{A}$, and ${}^0P$ is the set of all mappings ${}^0p$ such that for some $p \in P$,

$${}^0p = (p \upharpoonright N_n) \cup \{\langle 0, r \rangle\},$$

where $r = 1$ if $p(0) = 1/(1+e^{-g(m)})$, and $r = 0$ if $p(0) = 1/(1+e^{-h(m)})$. It is obvious that $F(\mathfrak{M}) \in K$ (i.e., $F(\mathfrak{M})$ is a standard model of Turing representation) and, conversely, for each $\mathfrak{M} \in K$, there is a model $\mathfrak{M}'$ in K' such that $F(\mathfrak{M}') = \mathfrak{M}$. Therefore, F is a mapping from K' onto K.

We may say now that $F(\mathfrak{M})$ is a standard approximation of $\mathfrak{M}$. It follows from condition (9.3.9) that in a model $\mathfrak{M}$ as above, where m is infinite, $g(m)$ is an infinite (nonstandard) positive real number and $h(m)$ is an infinite negative real number. Therefore, $p(0) \approx 1$ in the former case of (9.3.12), and since ${}^0p(0) = 1$, $p(0)$ is infinitesimally close to $^0p(0)$. In the latter case of (9.3.12), $p(0) \approx 0$, wherefore $p(0)$ is infinitesimally close to $^0p(0)$. Since F thus defines a limit procedure that is analogous to the one defined in Section 9.1, which transforms the models of K' into those of K, we can in an analogous way define a translation I, satisfying the relevant conditions presented in Chapter 5, of the $L_{\kappa\omega}(\tau)$-sentences (by means of which TR is characterized), into $L_{\kappa\omega}(\tau')$-sentences (characterizing NR). It follows that F and I determine a limiting case correspondence of TR to NR. In particular, the translation of (9.3.8) is (in the metalinguistic form):

> For all $p \in P$,
>
> If $p \!\upharpoonright \mathbb{N}_n \in E$, then $p(0) \approx 1$;
>
> If $p \!\upharpoonright \mathbb{N}_n \notin E$, then $p(0) \approx 0$

(expressible as an $L_{\kappa\omega}(\tau')$-sentence).

APPENDIX

DEFINABILITY[1]

Traditionally, the word 'definition' means something like explicit definition and, mainly in the philosophy of science, a very limited class of its generalizations. Traditional accounts of definability are often vague and obscure, however, so that it is not always clear what the word stands for. Hence it is instructive to place definitions in a more formal framework, as we shall see. Conversely, there are important pragmatic aspects of definition that cannot be dealt with by means of logical tools, but rather tools borrowed from the philosophy of language.

Section A.1 outlines some well-known views of traditional notions of definition and the distinction sometimes made between nominal and real definitions. In Section A.2, pragmatic aspects of definitions are looked at from a point of view of speech act theory. Sections A.3-5 are devoted to modern logical theories of definability, which are usually presented in model-theoretic terms, and in Section A.6 the traditional requirements concerning definitions are given a model-theoretic interpretation. Finally, Section A.7 studies the methodological and cognitive import of definitions. I shall not attempt any comprehensive survey of traditional or modern theories of definability, but rather to focus on some central points (many of which were needed earlier in this book) that I shall critically study.

A.1. TRADITIONAL ASPECTS

Whitehead and Russell argue in *Principia Mathematica* that a definition is wholly concerned with symbols and not with what they symbolize and that it is not true or false since it is an expression of a volition and not of a proposition. On the other hand, however, they define a definition as ". . . a declaration that a certain newly-introduced symbol or combination of symbols is to mean the same as a certain other combination of symbols of which the meaning is already known."[2] Hence they seem to say at the same time that a definition is wholly verbal and that it is not wholly verbal but about meanings as well. They again emphasize the latter aspect when they say that definitions often convey more important information than propositions in which they are used.

It should be clear, in any case, that if a definition is concerned with verbal expressions, it is also concerned with their meanings. A real difficulty that presents itself here is not, however, whether a definition is or is not concerned with meanings but the question of what is intended when it is said that a combination of symbols is to mean the same as a certain other combination of symbols. If meanings are contextual or relative, rather than objective, one has to specify the context or theoretical framework with respect to which meanings are considered. It is clear, of course, that in *Principia* this has been done, but that is not always the case in the traditional discussion of definability.

The notion of definition discussed in *Principia* is sometimes labelled as *nominal definition*, and similarly in Cohen and Nagel (1961). According to the latter, a nominal definition is ". . . an agreement or resolution concerning the use of verbal symbols. A new symbol called the *definiendum* is to be used for an already known group of words or symbols, the *definiens*. The definiendum is thus to have no meaning other than the definiens."[3] A standard example of explicit nominal definitions is the definition of classical implication in terms of negation and disjunction:

(A.1.1)   '*p* implies *q*' is by definition equivalent to 'not-*p* or *q*',

or symbolically, for example:

(A.1.2) $\mathbf{p} \to \mathbf{q} =_{\mathrm{df}} \neg\, \mathbf{p} \vee \mathbf{q}$.

Another scientific example mentioned is Comte's invention of the word 'sociology' for the study of human relations in organized group life. Furthermore, it is argued that all explicit definitions in the technique of modern mathematics are nominal.[4] According to Cohen and Nagel, too, a nominal definition is a resolution and not anything that would be true or false. Hence it cannot be a real premise of any argument and it does not extend our real knowledge – though it can be useful in scientific inquiry for technical reasons and may clarify our ideas.

In Robinson (1965), nominal definitions are also characterized as something concerning "words or signs or symbols."[5] The purpose of a nominal definition is to report or to establish the meaning of a symbol. This can be done either by saying that its meaning is the same as the meaning of another symbol (word-word definition) or by saying that it means a certain thing (word-thing definition). Since some nominal definitions are reports, they have a truth value. Thus Robinson's notion of nominal definition extends the respective notions mentioned in the above.

It is argued in Cohen and Nagel (1961) that in a *real definition* the definiens is an analysis of an idea, form, type, or universal symbolized by the definiendum.[6] It defines a word that possesses a meaning independently of the process of definition which equates it with the definiens indicating the structure of the defined entity. It follows that a real definition, unlike a nominal one, is a genuine proposition, true or false. They remark, on the other hand, that the distinction between nominal and real definitions is not as sharp as one could think on the basis of their characterization. Even in a verbal definition there is usually some reference to the analysis of the entity in question, and people may have emotional or other attitudes towards words defined that may confuse the character of definitions.

The distinction that Robinson makes between the two kinds of definition is more advanced. The distinction is for him intentional: 'nominal definition' and 'real definition' refer to purposes of definition rather than to methods.[7] One's immediate purpose determines to which of the two classes a definition belongs. While the immediate purpose of a nominal definition is to establish the meaning of a symbol, the immediate purpose of a real definition is to do with things (thing-thing definition), not symbols *qua* symbols. Nominal definitions are definitions of words, whereas real definitions are definitions of things. Robinson (tentatively) suggests, however, that from now on the word 'definition' should, perhaps, be restricted to nominal definitions, i.e., to processes concerning symbols, since the usage of the name 'real definition' has been ambiguous in the literature confusing activities of different kinds under one name.[8] He has found twelve such activities, so that, for instance, the analysis of things –

which is commonly associated with real definitions – is just one of them.

Robinson's view of mathematical definitions is slightly different from the view of Cohen and Nagel (1961). Thus he says, for instance, that definitions in mathematics are often real definitions in the sense of analyses of old ideas – whence they are often statements, true or false, and not just declarations concerning meanings of symbols.[9]

As we have seen, Whitehead and Russell (1973) and Cohen and Nagel (1961) regard a definition as being a syntactic expression in the first place, whereas for Robinson it is first of all an intellectual activity and its other possible senses are only secondary.[10] Thus, for example, whether a sentence – or, rather, an utterance – expresses a definition or just a report depends on the corresponding mental state of the speaker. This is another intentional feature in Robinson's notion of definition, and it follows that the question concerning whether something is a definition or which kind of definition is often contextual and the matter of interpretation. This suggests that pragmatic dimensions of definability can be studied by thinking of definitions as speech acts.

A.2. DEFINITIONS AS SPEECH ACTS

If definitions are speech acts, they can be classified accordingly. For example, definitions that are real in the sense discussed above would be assertives according to Searle's (1979) classification, and the different activities, or some of them, which Robinson reluctantly mentions as real definitions (as we saw in Section A.1) might be reclassified by studying their illocutionary force. More generally, if definitions are thought of as speech acts, then the whole theoretical machinery of speech acts is available. This would place the traditional discussion in a new perspective. Since the main purpose of the present article is to extend and criticize the traditional discussion by means of logical tools, I shall restrict myself to few observations concerning pragmatic aspect of definitions that are closely related to speech act theory.

According to Robinson (1965), there are nominal, word-thing definitions with which no truth value can be associated. Such is, for instance, stipulative definition, the act of assigning an object to a name or a name to an object.[11] The corresponding defining utterance, whatever its grammatical form is — there are different forms that are appropriate for the purpose — is not an assertion, but rather a proposal, request, or the like. It is obviously what Austin (1970) calls a performative utterance. One of Austin's examples of performative utterances is an act of naming:

(A.2.1)   I name this ship the *Queen Elizabeth.*

By uttering (A.2.1) one does not describe a christening ceremony, but actually performs the christening, and hence — according to Austin — it is not true nor false.

Though performative utterances are not true nor false, they are — Austin argues — satisfactory or unsatisfactory, they may succeed or fail. An utterance may suffer from various infelicities that arise if certain rules are broken. The rules constraining utterances are pragmatic rather than logical, and they are due to the conventional nature of language and the contextual and intentional character of utterances. First, the alleged social convention on which one relies when performing an act must exist and must be accepted by relevant institutions or communities. The performance of (A.2.1) fails to name the ship if it is part of a procedure that deviates from the accepted ceremony of christening ships. Second, the context in which an utterance is performed must be acceptable. (A.2.1) fails, for instance, if a person not authorized for the purpose performs the act according to the formally correct procedure.

Third, since performative utterances are often intentional — as, e.g., stipulative definitions, in particular — it is questionable whether a symbol will really become defined or a thing named if an utterance is not sincere or if there is a contradiction between the accepted, conventional interpretation of the utterance and the speaker's intention. If the utterance is not sincere and hence does not express the intentional state the speaker actually has, then it is obvious that there is a kind of infelicity in Austin's sense. On the other hand, however, it seems to depend on the strength of social constraints whether or not the definition is successful in such a case. If, for instance, I insincerely utter (A.2.1) in proper circumstances and in accordance with accepted rules, and as an authorized person, then it is obvious that I have named the ship the *Queen Elizabeth* as in this case the social constraints in question are powerful enough to overrule my insincerity.

All these cases of infelicity are somewhat inconvenient for the view advocated by Robinson that definitions are primarily mental activities. If an utterance is syntactically appropriate, then it seems to follow from what we have said above that in proper circumstances it is its syntactic form rather than the utterer's intention that makes it a definition, that is, it is the syntactic form on the basis of which it is generally accepted as a definition. Similar reservations concern Robinson's view that the distinction between nominal and real definition is intentional. In scientific contexts, an intention of defining something is often indicated by syntactic devices, as in (A.1.1) and (A.1.2), above, but in formal languages it is the syntactic form that is crucial.

A.3. THE LOGICAL ACCOUNT

The syntactic form of explicit definition, the most important kind of definition, is identity or equivalence in some sense. Since the notion of definition is very comprehensive in Robinson (1965), even explicit definitions have no fixed syntactic form and the question of their form is in fact inessential. Whether or not a verbal expression represents a definition or a definition of a given kind depends on the respective intentional state of the speaker and, as we saw, the context and social rules. This pragmatic conception is appropriate when definitions in natural language are concerned, but it is not appropriate for formal languages. Pragmatic dimensions become suppressed when formal ones are emphasized, and the syntactic form becomes essential. But it seems that some pragmatic aspects can be illuminated by using model-theoretic distinctions. More importantly, distinctions can be made that are not clear or possible without concepts of logical semantics. Consider, for example, the distinction between global and local definability. It is not always clear whether in traditional accounts a given definition is assumed to be contingent, to apply only to 'actual' objects, or whether they are thought of as being of a more necessary nature, applicable in 'all possible worlds', even though some attention has been paid to that distinction when real and nominal definitions have been discussed. Furthermore, logical semantics makes it possible to study semantic import of different kinds of definability.

We shall mainly restrict the discussion about definability to the familiar (one-sorted) first order predicate logic $L_{\omega\omega}$, that is, elementary logic, and only occasionally refer to other logics. Let τ be an admitted type of $L_{\omega\omega}$ and $\mathbf{P}$ a nonlogical constant not in τ. To simplify notation, we shall write 'L' for $L_{\omega\omega}(\tau)$ and '$L(\mathbf{s}_1,\ldots)$' for $L_{\omega\omega}(\tau \cup \{\mathbf{s}_1,\ldots\})$, where $\mathbf{s}_1,\ldots$ are symbols not in τ. Thus, for instance, '$L(\mathbf{P})$' means $L_{\omega\omega}(\tau \cup \{\mathbf{P}\})$. In what follows, $\mathsf{T}(\mathbf{P})$ will be a theory that is formulated by means of $L(\mathbf{P})$-sentences. T is the set of all L-sentences deducible from $\mathsf{T}(\mathbf{P})$ in L.

Here is some further notation. For convenience, the following abbreviations will be used: $\vec{a}$, $\vec{\mathbf{a}}$, and $\vec{\mathbf{x}}$ for '$a_1,\ldots,a_m$', '$\mathbf{a}_1,\ldots,\mathbf{a}_m$', and '$\mathbf{x}_1,\ldots,\mathbf{x}_m$', respectively; $\vec{\mathbf{z}}$ for '$\mathbf{z}_1,\ldots,\mathbf{z}_k$'; and $\exists\vec{\mathbf{x}}$ and $\forall\vec{\mathbf{z}}$ for '$\exists\mathbf{x}_1\ldots\exists\mathbf{x}_m$' and '$\forall\mathbf{z}_1\ldots\forall\mathbf{z}_k$', respectively. If the free variables of a formula φ are among $\vec{\mathbf{x}}$, we may indicate it by writing '$\varphi(\vec{\mathbf{x}})$', and similarly for other sets of variables and individual constants. For additional notation to be needed in what follows, the reader should consult Section 8.2.

I shall be considering the question of what it means to say that $\mathbf{P}$ is definable in terms of τ. Hence we are interested in relations between L-formulas and $L(\mathbf{P})$-formulas, and in the semantic counterparts of these relations. For simplicity, I shall assume that $\mathbf{P}$ is a unary predicate symbol. The assumption could be dispensed with; it would mainly cause some notational inconvenience, and possibly some additional conditions concerning definitions of function symbols and individual constants.[12]

Various formal definability concepts can be introduced. I shall consider explicit definability and some weaker notions and only summarize, with no proofs, some known results of elementary logic that relate syntactic and semantic aspects of these notions to each other. (For a more detailed discussion and proofs, see the works referred to below and Rantala, 1977a.) A syntactic definition means here a sentence of a certain form in the formal object language, that is, in this section we do not consider syntactic definitions in the metalanguage. Furthermore, since it does not make much sense to consider definitions *per se*, we shall consider such notions as definability in a theory and definability in a model.

Let us first consider the notions of explicit and implicit definability and their semantic counterpart. One of the most important model-theoretic results of elementary logic is the following, Beth's Theorem (which generalizes a result of Tarski, 1956):

THEOREM A.3.1 (Beth, 1953). The following conditions are equivalent:

    (i)   There is an L-formula $\varphi(\mathbf{x})$ such that

$$\mathsf{T}(\mathbf{P}) \vdash \forall \mathbf{x}(\mathbf{P}(\mathbf{x}) \leftrightarrow \varphi(\mathbf{x}));$$

    (ii) If $\mathbf{Q}$ is a unary predicate symbol not in τ and $\mathsf{T}(\mathbf{Q})$ the theory in $L(\mathbf{Q})$ obtained from $\mathsf{T}(\mathbf{P})$ by substituting $\mathbf{Q}$ for $\mathbf{P}$ in $\mathsf{T}(\mathbf{P})$, then

$$\mathsf{T}(\mathbf{P}) \cup \mathsf{T}(\mathbf{Q}) \vdash \forall \mathbf{x}(\mathbf{P}(\mathbf{x}) \leftrightarrow \mathbf{Q}(\mathbf{x}));$$

    (iii) Every model $\mathfrak{M}$ for L has at most one expansion $(\mathfrak{M}, P)$ for $L(\mathbf{P})$ such that $(\mathfrak{M}, P)$ is a model of $\mathsf{T}(\mathbf{P})$.

Here P is a subset of the domain of $\mathfrak{M}$. Condition (i) says that $\mathbf{P}$ is *explicitly definable* in $\mathsf{T}(\mathbf{P})$ (in terms of τ), i.e., that an explicit definition of $\mathbf{P}$ is derivable from the theory. This notion covers both of the two methodological situations: that the theory has been developed by adding an explicit definition to previous axioms (and thus expanding the language) and that the theory is given and an explicit definition is derived only afterwards. (ii) expresses the fact that $\mathbf{P}$ is *implicitly definable* in $\mathsf{T}(\mathbf{P})$. (iii) is the semantic counterpart of (ii), that is, their equivalence is immediate and independent of the assumption that we consider elementary logic here. If (iii) holds, $\mathbf{P}$ will be here called *identifiable* in $\mathsf{T}(\mathbf{P})$.[13] The equivalence of explicit and implicit definability in elementary logic − or, rather, that implicit definability implies explicit definability − is a deep result that is a consequence of Craig's Interpolation Theorem, and it does not hold for all logics.[14] It is one of the characteristic properties of elementary logic, and it has been pointed out that well-known extensions and deviations of elementary logic do not generally have the property.[15] The same obviously holds for many of the results below.

    Condition (iii) is model-theoretic, and it is quantitative in the sense that it states a restriction on the number (cardinality) of acceptable interpretations of $\mathbf{P}$ once the interpretations of the symbols in τ are fixed: there cannot be more than one such interpretation. The next two definability notions also impose such quantitative restrictions, but in a generalized sense. As we shall see, it is not obvious that the term 'definability' can be justifiably applied to these notions, but one justification will be provided later, in Section A.6.

THEOREM A.3.2 (Kueker, 1970). The following conditions are equivalent:

(i)   There are L-formulas $\sigma(\vec{x})$ and $\varphi_i(\vec{x}, y)$ $(i = 1, \ldots, n)$ such that

(a) $T(\mathbf{P}) \vdash \exists \vec{x} \sigma(\vec{x})$

(b) $T(\mathbf{P}) \vdash \forall \vec{x}(\sigma(\vec{x}) \to \bigvee_{1 \leq i \leq n} \forall y(\mathbf{P}(y) \leftrightarrow \varphi_i(\vec{x}, y)))$;

(ii) Every model $\mathfrak{M}$ for L has at most n expansions $(\mathfrak{M}, P)$ for $L(\mathbf{P})$ such that $(\mathfrak{M}, P)$ is a model of $T(\mathbf{P})$.

In Hintikka (1972), $\mathbf{P}$ is said to be *n-foldly identifiable* in $T(\mathbf{P})$ (in terms of τ) if (ii) holds, and *finitely identifiable* in $T(\mathbf{P})$ if it is n-foldly identifiable for some (finite) n.[16] We shall here call $\mathbf{P}$ *finitely definable* in $T(\mathbf{P})$ if (i) holds for some n. Thus, in elementary logic finite identifiability and finite definability are equivalent notions. If, in particular, $n = 1$, then $\mathbf{P}$ is explicitly (and implicitly) definable. The explicit definition that is then obtained from the conditions (a) and (b) is as follows:

$$\forall y(\mathbf{P}(y) \leftrightarrow \exists \vec{x}(\sigma(\vec{x}) \wedge \varphi_1(\vec{x}, y))).$$

In the next theorem, the number of acceptable expansions can be infinite. Let $|\mathfrak{M}|$ be the cardinality of $\mathfrak{M}$:

THEOREM A.3.3 (Chang, 1964; Makkai, 1964). The following conditions are equivalent:

(i)   There are L-formulas $\varphi_i(\vec{x}, y)$ $(i = 1, \ldots, n)$ such that

$$T(\mathbf{P}) \vdash \bigvee_{1 \leq i \leq n} \exists \vec{x} \forall y(\mathbf{P}(y) \leftrightarrow \varphi_i(\vec{x}, y));$$

(ii) Every infinite model $\mathfrak{M}$ for L has at most $|\mathfrak{M}|$ expansions $(\mathfrak{M}, P)$ for $L(\mathbf{P})$ such that $(\mathfrak{M}, P)$ is a model of $T(\mathbf{P})$.

Hintikka (1972) calls $\mathbf{P}$ *restrictedly identifiable* in $T(\mathbf{P})$ (in terms of τ) if (ii) holds. In that case, a fixed interpretation of the symbols in τ in an infinite domain $\mathfrak{M}$ leaves at most $|\mathfrak{M}|$ choices open for interpreting $\mathbf{P}$ in order to obtain a model of $T(\mathbf{P})$ — even though the number of all possible interpretations is $2^{|\mathfrak{m}|}$. **P** is *restrictedly definable* in T(**P**) if (i) holds.

The next theorem states a criterion for a predicate being definable in the models of a theory. We say that **P** is *definable in a model* for $L(\mathbf{P})$ (in terms of τ) if an explicit definition $\forall x(\mathbf{P}(x) \leftrightarrow \varphi(x))$, where $\varphi(x)$ is an L-formula, is true in the model.

THEOREM 3.4 (Hintikka and Tuomela, 1970). The following conditions are equivalent;

(i)   There are L-formulas $\varphi_i(x)$ $(i = 1, \ldots, n)$ such that

$$T(\mathbf{P}) \vdash \bigvee_{1 \le i \le n} \forall x(\mathbf{P}(x) \leftrightarrow \varphi_i(x));$$

(ii)   **P** is definable in every model of T(**P**);

(iii) **P** is explicitly definable in every complete consistent extension of T(**P**).

It is said that **P** is *piecewise definable* in T(**P**) (in terms of τ) if (i) holds. If **P** is piecewise definable, then **P** is definable in every model of the theory, but the definiens may be different in different models. This notion is again a generalization of the notion of explicit definability, which results if $n = 1$. On the other hand, it is implied by finite definability in the special case that the 'parameters' $\bar{x}$ can be dispensed with.[17] The conditions (i)-(ii) also have an important algebraic equivalent, as shown by Svenonius (1959).

In actual theory constructions, *conditional definitions* are often employed rather than explicit ones. This means that defined notions are not assumed to be uniquely applicable to all elements of the relevant domains, so some preconditions are needed. Consider, for example, the ordinary definition of division in arithmetic where the condition excluding division by zero is stated. Other examples are not hard to come by, although required conditions are not always explicitly presented.

Conditional definitions play an important historical role in the philosophy of science, in logical empiricists' discussion concerning the relation of 'theoretical terms' and 'observational terms' in scientific theories. For example, Carnap (1936-37) observed that certain theoretical terms, such as dispositional predicates, are not explicitly definable – explicit definablity was initially required by logical empiricists – and generally not even conditionally definable in terms of an observational language in an intuitively adequate way, if theories are assumed to be reconstructed in elementary logic. This drawback can be taken care of if elemetary logic is extended by adding subjunctive conditionals, which, however, results in difficult problems concerning formal semantics.[18] This characterization problem is just one indication of the fact that the expressive power of elementary logic is too weak for many metascientific purposes. Other examples, which are rather different, will be offered below.

In our present, slightly simplified logical framework, it is said that **P** is *conditionally definable* in $T(\mathbf{P})$ (in terms of τ) if a conditional definition is derivable from $T(\mathbf{P})$:

$$T(\mathbf{P}) \vdash \forall \mathbf{x}(\sigma(\mathbf{x}) \to (\mathbf{P}(\mathbf{x}) \leftrightarrow \phi(\mathbf{x}))).$$

Here σ and ϕ are L-formulas. The semantic import of conditional definability can be readily understood. For the cases of conditional definability where our simplification will not do — most mathematical definitions, for instance, are such — that is, where instead of a unary predicate symbol we need nonlogical constants of other kinds, some conditions for proper definitions must be added.[19]

It is not possible here to give any exhaustive overview of all different notions of definition that are mathematically, logically, or methodologically important and representable in formal logic. For the other notions, I can only refer to standard textbooks.

A.4. UNDEFINABILITY AND UNCERTAINTY

As we have seen, if **P** is explicitly definable in $T(\mathbf{P})$, then each interpretation of t uniquely determines the interpretation of **P** in the sense that every model for L has at most one expansion to a model of $T(\mathbf{P})$. Conversely, if **P** is not explicitly definable in $T(\mathbf{P})$, the theory leaves some freedom to interpret **P**. This freedom can also be construed as a kind of uncertainty concerning the interpretation of **P**. It means a kind of uncertainty for a person who is studying the theory and its models if he only knows how the constants in τ have been interpreted. Only if **P** is explicitly defined in the theory, the person may know how **P** has been interpreted in an arbitrary model of the theory *on the basis* of knowing how these other constants have been interpreted. The other kinds of definability do not completely remove this uncertainty, but some of them reduce it, as we can see from Theorems A.3.2-A.3.4.

It depends on the methodological standpoint whether undefinability is called freedom or uncertainty. The latter name is motivated, for example, if one is studying a theory and its models by stepwise processes during which one is gathering more and more information concerning how $\mathbf{P}$ and τ are syntactically or semantically related. Hintikka and Tuomela (1970) and Hintikka (1972) point out that by using an appropriate method of elementary logic, viz. the method of distributive normal forms, a stepwise syntactic process can be defined by means of which *global uncertainty* can be studied, i.e., uncertainty concerning the interpretation of $\mathbf{P}$ in all models of the theory. One's initial uncertainty concerning $\mathbf{P}$ may be reduced during the process, and at some point completely removed if $\mathbf{P}$ is explixitly definable in $\mathsf{T}(\mathbf{P})$. If, on the other hand, $\mathbf{P}$ is definable in one of the senses indicated in Theorems A.3.2-A.3.4, it will also become apparent at some point of the process. Conversely, if certain distinctive syntactic features are displayed in the course of the process, one can infer that $\mathbf{P}$ is definable in $\mathsf{T}(\mathbf{P})$ in one of these senses. Thus, global definability and the reduction of global uncertainty go hand in hand.

The method in question is based on Hintikka's notion of constituent, a special kind of formula, which can be used to represent various properties of elementary theories. In Hintikka and Rantala (1975) and Rantala (1977a), this syntactic method is combined with model-theoretic ones to study *local uncertainty*, that is, uncertainty concerning interpretation in single models. By these methods local definability results are obtained, that is, results concerning (finitary and infinitary) definability in a model. The notion of definability in a model is generalized for the other forms of definition with which we have become acquainted above, and in some of the mentioned results formulas occurring in definitions are infinitary, containing infinite conjunctions or disjunctions, and hence do not belong to elementary logic.

A.5. MODEL-THEORETIC DEFINABILITY

In model theory, questions concerning the expressive power of various logics extending elementary logic have been extensively studied. It is often important for metamathematical purposes to know whether given classes of mathematical structures, that is, of models, can be characterized in some logic in some well-defined sense. If L is any logic and K is a class of models for $L(\tau)$, then to say that K is characterizable in L means in the most straightforward sense that there is an $L(\tau)$-sentence φ (or possibly an appropriate set of such sentences) such that K is the class of all models in which φ is true, i.e., such that the following holds for all models $\mathfrak{M}$:

$$\mathfrak{M} \in K \Leftrightarrow \mathfrak{M} \models_L \varphi.$$

This equivalence expresses an explicit definition (derivable in the metatheory of L) in the metalanguage of L, and hence it is natural to say that K is *definable in L*, or *L-definable*, and φ *defines K in L*.[20]

The following generalization was also needed earlier. If $\mathfrak{M}$ is a model in $\mathrm{Mod}_L(\tau)$ and R an n-ary relation holding between individuals of $\mathfrak{M}$, R is L-definable in $\mathfrak{M}$ if for some $L(\tau)$-formula φ,

$$\langle a_1, \ldots, a_n \rangle \in R \Leftrightarrow \mathfrak{M} \models_L \varphi[a_1, \ldots, a_n];\text{[21]}$$

As recalled in Section 8.1, the expressive power of elementary logic is weak for many metamathematical purposes in the sense that not all important classes of mathematical structures, hence not all important mathematical notions, are definable in elementary logic. Such are, for example, the class of all finite sets (that is, the notion of finiteness), the class of all structures isomorphic to the standard model of arithmetics of natural numbers, and the class of all well-ordered sets (that is, the notion of well-ordering), to mention some of the most important cases.[22]

Questions concerning the characterization of classes of models in different logics belong to what in mathematical logic is called *definability theory* – which covers large and divergent areas of mathematical logic.[23] Our earlier discussion concerning the definability of **P** in $\mathrm{T}(\mathbf{P})$ is of course part of that general definability theory, and such results as Theorems A.3.1-A.3.4 have often been construed as indicating that there obtains a balance between expressive power and mathematical elegance of elementary logic. Similar results need not hold for extensions or deviations of elementary logic, as we already remarked in the above.

We may now see from this and preceding sections that as soon as definitions are studied in the model-theoretic framework, distinctions are made that are completely beside the question when they are studied in natural or informal scientific language. On the other hand,it is evident why such pragmatic considerations as the ones discussed in Section A.2, above, are not very relevant in the logical theory of definability. For instance, whether a formal expression indicates a definition is not so much dependent on the intentional state of the speaker; it is determined by logical conventions and rules.

A.6. CONDITIONS ON DEFINITIONS

Definitions are traditionally assumed to satisfy various requirements in order to be acceptable. Some of them are methodological, such as the rules that a definition must state the essence of the defined object and be given *per genus et differentiam* and that a definition must not be expressed in an obscure way, or logical, such as the rules that a definition must not be expressed negatively if it can be expressed positively and that the definiendum must be equivalent to the definiens. Aristotle's work is usually considered as the historical origin of these rules. In the literature there also exist other conditions – some of which are of pragmatic character – that are of later origin.[24]

In this section the requirements of eliminability and noncreativity are studied and generalized. These requirements can be formulated in the context of formal and mathematical languages, rather than of natural language. It is argued in *Principia Mathematica* that a definition is theoretically unnecessary in the sense that one might always use the definiens and dispense with the definiendum.[25] This condition seems to amount to eliminability, which in our formal framework is formulated as follows. $\mathbf{P}$ is *eliminable in* $\mathsf{T}(\mathbf{P})$ (in terms of τ) if for each $L(\mathbf{P})$-formula φ, there is an L-formula ψ (i.e., a formula not containing $\mathbf{P}$) such that

(A.6.1) $\mathsf{T}(\mathbf{P}) \vdash \varphi \leftrightarrow \psi$.

This notion is typically applied to cases in which a new term is introduced by means of a definition when a theory is being constructed, that is, in which $\mathsf{T}(\mathbf{P})$ is the set of all $L(\mathbf{P})$-sentences deducible from a set $\Sigma \cup \{\delta\}$, where Σ is a set of L-sentences (e.g., axioms and definitions of $\mathsf{T}(\mathbf{P})$ or another theory) and δ is a definition (of some kind) of $\mathbf{P}$ in terms of τ:

(A.6.2) $\mathsf{T}(\mathbf{P}) = \mathrm{Cn}(\Sigma \cup \{\delta\})$.

Then it is often required that the definition δ must be such that **P** is eliminable in T(**P**). The notion of eliminability was here defined more generally, however, so that it applies to arbitrary methodological situations.

The above notion of eliminability can be defined for any logic L, but in what follows, we again assume that L is elementary logic. If **P** is eliminable in T(**P**), then for each L(**P**)-formula there is an L-formula such that the two formulas are equivalent in T(**P**) – but they are not in general logically equivalent. In this sense the eliminability of **P** is only contextual. As can be expected, eliminability and explicit definability are equivalent notions. If **P** is explicitly definable in T(**P**), then ψ is obtained by replacing each subformula **P(t)** of φ (where **t** is an arbitrary term of the language) by the respective definiens. Conversely, if **P** is eliminable, then an explicit definition of **P** is obtained by choosing **P(x)** for φ in (A.6.1). It follows, then, by Beth's Theorem, that (for elementary logic L) the eliminability of **P** is equivalent to the condition that each model for L has at most one expansion for L(**P**) that is a model of T(**P**):[26]

THEOREM A.6.1. The following conditions are equivalent:
  (i)   **P** is eliminable in T(**P**);
  (ii) **P** is explicitly definable in T(**P**);
  (iii) Every model $\mathfrak{M}$ for L has at most one expansion $(\mathfrak{M}, P)$ for L(**P**) such that $(\mathfrak{M}, P)$ is a model of T(**P**).

That explicit definability implies eliminability is in fact a precise way to say that explicit definability has the following important methodological consequence, which is much discussed in the literature, but not always precisely expressed: T(**P**) is reducible to a theory in L (whose vocabulary is simpler)[27] – in the constructive case (A.6.2) to the theory Cn(Σ).

Is a similar concept of eliminability possible for definitions in natural language, that is, if we consider common speech instead of scientific and logical languages? Since it is difficult to specify different levels of language and to define the notion of linguistic context in natural language, the best thing one can do there is to define the eliminability of a symbol s as the feature that for each expression α in which s occurs, there is another one, β, in which it does not occur and which has the same meaning as α or is synonymous to α. The concepts of meaning and synonymy are, however, notoriously vague and confused, and they have been criticized by several philosophers, Quine (1953) in the first place.

Furthermore, the question how β is obtained from α is perplexing. If it is assumed, for example, that s is (explicitly) definable by means of a word-word definition[28] and that β is obtained by replacing s by the *definiens,* we get into well-known troubles with intensional contexts, e.g., if α is a propositional attitude phrase, and, more generally, with linguistic contexts that are not appropriately chosen.[29] In other words, only if relativized to appropriate, sufficiently narrow fragments of natural language the notion of eliminability makes sense, but, on the other hand, the task of delineating relevant fragments of common speech is as problematic as the task of defining the notion of synonymy or the sameness of meaning. The notion is fully relevant only for languages that are more technical and extensional than natural language. The same holds for the notion of noncreativity, which we next study.

Noncreativity is defined for theories of the form (A.6.2) where δ is now any $L(\mathbf{P})$-sentence that is used to extend Σ, i.e., to construct $\mathsf{T}(\mathbf{P})$, but which we may here think of as a definition. It is said that the sentence δ is *noncreative* in $\mathsf{T}(\mathbf{P})$ if any theorem of $\mathsf{T}(\mathbf{P})$ not containing $\mathbf{P}$ can already be deduced from Σ alone, that is, if for any L-sentence φ such that $\mathsf{T}(\mathbf{P}) \vdash \varphi$, it likewise holds that $\Sigma \vdash \varphi$. In other words, δ is noncreative if $\mathsf{T}(\mathbf{P})$ is a conservative extension of Σ.

If δ is noncreative, no new consequences in L can be obtained by means of δ, but it does not mean that no new consequences at all could be created by adding δ to Σ. New consequences are, of course, obtained, but they contain $\mathbf{P}$, and they need not be reducible to L-sentences if $\mathbf{P}$ is not also eliminable, that is, explicitly definable.

Noncreativity can also be expressed in semantic terms. Let us say that $\mathbf{P}$ is *Ramsey-eliminable* in $\mathsf{T}(\mathbf{P})$ if every model of Σ (which is a model for L) is expandable to a model of $\mathsf{T}(\mathbf{P})$. There exist different formulations of Ramsey-eliminability, which has in fact played an important methodological role in recent philosophy of science.[30] The following theorem − whose converse does not hold in general − is a direct consequence of the completeness of elementary logic. Its proof is well known and will not be presented here:

THEOREM A.6.2. If **P** is Ramsey-eliminable in T(**P**), then δ is non-creative in T(**P**).

The following theorem shows that noncreativity is, in a sense, a weaker notion than eliminability. The theorem will be proved here because the result is not, perhaps, well-known. The theorem justifies, in some weak sense, our terminology in Section A.3, above, where we generalized the notion of definability.

THEOREM A.6.3. The global definitions mentioned in Section A.3 are noncreative.

Proof. It is sufficient to consider restricted definability. Thus we assume that T(**P**) is as in (A.6.2), where δ is now of the form

$$\bigvee_{1 \leq i \leq n} \exists \vec{\mathbf{x}} \forall \mathbf{y}(\mathbf{P}(\mathbf{y}) \leftrightarrow \varphi_i(\vec{\mathbf{x}}, \mathbf{y})).$$

We show that in this case **P** is Ramsey-eliminable in T(**P**), which implies noncreativity. Let $\mathfrak{M}$ be a model of Σ and let $\varphi_i(\vec{\mathbf{x}}, \mathbf{y})$ be any of the formulas indicated above. Let $\vec{a}$ be arbitrary elements of the domain of $\mathfrak{M}$ and let $\vec{\mathbf{a}}$ be new individual constants (not in τ) that will be interpreted as $\vec{a}$, respectively. Furthermore, let P be the set of all elements of the domain of $\mathfrak{M}$ that satisfy the formula $\varphi_i(\vec{\mathbf{a}}, \mathbf{y})$ in $(\mathfrak{M}, \vec{a})$ (P may be empty). It follows that

$$(\mathfrak{M}, P, \vec{a}) \models \forall \mathbf{y}(\mathbf{P}(\mathbf{y}) \leftrightarrow \varphi_i(\vec{\mathbf{a}}, \mathbf{y})),$$

and hence

$$(\mathfrak{M}, P) \models \exists \vec{\mathbf{x}} \forall \mathbf{y}(\mathbf{P}(\mathbf{y}) \leftrightarrow \varphi_i(\vec{\mathbf{x}}, \mathbf{y})),$$

and therefore
$$(\mathfrak{M}, P) \models \delta.$$

It follows that

$$(\mathfrak{M}, P) \models \Sigma \cup \{\delta\}.$$

Thus $(\mathfrak{M}, P)$ is also a model of T(**P**), and so **P** is Ramsey-eliminable in T(**P**).

Hence all the definitions considered in Section A.3 can be thought of as proper definitions in the weak sense that they do not admit new consequences in L. With the exception of explicit definition, they do not meet the criteria of eliminability, as we noticed, but their image as proper definitions can be strengthened by showing that they satisfy certain weaker eliminability criteria: they admit eliminability in some local or contextual senses.[31] We define first a notion of local eliminability in an obvious way. **P** is *eliminable in a model* $\mathfrak{M}$ for $L(\mathbf{P})$ if for every $L(\mathbf{P})$-formula $\varphi(\vec{z})$, there is an L-formula $\psi(\vec{z})$ such that

$$\mathfrak{M} \models \forall \vec{z}(\varphi(\vec{z}) \leftrightarrow \psi(\vec{z})).$$

For each of the above forms of definability there is a corresponding notion of weak eliminability. Let us start with restricted definability. We say that **P** is *restrictedly eliminable* in $\mathsf{T}(\mathbf{P})$ if for every $L(\mathbf{P})$-formula $\varphi(\vec{z})$, there are L-formulas $\varphi_i(\vec{x}, \vec{z})$ $(i = 1, \ldots, n)$ such that

$$\mathsf{T}(\mathbf{P}) \vdash \bigvee_{1 \leq i \leq n} \exists \vec{x} \forall \vec{z}(\varphi(\vec{z}) \leftrightarrow \varphi_i(\vec{x}, \vec{z})).$$

That this notion corresponds to restricted definability is shown by the following theorem:

THEOREM A.6.4. The following conditions are equivalent:
   (i)   **P** is restrictedly definable in $\mathsf{T}(\mathbf{P})$;
   (ii) **P** is restrictedly eliminable in $\mathsf{T}(\mathbf{P})$.

Proof. Assume that (i) holds. Then there are L-formulas $\varphi_i(\vec{x}, \mathbf{y})$ $(i = 1, \ldots, n)$ such that

$$\mathsf{T}(\mathbf{P}) \vdash \bigvee_{1 \leq i \leq n} \exists \vec{x} \forall \mathbf{y}(\mathbf{P}(\mathbf{y}) \leftrightarrow \varphi_i(\vec{x}, \mathbf{y})).$$

If $\mathfrak{M}$ is a model of $\mathsf{T}(\mathbf{P})$, then for some $i = 1, \ldots, n$ it holds that

$$(\text{A.6.3}) \quad \mathfrak{M} \models \exists \vec{x} \forall \mathbf{y}(\mathbf{P}(\mathbf{y}) \leftrightarrow \varphi_i(\vec{x}, \mathbf{y})).$$

Hence for some elements $\vec{a}$ of the domain of $\mathfrak{M}$,

$$(\mathfrak{M}, \vec{a}) \models \forall \mathbf{y}(\mathbf{P}(\mathbf{y}) \leftrightarrow \varphi_i(\vec{\mathbf{a}}, \mathbf{y})),$$

where $\vec{\mathbf{a}}$ are new individual constants, interpreted as $\vec{a}$, respectively. Thus **P** is definable in the model $(\mathfrak{M}, \vec{a})$ for $L(\vec{\mathbf{a}})$ (in terms of $\tau \cup \{\vec{\mathbf{a}}\}$). It follows that if $\varphi(\vec{z})$ is an arbitrary $L(\mathbf{P})$-formula, there is an $L(\vec{\mathbf{a}})$-formula $\psi_i(\vec{\mathbf{a}}, \vec{z})$ such that

$$(\mathfrak{M}, \vec{a}) \models \forall \vec{z}(\varphi(\vec{z}) \leftrightarrow \psi_i(\vec{\mathbf{a}}, \vec{z})),$$

and so

$$\mathfrak{M} \models \exists \vec{x} \forall \vec{z} (\varphi(\vec{z}) \leftrightarrow \psi_i(\vec{x}, \vec{z})).$$

Since for each model $\mathfrak{M}$ of $T(\mathbf{P})$, (A.6.3) holds for some formula $\varphi_i(\vec{x}, \mathbf{y})$ ($i = 1, \ldots, n$), it follows that there are n formulas $\psi_i(\vec{x}, \vec{z})$) ($i = 1, \ldots, n$) such that

$$T(\mathbf{P}) \vdash \bigvee_{1 \leq i \leq n} \exists \vec{x} \forall \vec{z} (\varphi(\vec{z}) \leftrightarrow \psi_i(\vec{x}, \vec{z})),$$

thus (ii) holds.

Conversely, it is obvious that (ii) implies (i).

Restricted eliminability is a kind of contextual eliminability. If $\mathbf{P}$ is restrictedly eliminable in $T(\mathbf{P})$, then — as can be readily seen from the above proof — for each $L(\mathbf{P})$-formula and for each model $\mathfrak{M}$ of $T(\mathbf{P})$, $\mathbf{P}$ can be eliminated in a relevant expansion of $\mathfrak{M}$ of the form $(\mathfrak{M}, \vec{a})$, in terms of τ plus some new *parameters* $\vec{a}$. But with piecewise definability a notion of contextual eliminability without parameters can be associated, as can be expected. $\mathbf{P}$ is called *piecewise eliminable* in $T(\mathbf{P})$ (in terms of τ) if for every $L(\mathbf{P})$-formula $\varphi(\vec{z})$, there are L-formulas $\varphi_i(\vec{z})$ ($i = 1, \ldots, n$) such that

$$T(\mathbf{P}) \vdash \bigvee_{1 \leq i \leq n} \forall \vec{z} (\varphi(\vec{z}) \leftrightarrow \varphi_i(\vec{z})).$$

The proof of the following theorem is similar to the proof of the previous one:

THEOREM A.6.5. The following conditions are equivalent:
   (i) $\mathbf{P}$ is piecewise definable in $T(\mathbf{P})$;
   (ii) $\mathbf{P}$ is piecewise eliminable in $T(\mathbf{P})$;
   (iii) $\mathbf{P}$ is eliminable in every model of $T(\mathbf{P})$;
   (iv) $\mathbf{P}$ is eliminable in every complete consistent extension of $T(\mathbf{P})$.

It should be obvious by now how the eliminability notions corresponding to finite and conditional definability are defined. Their exact definitions are left to the reader.

A.7. WHY ARE DEFINITIONS NEEDED?

If a definition is noncreative and the term it introduces eliminable, is the definition worth the effort? The question is more relevant about definitions in formal and scientific languages than in natural language since, as we saw, eliminability and noncreativity are notions that are not immediately applicable to definitions in common speech. Furthermore, it is more relevant about explicit definition than the other kinds of definition discussed above since eliminability in the strict sense is not true of the latter. Since eliminability in appropriate local and contextual senses holds for the other kinds, the question is of some relevance in their cases as well. On the other hand, it has been pointed out in the extensive literature of the philosophy of science that introducing new 'theoretical' terms to a scientific theory is in general of great methodological, pragmatic, and heuristic importance.[32]

Many authors have posed this question and given rather similar answers. Whitehead and Russell (1973) state the core of the answers when they say, firstly, that though definitions are "theoretically superfluous," they nevertheless ". . . often convey more important information than is contained in the propositions in which they are used." This follows from the fact that when a definition is chosen, it shows us that its definiens is considered important, and from the fact that the definiendum is often more or less familiar to us, so that the definition, being a result of an analysis of a common idea, ". . . may therefore express a notable advance." A definition may, for instance, make a vague idea precise, as, for example, Cantor's definition of continuum. Secondly, there is a practical and economical dimension: ". . . if we introduced no definitions, our formulas would very soon become so lengthy as to be unmanageable . . .."[33]

What is said in *Principia* about the nature of information definitions may convey is of course very true, but something must be added to it. According to Frege (1884), conclusions we draw from a fruitful definition extend our knowledge. This remark seems to refer to a kind of information that is different from the kind of information mentioned in *Principia*, and it is the exact opposite of the view held by Cohen and Nagel (1961) – as regards nominal definitions.

What does it mean to say that a definition is fruitful and conclusions drawn from it extend our knowledge? According to Frege, a definition is fruithful only if it draws boundary lines that were not previously given at all. If, for example, the term defined is a unary predicate expression, then the definiens describes a collection of individuals in each world where the definition is assumed to be true. In this sense, the definition draws boundary lines in each such world. Therefore, one way to construe Frege's notion of fruitfulness is to say that a definition is fruitful if it makes a new classification of objects; new, e.g., in the sense that it was not known before or in the sense that it was not made, or that it was not known that it was made, by the earlier theory to which the definition was added. This kind of knowledge seems to be close to the kind of knowledge that Robinson (1965) means when arguing that a word-thing definition ". . . may give us new knowledge of the thing in that it abstracts it and sets it off from the rest of the world in a way which we probably had not done before . . .." In this way a definition may ". . . teach the contents of the world and the variety of human ideas."[34] As an example he provides the definition of derivative in mathematics, whereby a student learns this mathematical notion.

This kind of fruitfulness can be model-theoretically characterized by applying to definability Hintikka's notions of distributive normal form and semantic information. The characterization seems to lead to the conclusion made in Hintikka and Tuomela (1970) that definitions can be fruitful in the Fregean sense even though they are noncreative and the terms they introduce eliminable.[35] It appears, furthermore, that the activities that are needed (in the framework of distributive normal forms) to come upon new boundaries and to make them significant to us are analogous to the activities needed in logical deductions, as they are characterized in Hintikka (1973).[36] Hence, it seems to follow from Hintikka's theory of semantic information that there is an analogy between the ways in which information is obtained from fruitful definitions and from fruitful logical argumentations. This analogy can be considered as one aspect of the parallelism, emphasized by Tarski (1956), between the theory of definability and the theory of deduction.

Let us observe in passing that a definition is often used to introduce new objects – not only new boundary lines – as, for example, when a domain of known individuals is extended in science or mathematics.[37] Various extensions of number domains provide well-known mathematical examples. However, in what sense one considers the defined objects new, e.g., whether one thinks of the definition as introducing only new boundary lines or even ontologically new objects, depends on one's philosophical attitude. Philosophically this means a choice between realism and constructivism, but the choice may have implications concerning what methodological and formal procedures one chooses. It is clear that the kinds of information obtained in the two cases are different.

As observed in *Principia*, the main economical gain obtained by introducing definitions to a theory is that they reduce complexity. This is a more obvious advantage of definitions. It is clear, for example, that if a complex expression of natural language or scientific language is replaced by a simple term, it may be of a considerable cognitive and practical value. This is what Hempel (1972) means when he says that a formulation of theories solely in terms of primitives may become so involved as to be unintelligible.[38] One aspect of the reduction of complexity is the following, as Hintikka and Tuomela (1970) point out by using Hintikka's notion of semantic information. Consider a first order theory $T(\mathbf{P})$ as in (A.6.2), above. Even if the definition δ is noncreative, so that the theorems of $T(\mathbf{P})$ that are L-sentences can be deduced from Σ already, it is possible that within $T(\mathbf{P})$ they can be deduced in a simpler way. The complexity of deductions may be reduced in this manner. Such an advantage can be expected if the definition ". . . exhibits a genuine quantificational analysis of the definiendum, in the sense that its definiens has one or more (irreducible) layers of quantifiers . . .."[39] Though this observation concerns the formal theory of definition, it suggests that a similar gain is also obtainable from informal definitions. It is not difficult to find examples of this kind of economical gain, e.g., by studying definitions in mathematics.

The distinction between nominal and real definitions can be seen in a different light if definitions are looked at from the point of view of their utility. The characterization of nominal definition made in Cohen and Nagel (1961) — similar charcterizations are made by other writers — that they are only stipulations concerning how verbal symbols are to be used, that they are not true nor false, that they cannot be premises of any arguments, and that they do not extend our real knowledge,[40] are not relevant for formal definitions in logic, as it is easily seen from what we have learned in this section and Section A.3. It would be somewhat misleading to say, for example, that a formal sentence of elementary logic has no truth value, and false to say that it cannot be a premise of any argument. The last-mentioned characteristic is not appropriate for nominal definitions, whether formal or not, in the sense in which Cohen and Nagel use this notion, for they say that all explicit definitions in mathematics are nominal, but we have seen that we have reasons to suppose that such definitions may extend our knowledge. Furthermore, the term 'nominal' is often used by Cohen and Nagel (1961) – and many others, e.g., Hempel (1972) and Robinson (1965) – about definitions whose stipulative character is only apparent or partial, their ultimate purpose being analysis or explication. For instance, the definitions (A.1.1) and (A.1.2) of Section A.1 are syntactic representations of the intuitive idea of how the meaning of a connective is thought of as being related to other connectives. There are stipulative elements in the definitions since there are different ways to construe the meaning of implication and since (A.1.2) introduces a notation, but, on the other hand, they are results of an analysis and explication of the notion of implication, and therefore they extend our knowledge. It is evident that most important definitions in fact exhibit both features which are considered characteristic of nominal definitions and features considered characteristic of real definitions. Hence, it is questionable whether the distinction is relevant at all – and similar doubts are expressed by Cohen and Nagel themselves, as we saw in Section A.1. The distinction could perhaps be defined so as to coincide with the distinction between definitions that are and definitions that are not fruitful, or on the basis of their complexity, or the like. A distinction of that kind is in fact suggested by Hintikka and Tuomela (1970). By and large, however, the whole distinction seems to be unnecessary.

NOTES

CHAPTER 1

[1] Bohr considers classical theories as limiting cases in two senses: that quantum numbers become very large (and approach infinity); that vibrations become very slow (and approach zero).

[2] General relativity is not included in the grouping since, in Heisenberg's mind, it may not have reached its final form.

[3] For an overview of possible instances of the principle in physics and elsewhere and for an extensive literature on the subject, see Krajewski (1977).

[4] Thus Post, unlike the physicists mentioned above, does not here say that the Correspondence Principle means limiting case correspondence. All, or most, actual instances of the principle in physics are cases of limiting case correspondence. Obviously Post thus generalizes the notion.

[5] The first condition would presuppose, of course, that the old problems can be identified in the framework of the new theory or paradigm.

[6] Irrespective of whether this last proposal is associated with the idea that old theories are rejected when new ones are adopted, it evidently presupposes some notion of commensurability between theories; otherwise it is difficult to understand what it means to say that one theory is closer to the truth than another.

[7] For an overview, see, e.g., Suppe (1974).

[8] E.g., Popper (1962); Laudan (1977); Niiniluoto (1984), (1987). See especially Niiniluoto's comprehensive and penetrating discussion concerning truthlikeness. See Pearce (1987) for problem solving.

[9] There are attempts to define the principle for a broader setting, for example, for theory ensembles; see Pearce and Rantala (1984c); Pearce (1987).

[10] Glymour (1970); Pearce and Rantala (1984ab), (1985); Rantala (1989) and (1996). This notion will be studied in Parts Two and Three of this book.

[11] Sneed (1971); Balzer, Moulines and Sneed (1987).

[12] Krajewski (1977); Nowak (1980).

[13] Mayr (1981a); Moulines (1980); Tuomela (1985); Niiniluoto (1987).

[14] Cf., however, Tuomela (1985).

[15] In Chapter 6, below, I study in some detail the question whether counterfactual reduction can play such a role.

[16] Kuhn (1962), pp. 101-103.

[17] Ibid., pp. 102-103.

[18] See Kuhn (1970).

[19] Ibid., p. 202.

[20] For some reason, Hoyningen-Huene (1993) sometimes calls Kuhn's actual translation "everyday translation" but sometimes uses this term for the interpretative component of actual translation (see p. 216). If actual translation is the same as everyday translation and consists of the two components or processes (or "mo-

ments"), viz. translation in the narrow sense and interpretation, and, therefore, if
an expression is not translatable in the narrow sense, it cannot be translatable in
the everyday sense either. What Hoyningen-Huene seems to say, however, is that
even though an expression is not translatable in the narrow sense, it may still be
translatable in the everyday sense.

[21] Though Kuhn does not discuss the matter in model-theoretic terms, we can
nevertheless illustrate this fact by model-theoretic notions as follows. Within each
context in which the term occurs, it admits of quite a number of interpretations
(referents) that are provided by various models for the language in question. If,
however, one wants to choose only models in which the premises are true, this
does not help much since the conclusion is true in every model of the premises
and possibly in other models in addition, which provide additional interpretations
(referents).

[22] Hoyningen-Huene (1993), p. 215.

[23] Kuhn (1993), p. 324.

[24] See Section 1.5, above.

[25] A similar position is held by David Pearce, according to whom it is obvious
(even though it is not easy to see what Kuhn's criteria of adequacy are really since
he is not precise enough) that Kuhn imposes implausible constraints on acceptable translation (Pearce, 1987, p. 235).

[26] In particular, Pearce (1987) presents a sharp, extensive, and valuable criticism of Kuhn's later views of translation and commensurability.

[27] Hacking (1993), pp. 296-297. It is of some interest to notice that Hacking's
phenomenological argument is congenial (even though devised for a different purpose) with Jorge Borges' (1962) much quoted, anachronistic story about Pierre
Menard, who undertook to faithfully rewrite (without copying) *Don Quixote* by
attempting to "live" Cervantes, that is, learn Spanish, resurrect his Catholic con-
fession, fight the Turks, remove three hundred years of the history of Europe, in
brief, to annihilate everything that separates him from Cervantes. The attempt is
more or less unsuccessful, as is the similar enterprise in Hacking's ex-ample. See
also Danto (1981); Rantala and Wiesenthal (1989).

[28] Kuhn (1970), p. 204.

[29] P. 336.

[30] When we in the sequel discuss different kinds and different aims of translation, we can see that some of the kinds are more spectator-oriented than others;
that feature seems to be related to their different methodological functions.

[31] See, e.g., Kuhn (1993), p. 324.

[32] However, Buchwald (1993) criticizes Kuhn's (1970) hermeneutic description (in the Postscript, p. 202) of what Kuhn then called translation by arguing that
the kind of "effective penetration into an alien vocabulary and grammar rarely occurs among practising, creative scientists" (Buchwald, 1993, p. 188). But whether
Buchwald's notion of translation is the same as Kuhn's notion at that time, which
Kuhn later calls learning, does not become quite clear, for Buchwald refers (on
page 190) to literal translation, which may not be what Kuhn then meant. If Buch-

wald is correct, then what Kuhn is describing is the historian's reconstruction (as Kuhn himself seems to admit), but Buchwald's discovery would in fact establish discontinuity and incommensurability better than Kuhn's own approach in his (1970). The same holds true, of course, about Hacking's discovery as well, if correct about Kuhn.

[33] Kuhn (1962), p. 103.

[34] Cf. Section 1.5, above.

[35] Earlier it was employed by David Pearce and me in a number of works on intertheoretic relations; see, Pearce and Rantala (1984ab); Pearce (1987); Rantala (1989). The present book is a generalization of that earlier work.

[36] If incommensurability is really what Hacking's phenomenological-kind conditions suggest, Kuhn's definition of incommensurability as untranslatability may not feasible, as Hacking seems to suggest. The view that untranslatability entails incommensurability in such a phenomenological sense is hardly defensible.

[37] I shall take up the question of understanding in later chapters.

[38] Pearce (1987), p. 12.

[39] Thagard (1992) presents a computational approach to conceptual change, in which he investigates concepts as members of hierarchical systems. According to his theory, too, there can be some degree of continuity in conceptual revolutions, since parts of a system may survive across a revolutionary change.

CHAPTER 2

[1] Cf. Hacking (1993), p. 297.

[2] Quine (1960), p. 32.

[3] Kuhn (1993), p. 324.

[4] Burge (1978), pp. 141-142.

[5] See, e.g., Barthes (1967); Beardsley (1958).

[6] See Chapter 8, below. For an exact notion of situation, see, e.g., Barwise and Perry's (1983) treatment of situation semantics. In that book, the notions 'state of affairs' and 'event' are distinguished from each other. I shall not consider the distinction here.

[7] See Searle (1969) for the nonrelativized notion.

[8] Cf. Searle (1981), pp. 10-11.

[9] But the above 'more or less' qualification (in the second passage) is added because the unrestricted independence of linguistic or other cultural rules does not always make sense or is not interesting.

[10] The examples to come can, of course, be directly modified so as to apply to other kinds of illocutionary force.

[11] These are, however, more recalcitrant problems in connection with literary texts and scientific theories, as we shall see later. Interpretational multiplicity is often regarded as an aesthetically and cognitively valuable property of literary texts. The same holds for scientific theories, but in a different sense. In the former case, complexity is valuable, but in the latter, universal applicability.

[12] See Enderton (1972).

[13] See Nagel (1961); Hempel (1965).

[14] See Adams (1955); Sneed (1971); Stegmüller (1979); Mayr (1976), (1981a); Balzer, Moulines and Sneed (1987).

[15] See also Pearce and Rantala (1984b); Rantala and Vadén (1994). For an extensive discussion, see Pearce (1987).

[16] See Rantala (1989), (1996); Chapter 6, below.

[17] Some applications below even make sense only on the assumption that the hearer is *beforehand* aware of how the things are, that is, has an opinion about the subject matter to which the speaker is referring by his utterance, before the speaker's speech act.

[18] In the diagrams, the many-one nature of correlations, and thus the role of the refinement principle, is not presented for typographical reasons, but it will become evident in connection with the notion of corespondence. See Chapter 5.

[19] But the principle is not associated by the structuralists with translation, since they tend to dispense with the notion of object language.

[20] See Sneed (1971), p. 219.

[21] See Enderton (1967).

CHAPTER 3

[1] Danto (1981), p. 119.

[2] This should in fact read 'In the painting, the legs belong to a pearl diver or an oysterman'; and similarly for the other sentences in this section referring to the painting in question. With this understanding no philosophical ambiguity should arise, e.g., between what is fictional and what is not.

[3] See, e.g., Keenan (1978).

[4] Since in the actual situation the title of the painting is *Landscape with the Fall of Icarus*.

[5] This is analogous to what Glymour (1970) suggests to be asked in connection with scientific reductions where some special assumption is contrary-to-fact. See Section 6.3, below.

[6] This might be the case if the speaker thinks that the title is this.

[7] It is a somewhat controversial question among the philosophers of language whether linguistic expressions have any general, or grammatical, meanings, but we assume here that (3.1.1) is obvious enough to have one.

[8] See Section 8.4, below.

[9] This example was suggested by Arto Haapala in private communication.

[10] Though it is not always quite evident what is literal and what is metaphorical, it is simply assumed here, for the sake of the argument, that the distinction can be made.

[11] How such worlds are chosen is also discussed by Grund (1988).

[12] See Mac Cormac (1985), p. 9.

[13] 'Juliet is a celestial object'.

[14] 'The title is *Industry on Land and Sea*'.

[15] Phenomenological philosophers may disagree.

[16] Even this description may be too strong about sense data.

[17] For the notions of extensionality and intensionality, see Section 8.2, below.

[18] In the early eighties Arthur Danto raised the question whether and in what sense historical developments of art can be seen as being progressive, and ever since the topic has been much discussed; for references, see, e.g., Rollins (1993). As acknowledged by many authors, Danto's problem is comparable to Kuhn's, but their conclusions may not be similar; see my (1997b). Earlier, progress was considered by Gombrich (1960) to whom I shall refer in what follows.

[19] Gombrich (1960), p. 330.

[20] Naturally, σ and θ are now the syntactic parts of the respective works.

[21] This idea would be, however, as intricate as, and related to, the ideas discussed above about the sense-data theory of perception and the protocol language, and it is not commonly accepted anymore.

CHAPTER 4

[1] See, e.g., Lewis (1978); Pavel (1978); Rantala and Wiesenthal (1989); Rosen (1994).

[2] When it is said, for example, that an expression (with whatever illocutionary force) is satisfied at a world, it means roughly the same as before. See Section 2.3 for the discussion of the difficulties involved in the definition of the speaker's situations. For example, in some contexts, we may even let the hearer choose the speaker's worlds if that is epistemologically grounded. To speak of actual worlds is, of course, also problematic; they are worlds which the reader thinks are correct, given the text, its translation, context, etc. Thus the notion does not necessarily mean here the same as in possible worlds semantics in general.

[3] Later on, in Section 4.4, when studying narrative texts as sequences of sentences, I return to situations, referring to them as local components of worlds corresponding to events described by individual sentences.

[4] For a thorough discussion of this issue, see, e.g., Lewis (1978); Rantala and Wiesenthal (1989).

[5] See my (1987) and (1988b).

[6] For problems involved, see Oesch and Rantala (1999). One way to try to distinguish between what literary theorists call *reception* and *interpretation* is to argue that a reception is the emergence of the reader's courses of events when he or she is reading the text, but an interpretation includes, but is not exhausted by, the construction of worlds. Then we can see a little more exactly in what way reception is the logical basis of interpretation.

[7] It is not assumed here, as Goodman (1968) does, that only things that exist can be denoted. This means that the term is here assumed to apply to narrative fictions as well.

[8] For the principle, see Section 2.1.

[9] But they do not accept such evidence.

[10] According to a common view, literary works do not usually have a unique meaning. It would not even be desirable since the existence of many meanings is considered aesthetically and cognitively valuable in artworks.

[11] See Rantala and Wiesenthal (1989) for a more comprehensive discussion.

[12] *Op. cit.*, pp. 64, 66-68. It is a simple model-theoretic fact that, generally, the less there are worlds (models) the more truths they admit of.

[13] See Lewis (1978), p. 45, for a more detailed analysis of truth in fiction.

[14] Wolterstorff (1980), p. 121.

[15] There is one in Rantala and Wiesenthal (1989), pp. 77-79.

[16] For arguments defending the import of authors' intentions and for an overview, see also Haapala (1989).

[17] See Rantala (1987); and from a semiotic point of view, Eco (1979).

[18] Borges (1962). See my (1987).

[19] Hence the ideal reader could be considered as a 'hermeneutic limit' of an ordinary reader.

[20] Here we would then have another example of the back and forth character of reading.

CHAPTER 5

[1] It is immediately seen what these notions mean in the global case on the basis of their earlier characterization for the local case.

[2] This is tentatively sugested in my (1989), however. It extends earlier work by David Pearce and me; see, e.g., Pearce and Rantala (1983c), (1984a); and Pearce (1987).

[3] 'Possible world' is here used as a general term that, depending on the kind of theory, may mean such things as systems, structures, situations, courses of events, or models. Thus physical theories are often regarded as speaking of systems, mathematical ones of structures, historical ones of situations and courses of events, and formalized ones of models in the model-theoretic sense of the word.

[4] In the contexts of reduction and explanation, the terms 'primary theory' and 'secondary theory' are sometimes used in the literature, as, e.g., in Nagel (1961) and Glymour (1970).

[5] See Section 5.5, below. As noted earlier, there is no reason why the minimizing transformation should occur explicitly in the logical condition (4.1.1) of correspondence.

[6] It is of some interest to notice at this juncture that if Kuhn is correct when he says, in his (1977) and elsewhere, that the laws of a theory are *not* independent, since theories are integrated wholes, then this and the forthcoming definitions of correspondence may not be entirely adequate, since they at least involve a tacit assumption to the effect that a translation operates on independent sentences. This suggests that a global translation of the laws and other sentences of a theory would have features of a translation of an integrated text. What this idea would ultimate-

ly mean in practice is not very clear at the moment.

[7] It can be seen that the refinement principle is valid here, at least formally. We shall see later that the minimization principle is in force in the most important specific applications of correspondence.

[8] In this book, I shall not present any such case studies of historical theories, or other theories in the human or social sciences.

[9] See, e.g., Enderton (1972); Barwise (1975).

[10] These questions will be studied in Section 5.9.

[11] For this assumption, see Chapter 8.

[12] If it is required that F itself is definable, it is sufficient for certain purposes that it is definable in the (weaker) sense of being a *projective class*. If F itself is definable (in some sense) in a logic, it is an algebraic operation rather than a function in the proper sense; see Feferman (1974b); Gaifman (1974).

[13] Balzer, Moulines and Sneed (1987) also consider it too weak to represent a translation proper, since it does not preserve meaning.

[14] See Rantala (1988). In Section 6.9, below, the import of the notion is briefly studied.

[15] See (3.1.3).

[16] Recall Sections 1.1 and 1.4, above.

[17] See the respective quotation in Section 1.4.

[18] I shall here recapitulate the theory that can be found in earlier writings by Pearce and Rantala. The notions of nonstandard analysis that are needed are discussed and formally defined in Section 8.6.

[19] The notion of standard approximation of a model, employed below, varies a little bit according to the form of a model.

[20] See Pearce and Rantala (1983ab), (1984a).

[21] If T is axiomatized in some logic, H is the class of all models in which the axioms are true (in this logic). In this sense, H can be thought of as representing the axioms of T.

[22] Most often actual transformations are monomorphisms or other mappings between domains of models.

[23] See Pearce and Rantala (1983b), (1984a).

[24] See Section 8.5.

[25] In Pearce and Rantala (1984a), the notion of symmetry is even generalized, and Curie's Principle can be generalized accordingly. The former generalization is only defined there and in the present book, in Chapter 8, but not studied, since their relevance for actual scientific theories is not clear.

[26] See Section 8.2 for the concepts and notation used here.

[27] Such an expansion exists because $\mathfrak{M}$ is a model of T.

[28] In cases where the minimization principle is functioning.

[29] By (5.3.1). Therefore, K' is a subclass of $\mathrm{Mod}_L \cdot I(\theta)$.

[30] Assume just one.

[31] For supervenience, see, e.g, Kim (1990); for idealization, Nowak (1980); Niiniluoto (1990); and Lahti (1984) for idealization in quantum logic. Niiniluoto

shows that idealization involves counterfactual conditionals. Collier and Muller (1998) argue that emergence in science is a special case of supervenience.

[32] They may even argue that emergent properties can be eliminated in terms of physical ones.

[33] Nagel (1961), p. 372.

[34] It was briefly discussed above, in Section 5.9.

[35] Cf. the criticism in the Appendix of the distinction between real and nominal definitions.

[36] Primas (1998), p. 84.

[37] Danto (1981), pp. 119-120.

CHAPTER 6

[1] See, e.g., Salmon (1989) for such an overview.

[2] In particular, the forthcoming references to models must then be replaced by references to worlds or situations.

[3] For a more thorough discussion of pragmatic aspects of explanation see, e.g., Scriven (1962); Tuomela (1973). Sintonen (1984) employs speech act theory and Hintikka's (1976) question theory in studying pragmatic features of explanation.

[4] That is, $K \subseteq \mathrm{Mod}_L(\psi)$. Recall the definition of correspondence in Section 5.3.

[5] The constraint actually means that ψ is not logically 'too strong'.

[6] (6.2.2) and (6.2.3) can also be thought of as schemata of *problem solving*; see Pearce and Rantala (1984c); Pearce (1987).

[7] I neglect here the question whether the transformed theory can be really regarded as a theory in the same sense as T itself, or the transformed law as a law.

[8] The explanatory role of the correspondence relation can be better seen by investigating the case studies to come.

[9] For similar suggestions, see Rott (1987).

[10] See Section 8.4, below.

[11] This is also done in Rantala (1989), (1996).

[12] As we observed in Section 5.5, a limit procedure can be replaced by a limit condition by means of nonstandard analysis.

[13] This view is in fact somewhat contrary to what is maintained in Pearce and Rantala (1985) and Pearce (1987).

[14] See Section 8.4, below.

[15] And it is obvious that the same holds within any other theory of counterfactuals, due to their intensional nature.

[16] It is argued by Lewis (1973) that laws tend to be cotenable since they are so important to us.

[17] That there is follows, in fact, from Section 9.1.

[18] See Section 9.6.

[19] See Section 8.4.

[20] Notice that these assumptions imply that $v = \infty$.

[21] See also Sintonen (1984).

[22] This will become evident from the formal treatment of the correspondence in question, in Section 9.1.

[23] For this feature, see Pearce and Rantala (1984a).

[24] What is said here about the explanatory import of correspondence is, *mutatis mutandis*, applicable to cases where only correspondence sketches can be established between two theories or narratives and to cases where there exist global translations between literary works. Then the first step does not have similar logical difficulties; and if the correspondence (sketch) or translation does not contain counterfactual elements, problems connected with the third step will be lacking.

[25] See Chapter 1.

[26] Recall that they require that meanings be corrected rather than saved.

[27] See, in particular, Section 9.1.

[28] For instance, in a case where T' supplants T.

[29] In this book, however, I shall mainly restrict myself to the context of justification.

[30] See Scheibe (1973a), containing an excellent discussion of Bohr's views.

[31] Kitcher (1983), p. 227.

[32] *Ibid.*, p. 209.

[33] *Ibid.*, p. 218.

[34] Bonevac (1982), pp. 8-9.

[35] See, e.g., Suppe (1974).

[36] See, e.g., Quine (1970); Haack (1974); Briskman (1982).

[37] 'Interpretation' is often used in logic instead of 'reduction'.

[38] See Section 6.2.

CHAPTER 7

[1] This case was worked out in detail in Pearce and Rantala (1984a); see also Pearce (1987).

[2] See Chapters 8-9. The definitions of certain notions in Sections 8.6 and 9.1 require $L_{\kappa\omega}$ (where $\kappa \leq$ beth-ω), a many-sorted infinitary logic admitting of disjunctions and conjunctions of length less than κ but only finite strings of quantifiers. As pointed out by J. van Bentham, $L_{\omega_1\omega}$ is not sufficient, contrary to what was claimed in Pearce and Rantala (1984a). For simplicity, we shall try to avoid extensions that are unnecessarily strong in expressive power, such as second order logic.

[3] See Section 9.1.

[4] By 'velocity', 'acceleration', 'particle', etc., is here meant the informal interpretations of respective formal terms occurring in axioms. They have their natural meanings in intended models, but in some models they mean velocity, etc., only in a pathological sense. Similarly for the scientific terms occurring in the other case studies below.

[5] Recall Section 5.1.

[6] For the notion, see Section 9.1.

[7] For the principle, recall Section 5.7; see also Pearce and Rantala (1983b).

[8] For more exact effects of that assumption, see Section 9.1.

[9] Example 1 in Section 6.5.

[10] That is, interpreted as a classical mass of a particle in an intended or standard model, which means that $\mathbf{m} = \mathbf{m}_0$; and similarly for the other symbols here.

[11] Cf. Sections 6.5 and 9.1.

[12] More precisely, the term denoting the derivative of the position.

[13] For a detailed treatment, see that article.

[14] See Note 2, above.

[15] For the meanings of the terms 'mass' and 'planet' in this context, see Note 4, above.

[16] Recall Chapter 6.

[17] Consult, e.g., Matthews (1968).

[18] We assume that only one state function is present.

[19] E.g., the momentum and energy operators and Planck's constant.

[20] As can be seen from Section 9.1.

[21] Gasiorowicz (1996), pp. 125-128. Notation is slightly different here.

[22] See the relevant formal considerations and proofs in Chapters 8-9.

[23] Assuming similar conditions on the second derivative and on the other functions involved.

[24] Matthews (1968), pp. 157-158.

[25] Recall Kuhn's (1983) strict notion of translation; Chapter 1.

[26] Recall Chapters 2-3 and 6.

[27] Originated by Suppes (1957) and advanced, e.g., by Adams (1959); Sneed (1971); Stegmüller (1973), (1976); Moulines (1975); Balzer (1982); see especially Balzer, Moulines and Sneed (1987).

[28] This is, of course, a counterfactual assumption, but this time on the side of the reduced theory and not of the correspondence relation itself.

[29] There are more complicated and involved forms of the structuralist reduction to which Pearce's result does not necessarily apply.

[30] See Rantala and Vadén (1997).

[31] Their semantics is sometimes given from the outside, as in the case of digital machines. Then a system is not what Fetzer (1990) calls a semiotic system.

[32] See also Vadén (1995) and (1996) for a clarification of Smolensky's position. It is not always very clear what Smolensky means by his various notions (to be briefly discussed below).

[33] Cf. Smolensky (1988), pp. 10-11. (7.5.1.1) and (7.5.1.2) are not direct quotations.

[34] This interpretation is presented in Rantala and Vadén (1994).

[35] E.g. in Smolensky (1988), p. 12 (end of Section 5). There seems to be ambiguity here of the kind I referred to above.

[36] A detailed and formal investigation is presented in Section 9.3, below.

[37] Cf. (9.3.1) and (9.3.8), Section 9.3, for more precise formulations.

[38] Cf. (9.3.2) and (9.3.12), where the input argument is omitted, however, to

simplify the exposition.

[39] Self-organization means that given an input, N will adjust its connection weights so that an appropriate output follows.

[40] In actual reasearch on neural networks researchers are often aiming at smaller nets, not bigger. But this practical work has nothing to do with the problem concerning the conceptual relationship between the symbolic and subsymbolic.

[41] The conditions were presented in Section 6.4.

[42] P. 112.

[43] P. 282.

[44] P. 59.

[45] See also Section 1.2.

[46] See Rantala (1997).

[47] See Churchland (1992); Bechtel and Abrahamsen (1991).

[48] In Smolensky's terminology, this is a higher level than the neural level, but lower than the conceptual one.

[49] However, Smolensky says that subsymbolic models that are more likely to be reducible to the neural level should be favored in the coming research.

[50] Recall our discussions in Section 1.2 and Chapter 6.

[51] The nature of the correlation F makes this theoretically obvious in any case.

[52] I shall use the term 'knowledge' here (in agreement with computer scientists and many cognitive scientists) even though 'information' migth often be more appropriate, since philosophically more neutral.

[53] See, e.g., Bechtel and Abrahamsen (1991), p. 290.

[54] If this assumption is given up, the following considerations are relevant to artificial networks only.

[55] For the latter distinction, see below.

[56] Smolensky (1988), pp. 12 and 20.

[57] What relation he means is not quite clear.

[58] This would mean that S licenses θ' in something like the sense discussed in Section 8.3, below.

[59] Bechtel and Abrahamsen (1991), p. 163.

[60] See Bechtel and Abrahamsen (1991) for interpreting the kinds of knowledge in this way.

[61] Smolensky (1988), p. 5.

[62] It is obvious that here φ must be an appropriate translation (e.g., of the form 'the output is close to 1') of the sentence (e.g., of the form 'such-and-such a sequence of symbols is an expression of such-and-such a language') describing the propositional knowledge in question into the (mathematical) language to which S and ψ belong.

[63] Smolensky (1988), pp. 4-5.

[64] For a more exact treatment of N_∞, see Section 9.3.

[65] See Bechtel and Abrahamsen (1991).

[66] Quine's arguments are evidently the best known.

[67] The above studies give rise to further distinctions. For instance, they can be

used to explore a subsymbolic distinction between *competence* and *performance* in Smolensky's sense. He invites us to analyze a subsymbolic system in two ways, first at the subconceptual level and then at the conceptual level with and without suitable idealizations. If its processing is characterized at the former level (e.g., by means of S), a description of its performance is obtained, whereas desriptions of the latter kinds (like θ' and θ'+m *is infinite*) are about competence. Thus the distinction seems to be roughly the same as the distinction between nonpropositional and proposiotional knowledge of a connectionist system.

[68] This simplification should not influence the results below, but, again, make the possible relation of M to a Turing machine more obvious.

[69] Cf. θ' in Section 7.5.1. E would now be replaced by a set of true sentences.

[70] Cf. the theory S mentioned in Section 7.5.3, above.

[71] If more than one output unit were needed here, it would slightly modify (7.5.1.5).

[72] See Hintikka (1975).

[73] It is an idealization to assume that Ω is nonempty, but for logical purposes such idealizations are unavoidable. There are others here, of course.

[74] The present Kripke model can be considered a special case of the model defined there.

[75] Recall Section 5.10.

[76] Section 5.10.

[77] Here, of course, implication is defined in terms of negation and conjunction as usual.

[78] See Hughess and Cresswell (1984). 'Correspondence' has there a meaning that is different from its meaning in this book.

[79] The latter information I owe to Ken Manders.

CHAPTER 8

[1] Nonstandard arithmetic and analysis are cases in point. See Section 8.6, below.

[2] See, e.g., Section 9.1.

[3] See, e.g., Mostowski (1957); Lindström (1966) and (1969); Barwise (1974); Feferman (1974ab); Makowsky, Shelah and Stavi (1976).

[4] For some such requirements, see the next section.

[5] See the next section.

[6] This is what Quine has been arguing.

[7] This is sometimes written simply as '$\models$' if no confusion arises.

[8] As Feferman points out, one can as well take 'sentence' as the more basic expression and define 'formula' by its means.

[9] The earliest general definitions in the current sense are obviously those by Mostowski (1957) and Lindström (1966), (1969). Given Feferman's definition or any one of the definitions presented in abstract logic, important model-theoretic notions can be defined and general results proved; in particular, results that con-

cern interrelations of different logics, their model-theoretic properties, and definability of mathematical notions.

[10] However, this construal may stretch the original idea of type a little bit.

[11] As usual, W is a set representing possible worlds, R is an accessibility relation in W, and V is a valuation of propositional variables. Here it is assumed, of course, that the $\mathbf{p}_k$ are propositional variables of the logic, R does not occur in sentences of the logic, and $V(\mathbf{p}_k) = \mathfrak{M}(\mathbf{p}_k)$.

[12] For some remarks on this matter, see Rantala (1982b). See also Chapter 7, above.

[13] The same would evidently hold if one could develop a formal logic for narratives, as suggested in Chapter 4.

[14] Some such consequences and references to them can be found elsewhere, as, for instance, in Feferman (1974b) and Pearce (1987). Likewise, some passages in the present book concerning definability are of that kind.

[15] Carnap (1934); Carnap (1937), p. xiii.

[16] *Ibid.*, p. 180.

[17] *Ibid.*, p. 180.

[18] Sellars (1953), p. 322.

[19] P. 185. When Carnap speaks of ordinary meaning, he obviously dispenses with the problems involved in such a notion.

[20] Sellars himself argues that there might in fact be a kind of linguistic activity in which P-rules are indispensable: they often perform (in the metalanguage) the function of subjunctive conditionals; and the latter may not be permitted by the object language. For instance, the relation of inference (8.3.2) and conditional 'If a were a raven, a would be black' reflects this role of P-rules. As we know, this and related themes have been extensively discussed later on in logic and the philosophy of science.

[21] P. xiv.

[22] *Ibid.*

[23] P. xv.

[24] In this respect Carnap of course had predecessors, but the degree of generality was obviously new.

[25] P. xv.

[26] In what follows, I shall use notation that is a bit different from Carnap's notation.

[27] P. 169.

[28] My notation here is slightly different from that in Barwise (1972).

[29] See above. I shall not here study nonformal premises or interpreted ones. Their inclusion in a system leads to logical problems of other kinds. It is evident that in his general definition Carnap restricts himself to formal and extensional systems.

[30] See Rantala (1978).

[31] See, e.g., Pearce and Rantala (1983), (1984); Pearce (1987). But it has also turned out that such logical tools must be amended by appropriate pragmatic con-

siderations; see my (1989).

[32] Carnap (1937), p. xiii.

[33] Ibid., p. iv.

[34] Barwise (1972), p. 309.

[35] As can be seen, I am simplifying things by only considering Kripke models at the level of propositional logic. But the essential matters of counterfactual explanation are not affected by this assumption, as we saw in Chapter 6.

[36] See Lewis (1973).

[37] For the purposes of the present essay, we need not consider other conditions.

[38] If one wants to consider a specified 'actual' world to which truth and other notions are related, a distinguished world can be added to play the role of such a world. But it would not make much difference here.

[39] Logical inference is of course relative to the logic used; but in what follows, this should not cause any confusion.

[40] See, e.g., Lewis (1973) for a discussion of the notions; Niiniluoto (1987) for 'closeness' in the context of verisimilitude; Griffin (1977) for relative identity.

[41] See the case study in Section 9.2.

[42] For a further study, see, e.g., Goldblatt (1979); Maclane (1971).

[43] Notation '$f: a \to b$' means that the object a is the *domain* of f and b its *codomain*.

[44] Especially those that are considered characteristic (in some sense) of the theory. Such relations are, for example, the so-called constraints as introduced by Sneed (1971).

[45] See Pearce and Rantala (1984a).

[46] In, e.g., Robinson (1961), (1966).

[47] To be found, e.g., in Robinson and Zakon (1969); Stroyan and Luxemburg (1976).

[48] See Robinson and Zakon (1969).

[49] Notation '$\in$' will be used both for the membership relation and its restriction to A.

[50] Naturally, $*\!\in$ is the interpretation of $\in$ in $*\mathcal{A}$; i.e., $*\!\in \ = \in^{*\mathcal{A}}$, according to the familiar notation.

[51] For this notion of definability, see the Appendix.

[52] Naturally, $\models$ is here the satisfaction relation in $L_{\omega\omega}$.

[53] The following definitions may become trivial with respect to $*\mathcal{A}$ without certain additional requirements concerning it. Usually it is assumed that $*\mathcal{A}$ is an *enlargement* of $\mathcal{A}$; this guarantees that all the requisite nonstandard objects exist in $*\mathcal{A}$. In the applications of the present book, somewhat weaker existence assumptions will be sufficient.

[54] This generalization enables us to define some later concepts more generally.

[55] The following treatment can be generalized for a case where τ is not a subtype of τ'. For instance, a case where τ is embeddable in τ' (i.e., there is an appropriate mapping from τ to τ') may be of some intuitive interest. But to suppose that $\tau \subseteq \tau$' seems to more directly correspond to the idea of limiting case.

[56] In the former case, of course, a' is assumed to be in the domain X of st.

[57] It would be ontologically more economical to start with $U = \mathbb{N}$, thus to consider the natural numbers as the urelements, but in other respects it does not make any difference for our purposes here (contrary to what was claimed in Pearce and Rantala, 1984a).

[58] Th($\mathfrak{A}$) is the set of all $L_{\omega\omega}(\tau_0)$-sentences true in $\mathfrak{A}$. A model of analysis in his sense should not be confused with what is usually called by this name; the latter terminology refers to certain second order models. $\mathfrak{A}$ has an advantage of being a first order model and yet containing all mathematical objects needed in real and complex analysis in general (although in the specific applications of this paper we only need a small part of them).

[59] By the definition of $^\wedge a$ (in Section 8.6): $b \in {}^\wedge a \Leftrightarrow {}^*\mathfrak{A} \models \mathbf{x} \in \mathbf{a}[b]$.

[60] These sets are even *external* (see Bell and Machover, 1977).

[61] $\mathbb{E}$ and $\mathbb{I}$ can be defined, for instance, by the following formulas:

$\mathbb{E}$: $\bigvee_{a \in \mathbb{Z}^+} |x| < a$ ($\mathbb{Z}^+$ is the set of all positive integers in $\mathbb{R}$);

$\mathbb{I}$: $\bigwedge_{a \in \mathbb{Z}^+} |x| < 1/a$.

[62] The cardinal $\beth_\omega$ is one of the so-called beth numbers, viz. beth-ω. In Pearce and Rantala (1984a) it was erroneously claimed that $L\omega_1\omega$ would be enough, but, as I have mentioned earlier, it was pointed out by Johan van Benthem, in private communication, that it is not. This larger language is not, of course, as nice as $L\omega_1\omega$ but it makes no big difference in so far as just the correspondence relation is our concern, as here. The infinitely long disjunctions and conjunctions needed here only contain atomic formulas as disjuncts and conjuncts.

[63] It is obvious that both the 'structuralist' and the 'traditional' method, as they were employed in the philosophy of science, are too restricted for that purpose.

[64] For a more detailed study of these generalizations, see Robinson (1961).

[65] Expressed more completely: $^\wedge T = \{r \in {}^\wedge\mathbb{R} \mid a *< r *< b\}$.

[66] See the remarks earlier in this section concerning familiar mathematical notation.

[67] Extended, as usual, for (half-)closed intervals.

[68] This is a simplified version of Robinson's (1961) '*S*-continuity'.

[69] ^0f is not exactly the same function as its namesake in Robinson (1961).

CHAPTER 9

[1] The axiom systems of this section more or less closely imitate those presented in McKinsey, Sugar and Suppes (1953); Rubin and Suppes (1954); Sneed (1971). What is involved, in the physical sense, in these axiom systems and in the transformations to be discussed below is, of course, only a small part of what physicists call classical or relativistic particle mechanics.

[2] Notation for the familiar operations of real analysis is related to $\mathfrak{B}$, and they have the standard or nonstandard meanings according to whether $\mathfrak{B}$ is $\mathfrak{A}$ or one of its extensions $*\mathfrak{A}$; this was explained in Section 8.6.

[3] The general (three-dimensional) form of (9.1.4) is obtained by replacing $\mathbb{R}_\mathfrak{B}$

by $(\mathbb{R}^3)_{\mathfrak{B}}$. In a nonstandard case, $(\mathbb{R}^3)_{\mathfrak{B}} = {}^{\wedge}(\mathbb{R}^3) = \{ \ast\langle a,b,c\rangle \mid a,\ b,\ c \in {}^{\wedge}\mathbb{R} \}$. In (9.1.6)-(9.1.7) (and 9.1.8, below), the operations involved have to be generalized, in the well-known way, for vector calculus (see, e.g., Rantala, 1979).

[4] In the general case, $Ds(p,\ t)$ being a vector, '$(Ds(p,\ t)/c)^2$' has to be replaced by '$(|Ds(p,\ t)|/c)^2$'. See also Note 3 above. Naturally, a four-dimensional formulation of relativistic mechanics could also be handled by our method.

[5] (9.1.11) is a simplifying assumption. Its removal may lead to some complications.

[6] The effect of these conditions is that the functions in question resemble (as much as possible) natural extensions of 'smooth' standard functions. In view of the limit transformations (symmetries) defined later in this section, one cannot require, however, that they would actually be such extensions.

[7] The equation of (9.1.10) can be rewritten in the form

$$m(p)D^2 s(p,\ t) = (1-(Ds\ (p,\ t)/c\ )^2)^2 f(p,\ t).$$

[8] Recall that $\mathbb{U}$ is the set of all infinite real numbers; cf. Section 8.8.

[9] Now, only clauses (9.1.1) and (9.1.2) (that is, their generalizations for the full class H) require an infinitary logic; the others are expressible in elementary logic. (9.1.1.) is expressible by means of $\mathrm{Th}(\mathfrak{A})$ and (8.8.1), and (9.1.2) by the sentence

$$\bigvee_{1 \le n < \omega} \exists \mathbf{p}_1 \dots \mathbf{p}_n \forall \mathbf{p} \bigvee_{1 \le i < n} \mathbf{p} = \mathbf{p}_i.$$

[10] Remember that τ is the type of ${}^0\mathfrak{M}$ and τ' the type of $\mathfrak{M}$; see Section 9.1.

[11] Thus, for *each* individual a of ${}^0\mathfrak{M}_1$, $\mathbf{a}$ is an *individual constant* (and not any other term) whose denotation in ${}^0\mathfrak{M}_1$, and in $\mathfrak{M}_1$, too, is a.

[12] The only (primitive) relation symbols are $\in$ and $=$. In a formula of the form $\mathbf{u}_1 \in \mathbf{u}_2$ both $\mathbf{u}_1$ and $\mathbf{u}_2$ are of the sort of A (the first domain); but arbitrary sorts in a formula $\mathbf{u}_1 = \mathbf{u}_2$.

[13] '$\models$' is a shorthand for '$\models_L$'.

[14] Recall that ${}^0\mathbf{u}$ is the corresponding individual constant in τ_1; thus ${}^0\mathbf{u}$ and $\mathbf{u}$ have the same denotation in ${}^0\mathfrak{M}_1$.

[15] See, e.g., Krajewski (1977); Friedman (1982).

[16] E.g., Mayr (1981); Moulines (1980).

[17] For notation and operations in the nonstandard case, see, again, Sections 8.6-8.8. Naturally the usual notions of Galilean and Lorentz transformation are extendible for the nonstandard cases, too, as indicated earlier.

[18] This holds in the nonstandard cases as well as in the standard case, since the usual (formal) derivation rules and other rules hold in both cases.

[19] Notice that the induced mapping $T \mapsto \rho T$ is 1-1 and 'onto'.

[20] We shall not here discuss any properties of these categories; but even as they stand, this lemma and the next theorem seem to bring out something very essential of symmetries and their relationship.

[21] See Section 8.5, above; similarly for the next theorem.

[22] Again, we may first think of these transformations in terms of models in K_0 and K_0' and then generalize; similarly for the next theorem.

[23] $\mathbf{K}$ and $\mathbf{K}'$ are defined in Lemma 9.2.5.

[24] For this notion, see, e.g., Kleene (1964).

[25] Other labels could of course be used.

[26] More precisely, $\{\langle 1,s_1\rangle, \ldots, \langle n,s_n\rangle\}$.

[27] It is assumed here for simplicity that square 0 is originally blank.

[28] E.g., to belong to the open interval (0, 1).

[29] I.e., the sequence $\{\langle 1,p(1)\rangle, \ldots, \langle n,p(n)\rangle\}$.

[30] For the notation and concepts of what follows, see, in particular, Sections 8.6-9.1.

[31] As in Section 9.1, notation for the familiar operations of real analysis is related to $\mathfrak{B}$, and they have a standard or nonstandard meaning according to whether $\mathfrak{B}$ is $\mathfrak{A}$ or one of its extensions $*\mathfrak{A}$; see again Section 8.6.

[32] Cf., e.g., Barwise (1968).

[33] See Section 8.8.

[34] Here a notion of *intended model* could be justifiably and exactly defined. An intended model of TR − a model that more concretely corresponds to the above mathematical Turing representation of Exp$L$ − is a standard model of the above form such that $\mathfrak{B}$ is $\mathfrak{A}$ and $S = \text{Prim}L$, and $E = \text{Exp}L$.

[35] But not the collection of intended models, since it does not satisfy the isomorphism condition (8.2.3).

APPENDIX

[1] Excepting some minor changes and corrections, the Appendix is identical with my (1991).

[2] Whitehead and Russell (1973), p. 11.

[3] P. 228.

[4] *Op. cit.*, p. 232.

[5] P. 16.

[6] P. 230.

[7] Robinson, *op. cit.*, p. 16.

[8] Pp. 190-191.

[9] Robinson, *op. cit.*, p. 195.

[10] Robinson, *op. cit.*, p. 13.

[11] Pp. 59 ff.

[12] See Suppes (1957).

[13] Notice that the term 'identifiable' has different meanings in the literature; see my (1977b)

[14] For the Interpolation Theorem, see, e.g., Chang and Keisler (1973).

[15] See, e.g., Makowsky, Shelah and Stavi (1976).

[16] For somewhat different notions of identifiability, see, e.g., Hintikka (1991).

[17] See Kueker (1970).

[18] See, e.g., Przełecki (1969); Tuomela (1973); Suppe (1977) for the problem concerning the definability of theoretical terms; and Goodman (1947) and Lewis (1973) for subjunctive conditionals.

[19] See Suppes (1957).

[20] In other words, K is an EC_L-class. The notion can of course be generalized by allowing relevant sets of sentences.

[21] The right hand side means that the sequence $\langle a_1, \ldots, a_n \rangle$ satisfies (within L) φ in $\mathfrak{M}$.

[22] For model-theoretic definability, see, e.g., Barwise (1975).

[23] See Barwise (1975).

[24] In Robinson (1965), there is a comprehensive discussion of various historical conditions.

[25] Whitehead and Russell (1973), p. 11.

[26] Cf. Theorem A.3.1.

[27] See Chapter 5.

[28] Cf. Section A.1.

[29] See, e.g., Quine, *op. cit.*; Cresswell (1973); Saarinen (1979).

[30] See Tuomela (1973); Pearce and Rantala (1985b).

[31] See Rantala (1977a).

[32] See, e.g. Przelecki (1969); Tuomela (1973).

[33] Pp. 11-12.

[34] P. 33.

[35] See also Rantala (1977a).

[36] For a further analysis of the latter activities and for further references, see, for example, Rantala and Tselishchev (1987).

[37] See, e.g., Przelecki (1977).

[38] P. 17.

[39] *Op. cit.*, p. 317.

[40] Cf. Section A.1.

BIBLIOGRAPHY

Adams, E. W., *Axiomatic Foundations of Rigid Body Mechanics*, unpublished diss., Stanford Unversity, 1955.

Adams, E. W., 'The Foundations of Rigid Body Mechanics and the Derivation of Its Laws from Those of Particle Mechanics', in L. Henkin, P. Suppes and A. Tarski (eds.), *The Axiomatic Method*, North-Holland Publishing Co., Amsterdam, 1959, pp. 250-265.

Austin, J. L., 'Performative Utterances', in J. L. Austin, *Philosophical Papers* (ed. by J. O. Urmson and G. J. Warnock), Oxford University Press, Oxford, 1970, pp. 233-252.

Balzer, W., *Empirische Theorien: Modelle, Structures, Beispiele*, Braunschweig-Wiesbaden, 1982.

Balzer, W., Moulines, C. U. and Sneed, J. D., *An Architectonic for Science: The Structuralist Program*, D. Reidel Publishing Co., Dordrecht, 1987.

Barthes, R., Elements of Semiology, Jonathan Cape, London, 1967.

Barwise, J., 'Implicit Definability and Compactness in Infinitary Languages', in J. Barwise (ed.), *The Syntax and Semantics of Infinitary Languages*, Lecture Notes in Mathematics, vol. 72, Springer-Verlag, Berlin, Heidelberg, New York, 1968, pp. 1-35.

Barwise, J., 'Absolute Logics and $L_{\infty\omega}$', *Annals of Mathematical Logic* (1972), 309-340.

Barwise, J., 'Axioms for Abstract Model Theory', *Annals of Mathematical Logic* 4 (1972), 309-342.

Barwise, J., *Admissible Sets and Structures*, Springer-Verlag, Berlin, Heidelberg, New York, 1975.

Barwise, J. and Perry, J., *Situations and Attitudes*, The MIT Press, Cambridge, Mass., London, 1983.

Beardsley, M. C., *Aesthetics*, Harcourt, Brace, New York, 1958.

Bechtel, W., *Philosophy of Mind*, Lawrence Erlbaum Associates, Hillsdale, NJ, Hove, London, 1988.

Bechtel, W. and Abrahamsen, A., *Connectionism and the Mind*, Blackwell, Oxford, Cambridge, Mass., 1991.

Bell, J. and Machover, M., *A Course in Mathematical Logic*, North-Holland Publishing Co., Amsterdam, 1977.

Beth, E. W., 'On Padoa's Method in the Theory of Definition', *Indag. Math.* 15 (1953), 330-339.

Bohr, N., 'Über die Serienspektra der Elemente', *Z. Phys.* 2 (1920), 423.

Bonevac, D. A., *Reduction in the Abstract Sciences*, Hackett Publishing Co., Indianapolis, Cambridge, 1982.

Born, M., *Physics in My Generation*, Springer-Verlag, New York, 1969.

Borges, J. L., *Labyrinths* (ed. by D. A. Yates and J. E. Irby), A New Directions

Book, New York, 1962.

Briskman, L., 'From Logic to Logics (and Back Again)', *The British Journal for the Philosophy of Science* 33 (1982), 95-111.

Buchwald, J. Z., 'Design for Experimenting', in P. Horwich (ed.), *World Changes. Thomas Kuhn and the Nature of Science*, The MIT Press, Cambridge, Mass., London, 1993, pp. 169-206.

Bunge, M., *Philosophy of Physics*, D. Reidel Publishing Co., Dordrecht, 1973.

Burge, T., 'Self-Reference and Translation', in F. Guenther and M. Guenther-Reutter (eds.), *Meaning and Translation. Philosophical and Linguistic Approaches*, Duckworth, London, 1978, pp. 137-153.

Carnap, R., *The Logical Syntax of Language*, Routledge & Kegan Paul, London, 1937. Contains the translation of *Logische Syntax der Sprache*, Springer, Wien, 1934.

Carnap, R., 'Testability and Meaning', *Philosophy of Science* 3 (1936), 420-468; 4 (1937), 1-40.

Chalmers, A. F., 'Curie's Principle', *British Journal for the Philosophy of Science* 21 (1970), 133-148.

Chang, C. C., 'Some New Results in Definability', *Bull. Am. Math. Soc.* 70 (1964), 808-813.

Chang, C. C. and Keisler, H. J., *Model Theory*, North-Holland Publishing Co., Amsterdam, 1973.

Churchland, P. S., 'Reductionism and Antireductionism in Functionalist Theories of Mind', in B. Beakley and P. Ludlow (eds.), *The Philosophy of Mind*, The MIT Press, Cambridge, Mass., London, 1992, pp. 59-67.

Cohen, M. R. and Nagel, E., *An Introduction to Logic and Scientific Method*, Routledge & Kegan Paul Ltd, London, 1961.

Collier, J. D. and Muller S. J., 'The Dynamical Basis of Emergence in Natural Hierarchies', in G. L. Farre and T. Oksala (eds.), *Emergence, Complexity, Hierarchy, Organization* (Acta Polytechnica Scandinavica, Ma 91), Espoo, 1998, pp. 169-178.

Cresswell, M. J., *Logics and Languages*, Methuen, London, 1973.

Curie, P., 'Symétrie dans les phénomènes physique, *Journal de Physique* 3 (1894), 393-425.

Danto, A. C., *Analytical Philosophy of History*, Cambridge University Press, Cambridge, 1968.

Danto, A. C., *The Transfiguration of the Commonplace*, Harvard University Press, Cambridge, Mass., 1981.

Davidson, D., 'Radical Interpretation', *Dialectica* 27 (1973). Reprinted in *Inquiries into Truth and Interpretation*, Clarendon Press, Oxford, 1984, pp. 125-139.

Dennett, D., *Brainstorms*, The MIT Press, Cambridge, Mass., London, 1978.

Eco, U., *The Role of the Reader: Explorations in the Semiotics of Texts*, Indiana University Press, Bloomington, 1979.

Enderton, H., *A Mathematical Introduction to Logic*, Academic Press, New York, London, 1972.

Feferman, S., 'Many-Sorted Interpolation Theorems and Applications', in *Proceedings of the Tarski Symposium*, A. M. S. Proc. Symp. in Pure Math. 25, A. M. S., 1974a, 205-223.

Feferman, S., 'Two Notes on Abstract Model Theory I', *Fundamenta Mathematicae* 82 (1974b), 153-165.

Fetzer, J. H., *Artificial Intelligence: Its Scope and Limits*, Kluwer Academic Publishers, Dordrecht, Boston, London, 1990.

Feyerabend, P., *Against Method*, New Left Books, London, 1975.

Fodor, J. and Pylyshyn, Z., 'Connectionism and Cognitive Architecture: A Critical Analysis', *Cognition* 28 (1988), 3-71.

Frege, G., *Die Grundlagen der Arithmetik*, Wilhelm Koebner, Breslau, 1884.

Friedman, S., 'Is Intertheoretic Reduction Feasible?', *British Journal for the Philosophy of Science* 33 (1982), 17-40.

Gadamer, H-G., *Wahrheit und Methode*, J. C. B. Mohr, Tübingen, 1960.

Gaifman, H., 'Operations on Relational Structures, Functors and Classes I', in L. Henkin *et al.* (eds.), *Proceedings of the Tarski Symposium,* A. M. S. Proc. Symp. in Pure Math. 25, A. M. S., Providence, R. I., 1974, pp. 21-39.

Gasiorowicz, S., *Quantum Physics*, John Wiley & Sons, Inc., New York, Chichester, Brisbane, Toronto, Singapore, 1996.

Glymour, C., 'On Some Patterns of Reduction', *Philosophy of Science* 37 (1970), 340-353.

Glynn, S., 'Towards a Unified Epistemology of the Human and Natural Sciences', *Journal of the British Society for Phenomenology* 24 (1993), 173-189.

Goldblatt, R., *Topoi: A Categorial Analysis of Logic*, North-Holland Publishing Co., Amsterdam, 1979.

Gombrich, E. H., *Art and Illusion*, Phaidon, London, 1960.

Goodman, N., 'The Problem of Counterfactual Conditionals', *Journal of Philosophy* 44 (1947), 113-128. Reprinted in Goodman (1979).

Goodman, N., *Languages of Art*, Bobbs-Merrill, New York, 1968.

Goodman, N., *Fact, Fiction, and Forecast*, Harvester Press, Sussex, 1979.

Griffin, N., *Relative Identity*, Oxford University Press, Oxford, 1977.

Grund, C., 'Metaphors, Counterfactuals and Music', in V. Rantala, L. Rowell and E. Tarasti (eds.), *Essays on the Philosophy of Music* (Acta Philosophica Fennica, vol. 43), Helsinki, 1988, pp. 28-53.

Haack, S., *Deviant Logic*, Cambridge University Press, Cambridge, 1974.

Haapala, A., *What Is a Work of Literature?* (Acta Philosophica Fennica, vol. 46), Helsinki, 1989.

Hacking, I., 'Working in a New World: The Taxonomic Solution', in P. Horwich (ed.), *World Changes. Thomas Kuhn and the Nature of Science*, The MIT Press, Cambridge, Mass., London, 1993, pp. 275-310.

Hanson, N. R., *Patterns of Discovery*, Cambridge University Press, Cambridge, 1958.

Heisenberg, W., *Physics and Philosophy*, Harper & Row, New York, 1958.

Hempel, C. G., *Aspects of Scientific Explanation and Other Essays in the Philos-*

ophy of Science, The Free Press, New York, 1965.

Hempel, C. G., *Fundamentals of Concept Formation in Empirical Science*, University of Chicago Press, Chicago, 1972.

Hintikka, J., *Models for Modalities*, D. Reidel Publishing Co., Dordrecht, 1969.

Hintikka, J., 'Constituents and Finite Identifiability', *Journal of Philosophical Logic* 1 (1972), 45-52.

Hintikka, J., *Logic, Language Games and Information*, Clarendon Press, Oxford, 1973.

Hintikka, J., 'Imposible Possible Worlds Vindicated', *Journal of Philosophical Logic* 4 (1975), 475-484.

Hintikka, J., *The Semantics of Questions and the Questions of Semantics* (Acta Philosophica Fennica, vol. 28), North-Holland Publishing Co., Amsterdam, 1976.

Hintikka, J., 'Towards a General Theory of Identifiability', in J. H. Fetzer, D. Shatz and N. Schlesinger (eds.), *Definitions and Definability: Philosophical Perspectives*, Kluwer Academic Publishers, Dordrecht, Boston, London, 1991, pp. 161-183.

Hintikka, J. and Rantala, V., 'Systematizing Definability Theory', in S. Kanger (ed.), Proceedings of the Third Scandinavian Logic Symposium, North-Holland, Amsterdam, 1975, pp. 40-62.

Hintikka J. and Tuomela, R., 'Towards a General Theory of Auxiliary Concepts and Definability in First-Order Theories, in J. Hintikka and P. Suppes (eds.), *Information and Inference*, D. Reidel, Publishing Co., Dordrecht, 1970, pp. 298-330.

Hirsch, E. D., *Validity in Interpretation,* Yale University Press, New Haven, Conn., 1967.

Hoyningen-Huene, P., *Reconstructing Scientific Revolutions. Thomas S. Kuhn's Philosophy of Science*, The University of Chicago Press, Chicago, London, 1993.

Hughes, G. E. and Cresswell M. J., *A Companion to Modal Logic*, Methuen, London, New York, 1984.

Juhl, P. D., *Interpretation*, Princeton University Press, Princeton, 1980.

Kamlah, A., 'A logical Investigation of the Phlogiston Case', in W. Balzer, D. A. Pearce and H.-J. Schmidt (eds.), *Reduction in Science. Structure, Examples, Philosophical Problems*, D. Reidel Publishing Co., Dordrecht, Boston, Lancaster, 1984, pp. 217-238.

Keenan, E., 'Some Logical Problems in Translation', in F. Guenther and M. Guenther-Reutter (eds.), *Meaning and Translation. Philosophical and Linguistic Approaches*, Duckworth, London, 978, pp. 157-189.

Kim, J., 'Supervenience As a Philosophical Concept', *Metaphilosophy* 21 (1990), 1-27.

Kitcher, P., 'Theories, Theorists, and Theoretical Change', *Philosophical Review* 87 (1978), 519-547.

Kitcher, P., *The Nature of Mathematical Knowledge*, Oxford University Press,

New York, Oxford, 1983.

Kleene, S. C., *Introduction to Metamathematics*, North-Holland Publishing Co., Amsterdam, 1964.

Krajewski, W., *Correspondence Principle and Growth of Knowledge*, D. Reidel Publishing Co., Dordrecht, 1977.

Kueker, K., 'Generalized Interpolation and Definability', *Annals of Mathematical Logic* 1 (1970), 423-468.

Kuhn, T. S., *The Structure of Scientific Revolutions*, The University of Chicago Press, Chicago, 1962. Second Edition, Enlarged, 1970.

Kuhn, T. S., *The Essential Tension: Selected Studies in Scientific Tradition and Change*, The University of Chicago Press, Chicago, 1977.

Kuhn, T. S., 'Commensurability, Comparability, Communicability', in P. D. Asquith and T. Nickles (eds.), *PSA* 1982, Philosophy of Science Association, East Lansing, 1983, pp. 669-688.

Kuhn, T. S., 'Afterwords', in P. Horwich (ed.), *World Changes. Thomas Kuhn and the Nature of Science*, The MIT Press, Cambridge, Mass., London, 1993, pp. 311-341.

Lahti, P. J., 'Quantum Theory as a Factualization of Classical Theory', in W. Balzer, D. A. Pearce and H.-J. Schmidt (eds.), *Reduction in Science,* D. Reidel Publishing Co., Dordrecht, 1984, pp. 381-396.

Laudan, L., *Progress and Its Problems*, University of California Press, Berkeley, Los Angeles, London, 1977.

Levin, S. R., *Metaphoric Worlds. Conceptions of a Romantic Nature*, Yale University Press, New Haven and London, 1988.

Lewis, D., *Counterfactuals*, Blackwell, Oxford, 1973.

Lewis, D., 'Truth in Fiction', *American Philosophical Quarterly* 15 (1978), 37-46.

Lindström, P., 'First Order Logic with Generalized Quantifiers', *Theoria* 32 (1966), 186-195.

Lindström, P., 'On Extensions of Elementary Logic', *Theoria* 35 (1969), 1-11.

Mac Cormac, E. R., *A Cognitive Theory of Metaphor*, A Bradford Book, The MIT Press, Cambridge, Mass., London, 1985.

Machover, M. and Hirschfeld, J., *Lectures on Non-Standard Analysis*, Lecture Notes in Mathematics 94, Springer-Verlag, Berlin, Heidelberg, New York, 1969.

Maclane, S., *Categories for the Working Mathematicians*, Springer-Verlag, Berlin, Heidelberg, New York, 1971.

Makkai, M., 'A Generalization of a Theorem of E. W. Beth', *Acta Math. Sci. Hungar.* 15 (1964), 227-235.

Makowsky, J. A., Shelah, S. and Stavi, J., 'Δ-Logics and Generalized Quantifiers', *Annals of Mathematical Logic* 10 (1976), 155-192.

Margolis, J., *Art and Philosophy*, The Harvester Press, Brighton, 1980.

Matthews, P. T., *Introduction to Quantum Mechanics*, McGraw-Hill, London, 1968.

Mayr, D., 'Investigations of the Concept of Reduction, I', *Erkenntnis* 10 (1976),

275-294.

Mayr, D., 'Investigations of the Concept of Reduction, II', *Erkenntnis* 16 (1981a), 109-129.

Mayr, D., 'Approximate Reduction by Completion of Empirical Uniformities', in A. Hartkämper and H.-J. Schmidt (eds.), *Structure and Approximation in Physical Theories*, Plenum Press, New York, 1981b, pp. 55-70.

McKinsey, J. C. C., Sugar A. C. and Suppes, P., 'Axiomatic Foundations of Classical Particle Mechanics', *Journal of Rational Mechanics and Analysis* 2 (1953), 253-272.

Montague, R., 'Deterministic Theories', in R. Washburne (ed.), *Decisions, Values and Groups* II, Pergamon Press, Oxford, 1961, pp. 325-370.

Mostowski, A., 'On a Generalization of Quantifiers', *Fundamenta Mathematicae* 44 (1957), 12-36.

Moulines, C. U., 'A Logical Reconstruction of Simple Equilibrium Thermodynamics', *Erkenntnis* 9 (1975), 101-130.

Moulines, C. U., 'Intertheoretic Approximation: The Kepler-Newton Case', *Synthese* 45 (1980), 387-412.

Moulines, C. U., 'A General Scheme for Intertheoretic Approximation', in A. Hartkämper and H.-J. Schmidt (eds.), *Structure and Approximation in Physical Theories*, Plenum Press, New York, 1981, pp. 123-146.

Nagel, E., *The Structure of Science*, Routledge & Kegan Paul, London, 1961.

Niiniluoto, I., *Is Science Progressive?*, D. Reidel Publishing Co., Dordrecht, 1984.

Niiniluoto, I., *Truthlikeness*, D. Reidel Publishing Co., Dordrecht, 1987.

Niiniluoto, I., 'Theories, Approximations, and Idealizations', in J. Brzezinski, F. Coniglione, T. A. F. Kuipers and L. Nowak (eds.), *Idealization I: General Problems*, Poznan Studies in the Philosophy of Sciences and the Humanities, vol. 16, Rodopi, Amsterdam, Atlanta, GA, 1989, pp. 9-57.

Nowak, L., *The Structure of Idealization*, D. Reidel Publishing Co., Dordrecht, 1980.

Nute, D., 'Conditional Logic', in D. Gabbay and F. Guenther (eds.), *Handbook of Philosophical Logic*, Vol. II, D. Reidel Publishing Co., Dordrecht, 1984, pp. 387-439.

Oesch, E., and Rantala, V., 'Interpretation, Reception, and Aesthetic Experience', in A. Haapala and O. Naukkarinen (eds.), *Interpretation and Its Boundaries*, Helsinki University Press, Helsinki, 1999, pp. 32-45.

Pavel, T. G., *Fictional Worlds*, Harvard University Press, Cambridge, Mass., 1986.

Pearce, D., *Translation, Reduction and Equivalence: Some Topics in Intertheory Relations*, University of Sussex, 1979.

Pearce, D., 'Logical Properties of the Structuralist Concept of Reduction', *Erkenntnis* 18 (1982), 307-333.

Pearce, D., *Roads to Commensurability*, D. Reidel Publishing Co., Dordrecht, 1987.

Pearce, D. and Rantala, V., 'New Foundations for Metascience', *Synthese* 56

(1983a), 1-26.

Pearce, D. and Rantala, V., 'The Logical Study of Symmetries in Scientific Change', in P. Weingartner and H. Czermak (eds.), *Epistemology and Philosophy of Science*, Hölder-Pichler-Tempsky, Vienna, 1983b, pp. 330-332.

Pearce, D. and Rantala, V., 'Logical Aspects of Scientific Reduction', in P. Weingartner and J. Czermak (eds.), *Epistemology and Philosophy of Science*, Hölder-Pichler-Tempsky, Vienna, 1983c.

Pearce, D. and Rantala, V., 'A Logical Study of the Correspondence Relation, *Journal of Philosophical Logic* 13 (1984a), 47-84.

Pearce, D. and Rantala, V., 'Limiting-Case Correspondence between Physical Theories', in W. Balzer, D. A. Pearce and H.-J. Schmidt (eds.), *Reduction in Science,* D. Reidel Publishing Co., Dordrecht, 1984b, pp. 153-185.

Pearce, D. and Rantala, V., 'Scientific Change, Continuity and Problem Solving', *Philosophia Naturalis* 21 (1984c), 389-399.

Pearce, D. and Rantala, V., 'Approximate Explanation is Deductive-Nomological', *Philosophy of Science* 52 (1985a), 126-140.

Pearce, D. and Rantala, V., 'Ramsey Eliminability Revisited', *Communication and Cognition* 18 (1985b), 157-176.

Popper, K. R., 'Some Comments on Truth and the Growth of Knowledge', in E. Nagel, P. Suppes and A. Tarski (eds.), *Logic, Methodology, and Philosophy of Science*, Stanford University Press, Stanford, 1962, pp. 285-292.

Post, H. R., 'Correspondence, Invariance and Heuristics', *Studies in History and Philosophy of Science* 2 (1971), 213-255.

Primas, H., 'Emergence in Exact Natural Sciences', in G. L. Farre and T. Oksala (eds.), *Emergence, Complexity, Hierarchy, Organization* (Acta Polytechnica Scandinavica, Ma 91), Espoo, 1998, pp. 83-98.

Przelecki, M., *The Logic of Empirical Theories*, Routledge and Kegan Paul, London, 1969.

Przelecki, M., 'On Identifiability in Extended Domains', in R. E. Butts and J. Hintikka (eds.), *Basic Problems in Methodology and Linguistics*, D. Reidel Publishing Co., Dordrecht, 1977, pp. 81-90.

Putnam, H., 'The Logic of Quantum Mechanics', in H. Putnam, *Mathematics, Matter and Method*, Cambridge University Press, Cambridge, 1975, pp. 174-197.

Quine, W. V. O., *From a Logical Point of View*, Harper & Row, New York, 1953.

Quine, W. V. O., *Word and Object*, The MIT Press, Cambridge, Mass., 1960.

Quine, W. V. O., *Philosophy of Logic*, Prentice-Hall, New Jersey, 1970.

Rantala, V., *Aspects of Definability* (Acta Philosophica Fennica, vol. 29), North-Holland, Amsterdam, 1977a.

Rantala, V., 'Prediction and Identifiability', in R. E. Butts and J. Hintikka (eds.), *Basic Problems in Methodology and Linguistics*, D. Reidel Publishing Co., Dordrecht, 1977b, pp. 91-102.

Rantala, V., 'The Old and the New Logic of Metascience', *Synthese* 39 (1978), 233-247.

Rantala, V., 'Correspondence and Non-Standard Models: A Case Study', in I. Niiniluoto and R. Tuomela (eds.) *The Logic and Epistemology of Scientific Change* (Acta Philosophica Fennica, vol. 30), North-Holland, Amsterdam, 1979, pp. 366-378.

Rantala, V., 'Impossible Worlds Semantics and Logical Omniscience', in I. Niiniluoto and E. Saarinen (eds.), *Intensional Logic: Theory and Applications* (Acta Philosophica Fennica, vol. 35), Helsinki, 1982a, pp. 106-115.

Rantala, V., 'Quantified Modal Logic: Non-Normal Worlds and Propositional Attitudes', *Studia Logica* 41 (1982b), 41-65.

Rantala, V., 'Interpreting Narratives. Some Logical Questions', in I. Ruzsa and A. Szabolcsi (eds.), *Logic and Language*, Akadémiai Kiadó, Budapest, 1987, pp. 121-132.

Rantala, V., 'Scientic Change and Change of Logic', in I. M. Bodnar, A. Maté and L. Polos (eds.), *Intensional Logic, History of Philosophy, and Methodology*, Budapest, 1988, pp. 247-252.

Rantala, V., 'Counterfactual Reduction', in K. Gavroglu, Y. Goudaroulis and P. Nicolacopoulos (eds), *Imre Lakatos and Theories of Scientific Change*, Kluwer Academic Publishers, Dordrecht, 1989, pp. 347-360.

Rantala, V., 'Definitions and Definability', in J. H. Fetzer, D. Shatz and N. Schlesinger (eds.), *Definitions and Definability: Philosophical Perspectives*, Kluwer Academic Publishers, Dordrecht, Boston, London, 1991, pp. 135-159.

Rantala, V., 'Reduction and Explanation: Science vs. Mathematics', in J. Echeverria, A. Ibarra and T. Mormann (eds.), *The Space of Mathematics*, De Gruyter, Berlin, New York, 1992, pp. 47-59.

Rantala, V., 'Understanding Scientific Change', in P. I. Bystrov and V. N. Sadovsky (eds.), *Philosophical Logic and Logical Philosophy*, Kluwer Academic Publishers, Dordrecht, Boston, London, 1996, pp. 3-15.

Rantala, V., 'Explanatory Translation. Beyond Functionalism and Reductionism', in M. L. Dalla Chiara, K. Doets, D. Mundici and J. van Benthem (eds.), *Logic and Scientific Methods*, Kluwer Academic Publishers, Dordrecht, Boston, London, 1997a, pp. 399-41.

Rantala, V., 'Theoreticity in Art', in A. Haapala, J. Levinson and V. Rantala (eds.), *The End of Art and Beyond. Essays after Danto*, Humanities Press, New Jersey, 1997b, pp. 110-121.

Rantala, V. and Tselishchev, V., 'Surface Information and Analyticity', in R. Bogdan (ed.), *Jaakko Hintikka*, D. Reidel, Dordrecht, 1987, pp. 77-90.

Rantala, V. and Vadén, T., 'Idealization in Cognitive Science. A Study in Counterfactual Correspondence', in M. Kuokkanen (ed.), *Idealization VII: Structuralism, Idealization and Approximation* (Poznan Studies in the Philosophy of the Sciences and the Humanities, Vol. 42), Rodopi, Amsterdam, Atlanta, GA, 1994, pp. 179-198.

Rantala, V. and Wiesenthal, L., 'The Worlds of Fiction and the Worlds of Science: A Comparative Study', *Synthese* 78 (1989), 53-86.

Redhead, M. L. G., 'Symmetry in Intertheory Relations', *Synthese* 32 (1975),

77-112.

Richardson, R. C., 'Functionalism and Reductionism', *Philosophy of Science* 46 (1979), 533-558.

Ricoeur, P. 'Life Quest of Narrative', in D. Wood (ed.), *On Paul Ricoeur. Narrative and Interpretation*, Routledge, London, New York, 1992, pp. 20-33.

Robinson, A., 'Non-Standard Analysis', *Proceedings of the Royal Academy of Sciences* (A) 64 (1961), 432-440.

Robinson, A., *Non-Standard Analysis*, North-Holland, Amsterdam, 1966.

Robinson, R., *Definition*, Clarendon Press, Oxford, 1965.

Robinson, A. and Zakon, E., 'A Set-Theoretical Characterization of Enlargements', in W. A. J. Luxemburg (ed.), *Applications of Model Theory to Algebra, Analysis, and Probability*, Holt, Rinehart and Winston, New York, 1969, pp. 109-122.

Rollins, M. (ed.), *Danto and His Critics*, Blackwell, Oxford, 1993.

Rosen, R., *Possible Worlds and Literary Theory*, Cambridge University Press, Cambridge, 1994.

Rott, H., 'On Relations between Successive Theories' in *Abstracts of the LMPS 87,* Vol.4., Part 2, Moscow, 1987, pp. 123-127.

Rubin, H. and Suppes, P., 'Transformations of Systems of Relativistic Particle Mechanics', *Pacific Journal of Mathematics* 4 (1954), 563-601.

Saarinen, E., 'Propositional Attitudes Are Not Attitudes towards Propositions', in I. Niiniluoto and E. Saarinen (eds.), *Intensional Logic: Theory and Applications* (Acta Philosophica Fennica, vol. 35), Helsinki, 1982, pp. 130-162.

Salmon, W. C., *Four Decades of Scientific Explanation*, University of Minnesota Press, Minneapolis, 1989.

Scheibe, E., *The Logical Analysis of Quantum Mechanics,* Pergamon Press, Oxford, 1973a.

Scheibe, E., 'The Approximate Explanation and the Development of Physics', in P. Suppes, L. Henkin, A. Joja and G. Moisil (eds.), *Logic, Methodology and Philosophy of Science IV*, North-Holland, Amsterdam, 1973b, pp. 930-942.

Scriven, M., 'Explanations Predictions, and Laws', in H. Feigl and G. Maxwell (eds), *Minnesota Studies in the Philosophy of Science,* Vol. III, University of Minnesota Press, Minneapolis, 1962, pp. 170-230.

Searle, J. R., *Speech Acts,* Cambridge University Press, Cambridge, 1969.

Searle, J. R., *Expression and Meaning*, Cambridge University Press, Cambridge, 1979.

Searle, J. R., *Intentionality*, Cambridge University Press, Cambridge, 1981.

Segerberg, K., *An Essay in Classical Modal Logic,* Filosofiska Studier, no. 13, Uppsala, 1971.

Sellars, W., 'Inference and Meaning', *Mind* 62 (1953), 313-330.

Sintonen, M., *The Pragmatics of Scientific Explanation* (Acta Philosophica Fennica, vol. 37), Helsinki, 1984.

Smolensky, P., 'On the Proper Treatment of Connectionism' (with open peer commentary and author's response), *Behavioral and Brain Sciences* 11 (1988), 1-74.

Sneed, J. D., *The Logical Structure of Mathematical Physics*, D. Reidel Publishing Co., Dordrecht, 1971.

Solomon, R. C. and Higgins, K. M., 'Atomism, Art, and Arthur: Danto's Hegelian Turn', in M. Rollins (ed.), *Danto and His Critics*, Blackwell, Oxford, 1993, pp. 107-126.

Stegmüller, W., *Theorie und Erfahrung* (Zweiter Halbband). *Theorienstructuren und Theoriendynamik*, Springer-Verlag, Berlin, Heidelberg, New York, 1973.

Stegmüller, W., *The Structure and Dynamics of Theories*, Springer-Verlag, Berlin, Heidelberg, New York, 1976.

Strawson, P. F., *Individuals*, Methuen, London, 1959.

Stroyan, K. and Luxemburg, W. A. J., *Introduction to the Theory of Infinitesimals*, Academic Press, New York, 1976.

Suppe, F., 'The Search for Philosophic Understanding of Scientific Theories', in F. Suppe (ed.), *The Structure of Scientific Theories*, University of Illinois Press, Urbana, Chicago, London, 1977, pp. 1-232.

Suppes, P., *Introduction to Logic*, Van Nostrand, New York, 1957.

Svenonius, L., 'A Theorem of Permutation in Models', *Theoria* 25 (1959), 173-178.

Tarski, A., *Logic, Semantics, Metamathematics: Papers from 1923 to 1938* (translated by J. H. Woodger), Clarendon Press, Oxford, 1956.

Thagard, P., *Conceptual Revolutions*, Princeton University Press, Princeton, 1992.

Tormey, A., 'Metaphors and Counterfactuals', in J. Fisher (ed.), *Essays in Aesthetics: Perspectives on the Work of Monroe C. Beardsley*, Temple University Press, Philadelphia, 1983, pp. 235-246.

Tuomela, R., *Theoretical Concepts,* Springer-Verlag, Wien, New York, 1973.

Tuomela, R., *Science, Action, and Reality*, D. Reidel Publishing Co., Dordrecht, 1985.

Vadén, T., 'From Limitivism to Correspondence: The Symbolic-Subsymbolic Relation', in L. Niklasson and M. Bodén (eds.), *Current Trends in Connectionism*, Lawrence Erlbaum, New Jersey, 1995.

Vadén, T., *The Symbolic and the Subsymbolic in Cognitive Science* (diss.), University of Tampere, Tampere, 1996.

Waerden, B. L. van der, 'Introduction', in B. L. van der Waerden (ed.), *Sources of Quantum Mechanics,* Dover Publications Inc., New York, 1967, pp. 1-59

Wheeler, S. C., 'Indeterminacy of Radical Interpretation and the Causal Theory of Reference', in F. Guenther and M. Guenther-Reutter (eds.), *Meaning and Translation. Philosophical and Linguistic Approaches*, Duckworth, London, 1978, pp. 83-94.

Whitehead, A. N. and Russell, B., *Principia Mathematica to *56* (from *Principia Mathematica*, 1910), Cambridge University Press, Cambridge, 1973.

Wimsatt Jr., W. K. ja Beardsley, M. C., 'The Intentional Fallacy', in J. Margolis (ed.), *Philosophy Looks at the Arts*, Temple University Press, Philadelphia, 1978, s. 293-306.

Wolterstorff, N., *Works and Worlds of Art,* Clarendon Press, Oxford, 1980.

NAME INDEX

Abrahamsen, A. 157; 161; 167-168; 179; 279
Adams, E. W. 150; 272; 278
Austin, J. L. 249

Balzer, W. 269; 272; 275; 278
Barthes, R. 64; 271
Barwise, J. 202-204; 271; 275; 280-282; 285-286
Beardsley, M. C. 66; 271
Bechtel, W. 157; 161; 167-168; 179; 279
Bell, J. 210-211; 283
Benthem, J. van 277; 283
Beth, E. W. 252
Bohr, N. 3-5; 89; 128; 269
Bonevac, D. A. 115; 124; 131; 277
Borges, J. L. 70; 270; 274
Born, M. 4
Brahe, T. 54; 56-57
Breughel, P. 39; 45-46; 87
Briskman, L. 277
Buchwald, J. Z. 270
Bunge, M. 8
Burge, T. 27; 271

Cantor, G. 129
Carnap, R. 187; 197; 199-205; 254; 281-282
Chalmers, A., F. 93
Chang, C. C. 253; 285
Church, A. 158
Churchland, P. S. 161; 279
Cohen, M. R. 246-248; 264; 267
Collier, J. D. 276
Cresswell, M. J. 280; 286
Curie, P. 93-94; 138-139; 141; 220; 230; 275

Danto, A. C. 38-39; 42; 83; 87; 101; 107; 270; 272-273; 276
Davidson, D. 22-23; 31
Dennett, D. 23

Eco, U. 274
Einstein, A. 4; 10; 13; 92

Enderton, H. 272; 275
Feferman, S. 192; 194-196; 202; 275; 280-281
Fetzer, J. H. 278
Feyerabend, P. 8; 16
Fodor, J. 151
Frege, G. 196; 264-265
Friedman, S. 284

Gadamer, H-G. 24-26
Gaifman, H. 275
Galilei, G. 138; 230; 232; 234
Gasiorowicz, S. 145; 278
Glymour, C. 114-115; 117; 119; 124; 269; 272; 274
Glynn, S. 124
Goldblatt, R. 282
Gombrich, E. H. 58; 60; 273
Goodman, N. 22; 51; 160; 205-207; 273; 285
Griffin, N. 282
Grund, C. 50; 272

Haack, S. 277
Haapala, A. 272; 274
Hacking, I. 16; 19; 25; 270-271
Hanson, N. R. 54-55; 57; 59
Heisenberg, W. 4; 8; 144; 146; 269
Hempel, C. G. 82; 102-104; 266-267; 272
Higgins, K. M. 83
Hintikka, J. 48; 253-254; 256; 265-267; 276; 280; 285
Hirsch, E. D. 66
Hirschfelt, J. 210
Hoyningen-Huene, P. 12; 269-270
Hughess, G. E. 280

Juhl, P. D. 66

Kamlah, A. 149-150
Keenan, E. 272
Keisler, H, J. 285
Kepler, J. 8; 54; 56-57; 141-142
Kim, J., 275

Svenonius, L. 254

Tarski, A. 252; 265
Tormey, A. 50-53
Tselishchev, V. 286
Tuomela, R. 254; 256; 265-267; 276; 285-286
Turing, A. 155; 157-159; 170-173; 235; 242; 285

Vadén, T. 151; 154; 220; 235; 238; 272; 278

Waerden, B. L. van der 3
Whitehead, A. N. 245; 264; 285-286
Wiesenthal, L. 69; 270; 273-274
Wigner, E. 91
Wimsatt Jr., W. K. 66
Wittgenstein, L. 54; 101; 197
Wolterstorff, N. 68; 274

Zakon, E. 282
Zermelo, E. 136

SUBJECT INDEX

art
> Danto on 273
> Gombrich on 58; 60; 273
> and perception 60
> progress of 58-60; 273

category theory 209-210; 282
classical mechanics 5; 25; 142
classical particle mechanics 127
cognition
> in Fodor and Pylyshyn 151
> subsymbolic 151
> symbolic 151
cognitive science
> counterfactual correspondence in 164
> limiting case correspondence in 154; 243
> Smolensky on 150; 153-154; 164
> subsymbolic representation in 150; 159; 163; 166
> symbolic representation in 150; 159; 163; 166
> translation in 158; 169; 243
conceptual change 3; 18
> in art 58
> in chemistry 147
> Kuhn on 11; 14
> in logic 128
> in mathematics 128
> in physics 128
> in science 4-6; 8-11
> Thagard on 270
> and translation 21
connectionist systems
> conceptual level in 152; 167-168; 280
> epistemic logic of 173; 177
> external (individual) logic of 176
> external logic of 173
> individual knowledge in 169
> internal logic of 172; 175
> and Kripke models 177
> learning in 173
> logic learning in 174
> pattern recognition in 173

and reduction 167
and semantics 175
Smolensky on 153
subconceptual level in 152; 167-169; 280
Correspondence Principle 3-5; 8; 17-18; 128; 143
> Bohr on 3; 269
> Born on 4
> Bunge on 8
> cognitive 151
> Einstein on 4
> extended 19
> Heisenberg on 4; 8; 269
> Kuhn on 10; 17
> Post on 4; 269
> van der Waerden on 3
correspondence relation 4-5; 79; 86
> approximate 90
> in celestial mechanics 141
> in chemistry 146
> classical reduction as 99-100
> in cognitive science 150-158; 169; 235-244
> and commensurability 86
> and conceptual change 80
> conditional definitions as 98
> correlation in 82
> correspondence sketch as 82
> counterfactual 9-10; 88; 116
> definition of 82
> definitions as 96
> example of 90-91
> and explanation 112, 124
> explanatory import of 109
> explicit definitions as 96
> finite definitions as 98
> of formal theories 85-86
> in historical sciences 82-83
> limiting case 3-4; 11; 25; 35; 88-89; 144
> and logic 85
> in logic 134
> of logics 87
> and Minkowski's force law 89

SYNTHESE LIBRARY

1. J. M. Bochénski, *A Precis of Mathematical Logic*. Translated from French and German by O. Bird. 1959 ISBN 90-277-0073-7
2. P. Guiraud, *Problèmes et méthodes de la statistique linguistique*. 1959 ISBN 90-277-0025-7
3. H. Freudenthal (ed.), *The Concept and the Role of the Model in Mathematics and Natural and Social Sciences*. 1961 ISBN 90-277-0017-6
4. E. W. Beth, *Formal Methods*. An Introduction to Symbolic Logic and to the Study of Effective Operations in Arithmetic and Logic. 1962 ISBN 90-277-0069-9
5. B. H. Kazemier and D. Vuysje (eds.), *Logic and Language*. Studies dedicated to Professor Rudolf Carnap on the Occasion of His 70th Birthday. 1962 ISBN 90-277-0019-2
6. M. W. Wartofsky (ed.), *Proceedings of the Boston Colloquium for the Philosophy of Science, 1961–1962*. [Boston Studies in the Philosophy of Science, Vol. I] 1963 ISBN 90-277-0021-4
7. A. A. Zinov'ev, *Philosophical Problems of Many-valued Logic*. A revised edition, edited and translated (from Russian) by G. Küng and D.D. Comey. 1963 ISBN 90-277-0091-5
8. G. Gurvitch, *The Spectrum of Social Time*. Translated from French and edited by M. Korenbaum and P. Bosserman. 1964 ISBN 90-277-0006-0
9. P. Lorenzen, *Formal Logic*. Translated from German by F.J. Crosson. 1965 ISBN 90-277-0080-X
10. R. S. Cohen and M. W. Wartofsky (eds.), *Proceedings of the Boston Colloquium for the Philosophy of Science, 1962–1964*. In Honor of Philipp Frank. [Boston Studies in the Philosophy of Science, Vol. II] 1965 ISBN 90-277-9004-0
11. E. W. Beth, *Mathematical Thought*. An Introduction to the Philosophy of Mathematics. 1965 ISBN 90-277-0070-2
12. E. W. Beth and J. Piaget, *Mathematical Epistemology and Psychology*. Translated from French by W. Mays. 1966 ISBN 90-277-0071-0
13. G. Küng, *Ontology and the Logistic Analysis of Language*. An Enquiry into the Contemporary Views on Universals. Revised ed., translated from German. 1967 ISBN 90-277-0028-1
14. R. S. Cohen and M. W. Wartofsky (eds.), *Proceedings of the Boston Colloquium for the Philosophy of Sciences, 1964–1966*. In Memory of Norwood Russell Hanson. [Boston Studies in the Philosophy of Science, Vol. III] 1967 ISBN 90-277-0013-3
15. C. D. Broad, *Induction, Probability, and Causation*. Selected Papers. 1968 ISBN 90-277-0012-5
16. G. Patzig, *Aristotle's Theory of the Syllogism*. A Logical-philosophical Study of *Book A* of the *Prior Analytics*. Translated from German by J. Barnes. 1968 ISBN 90-277-0030-3
17. N. Rescher, *Topics in Philosophical Logic*. 1968 ISBN 90-277-0084-2
18. R. S. Cohen and M. W. Wartofsky (eds.), *Proceedings of the Boston Colloquium for the Philosophy of Science, 1966–1968, Part I*. [Boston Studies in the Philosophy of Science, Vol. IV] 1969 ISBN 90-277-0014-1
19. R. S. Cohen and M. W. Wartofsky (eds.), *Proceedings of the Boston Colloquium for the Philosophy of Science, 1966–1968, Part II*. [Boston Studies in the Philosophy of Science, Vol. V] 1969 ISBN 90-277-0015-X
20. J. W. Davis, D. J. Hockney and W. K. Wilson (eds.), *Philosophical Logic*. 1969 ISBN 90-277-0075-3
21. D. Davidson and J. Hintikka (eds.), *Words and Objections*. Essays on the Work of W. V. Quine. 1969, rev. ed. 1975 ISBN 90-277-0074-5; Pb 90-277-0602-6
22. P. Suppes, *Studies in the Methodology and Foundations of Science. Selected Papers from 1951 to 1969*. 1969 ISBN 90-277-0020-6
23. J. Hintikka, *Models for Modalities*. Selected Essays. 1969 ISBN 90-277-0078-8; Pb 90-277-0598-4

SYNTHESE LIBRARY

24. N. Rescher *et al.* (eds.), *Essays in Honor of Carl G. Hempel.* A Tribute on the Occasion of His 65th Birthday. 1969   ISBN 90-277-0085-0
25. P. V. Tavanec (ed.), *Problems of the Logic of Scientific Knowledge.* Translated from Russian. 1970   ISBN 90-277-0087-7
26. M. Swain (ed.), *Induction, Acceptance, and Rational Belief.* 1970   ISBN 90-277-0086-9
27. R. S. Cohen and R. J. Seeger (eds.), *Ernst Mach: Physicist and Philosopher.* [Boston Studies in the Philosophy of Science, Vol. VI]. 1970   ISBN 90-277-0016-8
28. J. Hintikka and P. Suppes, *Information and Inference.* 1970   ISBN 90-277-0155-5
29. K. Lambert, *Philosophical Problems in Logic.* Some Recent Developments. 1970
                                                                              ISBN 90-277-0079-6
30. R. A. Eberle, *Nominalistic Systems.* 1970   ISBN 90-277-0161-X
31. P. Weingartner and G. Zecha (eds.), *Induction, Physics, and Ethics.* 1970 ISBN 90-277-0158-X
32. E. W. Beth, *Aspects of Modern Logic.* Translated from Dutch. 1970   ISBN 90-277-0173-3
33. R. Hilpinen (ed.), *Deontic Logic.* Introductory and Systematic Readings. 1971
     See also No. 152.   ISBN Pb (1981 rev.) 90-277-1302-2
34. J.-L. Krivine, *Introduction to Axiomatic Set Theory.* Translated from French. 1971
                                                          ISBN 90-277-0169-5; Pb 90-277-0411-2
35. J. D. Sneed, *The Logical Structure of Mathematical Physics.* 2nd rev. ed., 1979
                                                          ISBN 90-277-1056-2; Pb 90-277-1059-7
36. C. R. Kordig, *The Justification of Scientific Change.* 1971
                                                          ISBN 90-277-0181-4; Pb 90-277-0475-9
37. M. Čapek, *Bergson and Modern Physics.* A Reinterpretation and Re-evaluation. [Boston Studies in the Philosophy of Science, Vol. VII] 1971   ISBN 90-277-0186-5
38. N. R. Hanson, *What I Do Not Believe, and Other Essays.* Ed. by S. Toulmin and H. Woolf. 1971   ISBN 90-277-0191-1
39. R. C. Buck and R. S. Cohen (eds.), *PSA 1970.* Proceedings of the Second Biennial Meeting of the Philosophy of Science Association, Boston, Fall 1970. In Memory of Rudolf Carnap. [Boston Studies in the Philosophy of Science, Vol. VIII] 1971
                                                          ISBN 90-277-0187-3; Pb 90-277-0309-4
40. D. Davidson and G. Harman (eds.), *Semantics of Natural Language.* 1972
                                                          ISBN 90-277-0304-3; Pb 90-277-0310-8
41. Y. Bar-Hillel (ed.), *Pragmatics of Natural Languages.* 1971
                                                          ISBN 90-277-0194-6; Pb 90-277-0599-2
42. S. Stenlund, *Combinators, γ Terms and Proof Theory.* 1972   ISBN 90-277-0305-1
43. M. Strauss, *Modern Physics and Its Philosophy.* Selected Paper in the Logic, History, and Philosophy of Science. 1972   ISBN 90-277-0230-6
44. M. Bunge, *Method, Model and Matter.* 1973   ISBN 90-277-0252-7
45. M. Bunge, *Philosophy of Physics.* 1973   ISBN 90-277-0253-5
46. A. A. Zinov'ev, *Foundations of the Logical Theory of Scientific Knowledge (Complex Logic).* Revised and enlarged English edition with an appendix by G. A. Smirnov, E. A. Sidorenka, A. M. Fedina and L. A. Bobrova. [Boston Studies in the Philosophy of Science, Vol. IX] 1973
                                                          ISBN 90-277-0193-8; Pb 90-277-0324-8
47. L. Tondl, *Scientific Procedures.* A Contribution concerning the Methodological Problems of Scientific Concepts and Scientific Explanation. Translated from Czech by D. Short. Edited by R.S. Cohen and M.W. Wartofsky. [Boston Studies in the Philosophy of Science, Vol. X] 1973
                                                          ISBN 90-277-0147-4; Pb 90-277-0323-X
48. N. R. Hanson, *Constellations and Conjectures.* 1973   ISBN 90-277-0192-X

49. K. J. J. Hintikka, J. M. E. Moravcsik and P. Suppes (eds.), *Approaches to Natural Language.* 1973   ISBN 90-277-0220-9; Pb 90-277-0233-0

50. M. Bunge (ed.), *Exact Philosophy.* Problems, Tools and Goals. 1973   ISBN 90-277-0251-9

51. R. J. Bogdan and I. Niiniluoto (eds.), *Logic, Language and Probability.* 1973   ISBN 90-277-0312-4

52. G. Pearce and P. Maynard (eds.), *Conceptual Change.* 1973   ISBN 90-277-0287-X; Pb 90-277-0339-6

53. I. Niiniluoto and R. Tuomela, *Theoretical Concepts and Hypothetico-inductive Inference.* 1973   ISBN 90-277-0343-4

54. R. Fraissé, *Course of Mathematical Logic* – Volume 1: *Relation and Logical Formula.* Translated from French. 1973   ISBN 90-277-0268-3; Pb 90-277-0403-1 (For *Volume 2* see under No. 69).

55. A. Grünbaum, *Philosophical Problems of Space and Time.* Edited by R.S. Cohen and M.W. Wartofsky. 2nd enlarged ed. [Boston Studies in the Philosophy of Science, Vol. XII] 1973   ISBN 90-277-0357-4; Pb 90-277-0358-2

56. P. Suppes (ed.), *Space, Time and Geometry.* 1973   ISBN 90-277-0386-8; Pb 90-277-0442-2

57. H. Kelsen, *Essays in Legal and Moral Philosophy.* Selected and introduced by O. Weinberger. Translated from German by P. Heath. 1973   ISBN 90-277-0388-4

58. R. J. Seeger and R. S. Cohen (eds.), *Philosophical Foundations of Science.* [Boston Studies in the Philosophy of Science, Vol. XI] 1974   ISBN 90-277-0390-6; Pb 90-277-0376-0

59. R. S. Cohen and M. W. Wartofsky (eds.), *Logical and Epistemological Studies in Contemporary Physics.* [Boston Studies in the Philosophy of Science, Vol. XIII] 1973   ISBN 90-277-0391-4; Pb 90-277-0377-9

60. R. S. Cohen and M. W. Wartofsky (eds.), *Methodological and Historical Essays in the Natural and Social Sciences. Proceedings of the Boston Colloquium for the Philosophy of Science, 1969–1972.* [Boston Studies in the Philosophy of Science, Vol. XIV] 1974   ISBN 90-277-0392-2; Pb 90-277-0378-7

61. R. S. Cohen, J. J. Stachel and M. W. Wartofsky (eds.), *For Dirk Struik. Scientific, Historical and Political Essays.* [Boston Studies in the Philosophy of Science, Vol. XV] 1974   ISBN 90-277-0393-0; Pb 90-277-0379-5

62. K. Ajdukiewicz, *Pragmatic Logic.* Translated from Polish by O. Wojtasiewicz. 1974   ISBN 90-277-0326-4

63. S. Stenlund (ed.), *Logical Theory and Semantic Analysis.* Essays dedicated to Stig Kanger on His 50th Birthday. 1974   ISBN 90-277-0438-4

64. K. F. Schaffner and R. S. Cohen (eds.), *PSA 1972. Proceedings of the Third Biennial Meeting of the Philosophy of Science Association.* [Boston Studies in the Philosophy of Science, Vol. XX] 1974   ISBN 90-277-0408-2; Pb 90-277-0409-0

65. H. E. Kyburg, Jr., *The Logical Foundations of Statistical Inference.* 1974   ISBN 90-277-0330-2; Pb 90-277-0430-9

66. M. Grene, *The Understanding of Nature.* Essays in the Philosophy of Biology. [Boston Studies in the Philosophy of Science, Vol. XXIII] 1974   ISBN 90-277-0462-7; Pb 90-277-0463-5

67. J. M. Broekman, *Structuralism: Moscow, Prague, Paris.* Translated from German. 1974   ISBN 90-277-0478-3

68. N. Geschwind, *Selected Papers on Language and the Brain.* [Boston Studies in the Philosophy of Science, Vol. XVI] 1974   ISBN 90-277-0262-4; Pb 90-277-0263-2

69. R. Fraissé, *Course of Mathematical Logic* – Volume 2: *Model Theory.* Translated from French. 1974   ISBN 90-277-0269-1; Pb 90-277-0510-0 (For *Volume 1* see under No. 54)

70. A. Grzegorczyk, *An Outline of Mathematical Logic*. Fundamental Results and Notions explained with all Details. Translated from Polish. 1974   ISBN 90-277-0359-0; Pb 90-277-0447-3
71. F. von Kutschera, *Philosophy of Language*. 1975   ISBN 90-277-0591-7
72. J. Manninen and R. Tuomela (eds.), *Essays on Explanation and Understanding*. Studies in the Foundations of Humanities and Social Sciences. 1976   ISBN 90-277-0592-5
73. J. Hintikka (ed.), *Rudolf Carnap, Logical Empiricist*. Materials and Perspectives. 1975
      ISBN 90-277-0583-6
74. M. Čapek (ed.), *The Concepts of Space and Time*. Their Structure and Their Development. [Boston Studies in the Philosophy of Science, Vol. XXII] 1976
      ISBN 90-277-0355-8; Pb 90-277-0375-2
75. J. Hintikka and U. Remes, *The Method of Analysis*. Its Geometrical Origin and Its General Significance. [Boston Studies in the Philosophy of Science, Vol. XXV] 1974
      ISBN 90-277-0532-1; Pb 90-277-0543-7
76. J. E. Murdoch and E. D. Sylla (eds.), *The Cultural Context of Medieval Learning*. [Boston Studies in the Philosophy of Science, Vol. XXVI] 1975
      ISBN 90-277-0560-7; Pb 90-277-0587-9
77. S. Amsterdamski, *Between Experience and Metaphysics*. Philosophical Problems of the Evolution of Science. [Boston Studies in the Philosophy of Science, Vol. XXXV] 1975
      ISBN 90-277-0568-2; Pb 90-277-0580-1
78. P. Suppes (ed.), *Logic and Probability in Quantum Mechanics*. 1976
      ISBN 90-277-0570-4; Pb 90-277-1200-X
79. H. von Helmholtz: *Epistemological Writings. The Paul Hertz / Moritz Schlick Centenary Edition of 1921 with Notes and Commentary by the Editors*. Newly translated from German by M. F. Lowe. Edited, with an Introduction and Bibliography, by R. S. Cohen and Y. Elkana. [Boston Studies in the Philosophy of Science, Vol. XXXVII] 1975
      ISBN 90-277-0290-X; Pb 90-277-0582-8
80. J. Agassi, *Science in Flux*. [Boston Studies in the Philosophy of Science, Vol. XXVIII] 1975
      ISBN 90-277-0584-4; Pb 90-277-0612-2
81. S. G. Harding (ed.), *Can Theories Be Refuted?* Essays on the Duhem-Quine Thesis. 1976
      ISBN 90-277-0629-8; Pb 90-277-0630-1
82. S. Nowak, *Methodology of Sociological Research*. General Problems. 1977
      ISBN 90-277-0486-4
83. J. Piaget, J.-B. Grize, A. Szemińska and V. Bang, *Epistemology and Psychology of Functions*. Translated from French. 1977   ISBN 90-277-0804-5
84. M. Grene and E. Mendelsohn (eds.), *Topics in the Philosophy of Biology*. [Boston Studies in the Philosophy of Science, Vol. XXVII] 1976   ISBN 90-277-0595-X; Pb 90-277-0596-8
85. E. Fischbein, *The Intuitive Sources of Probabilistic Thinking in Children*. 1975
      ISBN 90-277-0626-3; Pb 90-277-1190-9
86. E. W. Adams, *The Logic of Conditionals*. An Application of Probability to Deductive Logic. 1975   ISBN 90-277-0631-X
87. M. Przełęcki and R. Wójcicki (eds.), *Twenty-Five Years of Logical Methodology in Poland*. Translated from Polish. 1976   ISBN 90-277-0601-8
88. J. Topolski, *The Methodology of History*. Translated from Polish by O. Wojtasiewicz. 1976
      ISBN 90-277-0550-X
89. A. Kasher (ed.), *Language in Focus: Foundations, Methods and Systems*. Essays dedicated to Yehoshua Bar-Hillel. [Boston Studies in the Philosophy of Science, Vol. XLIII] 1976
      ISBN 90-277-0644-1; Pb 90-277-0645-X

90. J. Hintikka, *The Intentions of Intentionality and Other New Models for Modalities.* 1975
                                                    ISBN 90-277-0633-6; Pb 90-277-0634-4
91. W. Stegmüller, *Collected Papers on Epistemology, Philosophy of Science and History of Philosophy.* 2 Volumes. 1977   Set ISBN 90-277-0767-7
92. D. M. Gabbay, *Investigations in Modal and Tense Logics with Applications to Problems in Philosophy and Linguistics.* 1976   ISBN 90-277-0656-5
93. R. J. Bogdan, *Local Induction.* 1976   ISBN 90-277-0649-2
94. S. Nowak, *Understanding and Prediction.* Essays in the Methodology of Social and Behavioral Theories. 1976   ISBN 90-277-0558-5; Pb 90-277-1199-2
95. P. Mittelstaedt, *Philosophical Problems of Modern Physics.* [Boston Studies in the Philosophy of Science, Vol. XVIII] 1976   ISBN 90-277-0285-3; Pb 90-277-0506-2
96. G. Holton and W. A. Blanpied (eds.), *Science and Its Public: The Changing Relationship.* [Boston Studies in the Philosophy of Science, Vol. XXXIII] 1976
                                                    ISBN 90-277-0657-3; Pb 90-277-0658-1
97. M. Brand and D. Walton (eds.), *Action Theory.* 1976   ISBN 90-277-0671-9
98. P. Gochet, *Outline of a Nominalist Theory of Propositions.* An Essay in the Theory of Meaning and in the Philosophy of Logic. 1980   ISBN 90-277-1031-7
99. R. S. Cohen, P. K. Feyerabend, and M. W. Wartofsky (eds.), *Essays in Memory of Imre Lakatos.* [Boston Studies in the Philosophy of Science, Vol. XXXIX] 1976
                                                    ISBN 90-277-0654-9; Pb 90-277-0655-7
100. R. S. Cohen and J. J. Stachel (eds.), *Selected Papers of Léon Rosenfeld.* [Boston Studies in the Philosophy of Science, Vol. XXI] 1979   ISBN 90-277-0651-4; Pb 90-277-0652-2
101. R. S. Cohen, C. A. Hooker, A. C. Michalos and J. W. van Evra (eds.), *PSA 1974. Proceedings of the 1974 Biennial Meeting of the Philosophy of Science Association.* [Boston Studies in the Philosophy of Science, Vol. XXXII] 1976   ISBN 90-277-0647-6; Pb 90-277-0648-4
102. Y. Fried and J. Agassi, *Paranoia.* A Study in Diagnosis. [Boston Studies in the Philosophy of Science, Vol. L] 1976   ISBN 90-277-0704-9; Pb 90-277-0705-7
103. M. Przełęcki, K. Szaniawski and R. Wójcicki (eds.), *Formal Methods in the Methodology of Empirical Sciences.* 1976   ISBN 90-277-0698-0
104. J. M. Vickers, *Belief and Probability.* 1976   ISBN 90-277-0744-8
105. K. H. Wolff, *Surrender and Catch.* Experience and Inquiry Today. [Boston Studies in the Philosophy of Science, Vol. LI] 1976   ISBN 90-277-0758-8; Pb 90-277-0765-0
106. K. Kosík, *Dialectics of the Concrete.* A Study on Problems of Man and World. [Boston Studies in the Philosophy of Science, Vol. LII] 1976   ISBN 90-277-0761-8; Pb 90-277-0764-2
107. N. Goodman, *The Structure of Appearance.* 3rd ed. with an Introduction by G. Hellman. [Boston Studies in the Philosophy of Science, Vol. LIII] 1977
                                                    ISBN 90-277-0773-1; Pb 90-277-0774-X
108. K. Ajdukiewicz, *The Scientific World-Perspective and Other Essays, 1931-1963.* Translated from Polish. Edited and with an Introduction by J. Giedymin. 1978   ISBN 90-277-0527-5
109. R. L. Causey, *Unity of Science.* 1977   ISBN 90-277-0779-0
110. R. E. Grandy, *Advanced Logic for Applications.* 1977   ISBN 90-277-0781-2
111. R. P. McArthur, *Tense Logic.* 1976   ISBN 90-277-0697-2
112. L. Lindahl, *Position and Change.* A Study in Law and Logic. Translated from Swedish by P. Needham. 1977   ISBN 90-277-0787-1
113. R. Tuomela, *Dispositions.* 1978   ISBN 90-277-0810-X
114. H. A. Simon, *Models of Discovery and Other Topics in the Methods of Science.* [Boston Studies in the Philosophy of Science, Vol. LIV] 1977   ISBN 90-277-0812-6; Pb 90-277-0858-4

115. R. D. Rosenkrantz, *Inference, Method and Decision*. Towards a Bayesian Philosophy of Science. 1977   ISBN 90-277-0817-7; Pb 90-277-0818-5

116. R. Tuomela, *Human Action and Its Explanation*. A Study on the Philosophical Foundations of Psychology. 1977   ISBN 90-277-0824-X

117. M. Lazerowitz, *The Language of Philosophy*. Freud and Wittgenstein. [Boston Studies in the Philosophy of Science, Vol. LV] 1977   ISBN 90-277-0826-6; Pb 90-277-0862-2

118. Not published 119. J. Pelc (ed.), *Semiotics in Poland, 1894–1969*. Translated from Polish. 1979   ISBN 90-277-0811-8

120. I. Pörn, *Action Theory and Social Science*. Some Formal Models. 1977   ISBN 90-277-0846-0

121. J. Margolis, *Persons and Mind*. The Prospects of Nonreductive Materialism. [Boston Studies in the Philosophy of Science, Vol. LVII] 1977   ISBN 90-277-0854-1; Pb 90-277-0863-0

122. J. Hintikka, I. Niiniluoto, and E. Saarinen (eds.), *Essays on Mathematical and Philosophical Logic*. 1979   ISBN 90-277-0879-7

123. T. A. F. Kuipers, *Studies in Inductive Probability and Rational Expectation*. 1978   ISBN 90-277-0882-7

124. E. Saarinen, R. Hilpinen, I. Niiniluoto and M. P. Hintikka (eds.), *Essays in Honour of Jaakko Hintikka on the Occasion of His 50th Birthday*. 1979   ISBN 90-277-0916-5

125. G. Radnitzky and G. Andersson (eds.), *Progress and Rationality in Science*. [Boston Studies in the Philosophy of Science, Vol. LVIII] 1978   ISBN 90-277-0921-1; Pb 90-277-0922-X

126. P. Mittelstaedt, *Quantum Logic*. 1978   ISBN 90-277-0925-4

127. K. A. Bowen, *Model Theory for Modal Logic*. Kripke Models for Modal Predicate Calculi. 1979   ISBN 90-277-0929-7

128. H. A. Bursen, *Dismantling the Memory Machine*. A Philosophical Investigation of Machine Theories of Memory. 1978   ISBN 90-277-0933-5

129. M. W. Wartofsky, *Models*. Representation and the Scientific Understanding. [Boston Studies in the Philosophy of Science, Vol. XLVIII] 1979   ISBN 90-277-0736-7; Pb 90-277-0947-5

130. D. Ihde, *Technics and Praxis*. A Philosophy of Technology. [Boston Studies in the Philosophy of Science, Vol. XXIV] 1979   ISBN 90-277-0953-X; Pb 90-277-0954-8

131. J. J. Wiatr (ed.), *Polish Essays in the Methodology of the Social Sciences*. [Boston Studies in the Philosophy of Science, Vol. XXIX] 1979   ISBN 90-277-0723-5; Pb 90-277-0956-4

132. W. C. Salmon (ed.), *Hans Reichenbach: Logical Empiricist*. 1979   ISBN 90-277-0958-0

133. P. Bieri, R.-P. Horstmann and L. Krüger (eds.), *Transcendental Arguments in Science*. Essays in Epistemology. 1979   ISBN 90-277-0963-7; Pb 90-277-0964-5

134. M. Marković and G. Petrović (eds.), *Praxis*. Yugoslav Essays in the Philosophy and Methodology of the Social Sciences. [Boston Studies in the Philosophy of Science, Vol. XXXVI] 1979   ISBN 90-277-0727-8; Pb 90-277-0968-8

135. R. Wójcicki, *Topics in the Formal Methodology of Empirical Sciences*. Translated from Polish. 1979   ISBN 90-277-1004-X

136. G. Radnitzky and G. Andersson (eds.), *The Structure and Development of Science*. [Boston Studies in the Philosophy of Science, Vol. LIX] 1979   ISBN 90-277-0994-7; Pb 90-277-0995-5

137. J. C. Webb, *Mechanism, Mentalism and Metamathematics*. An Essay on Finitism. 1980   ISBN 90-277-1046-5

138. D. F. Gustafson and B. L. Tapscott (eds.), *Body, Mind and Method*. Essays in Honor of Virgil C. Aldrich. 1979   ISBN 90-277-1013-9

139. L. Nowak, *The Structure of Idealization*. Towards a Systematic Interpretation of the Marxian Idea of Science. 1980   ISBN 90-277-1014-7

140. C. Perelman, *The New Rhetoric and the Humanities*. Essays on Rhetoric and Its Applications. Translated from French and German. With an Introduction by H. Zyskind. 1979
ISBN 90-277-1018-X; Pb 90-277-1019-8

141. W. Rabinowicz, *Universalizability*. A Study in Morals and Metaphysics. 1979
ISBN 90-277-1020-2

142. C. Perelman, *Justice, Law and Argument*. Essays on Moral and Legal Reasoning. Translated from French and German. With an Introduction by H.J. Berman. 1980
ISBN 90-277-1089-9; Pb 90-277-1090-2

143. S. Kanger and S. Öhman (eds.), *Philosophy and Grammar*. Papers on the Occasion of the Quincentennial of Uppsala University. 1981 ISBN 90-277-1091-0

144. T. Pawlowski, *Concept Formation in the Humanities and the Social Sciences*. 1980
ISBN 90-277-1096-1

145. J. Hintikka, D. Gruender and E. Agazzi (eds.), *Theory Change, Ancient Axiomatics and Galileo's Methodology*. Proceedings of the 1978 Pisa Conference on the History and Philosophy of Science, Volume I. 1981 ISBN 90-277-1126-7

146. J. Hintikka, D. Gruender and E. Agazzi (eds.), *Probabilistic Thinking, Thermodynamics, and the Interaction of the History and Philosophy of Science*. Proceedings of the 1978 Pisa Conference on the History and Philosophy of Science, Volume II. 1981 ISBN 90-277-1127-5

147. U. Mönnich (ed.), *Aspects of Philosophical Logic*. Some Logical Forays into Central Notions of Linguistics and Philosophy. 1981 ISBN 90-277-1201-8

148. D. M. Gabbay, *Semantical Investigations in Heyting's Intuitionistic Logic*. 1981
ISBN 90-277-1202-6

149. E. Agazzi (ed.), *Modern Logic – A Survey*. Historical, Philosophical, and Mathematical Aspects of Modern Logic and Its Applications. 1981 ISBN 90-277-1137-2

150. A. F. Parker-Rhodes, *The Theory of Indistinguishables*. A Search for Explanatory Principles below the Level of Physics. 1981 ISBN 90-277-1214-X

151. J. C. Pitt, *Pictures, Images, and Conceptual Change*. An Analysis of Wilfrid Sellars' Philosophy of Science. 1981 ISBN 90-277-1276-X; Pb 90-277-1277-8

152. R. Hilpinen (ed.), *New Studies in Deontic Logic*. Norms, Actions, and the Foundations of Ethics. 1981 ISBN 90-277-1278-6; Pb 90-277-1346-4

153. C. Dilworth, *Scientific Progress*. A Study Concerning the Nature of the Relation between Successive Scientific Theories. 3rd rev. ed., 1994 ISBN 0-7923-2487-0; Pb 0-7923-2488-9

154. D. Woodruff Smith and R. McIntyre, *Husserl and Intentionality*. A Study of Mind, Meaning, and Language. 1982 ISBN 90-277-1392-8; Pb 90-277-1730-3

155. R. J. Nelson, *The Logic of Mind*. 2nd. ed., 1989 ISBN 90-277-2819-4; Pb 90-277-2822-4

156. J. F. A. K. van Benthem, *The Logic of Time*. A Model-Theoretic Investigation into the Varieties of Temporal Ontology, and Temporal Discourse. 1983; 2nd ed., 1991 ISBN 0-7923-1081-0

157. R. Swinburne (ed.), *Space, Time and Causality*. 1983 ISBN 90-277-1437-1

158. E. T. Jaynes, *Papers on Probability, Statistics and Statistical Physics*. Ed. by R. D. Rozenkrantz. 1983 ISBN 90-277-1448-7; Pb (1989) 0-7923-0213-3

159. T. Chapman, *Time: A Philosophical Analysis*. 1982 ISBN 90-277-1465-7

160. E. N. Zalta, *Abstract Objects*. An Introduction to Axiomatic Metaphysics. 1983
ISBN 90-277-1474-6

161. S. Harding and M. B. Hintikka (eds.), *Discovering Reality*. Feminist Perspectives on Epistemology, Metaphysics, Methodology, and Philosophy of Science. 1983
ISBN 90-277-1496-7; Pb 90-277-1538-6

162. M. A. Stewart (ed.), *Law, Morality and Rights*. 1983 ISBN 90-277-1519-X

163. D. Mayr and G. Süssmann (eds.), *Space, Time, and Mechanics.* Basic Structures of a Physical Theory. 1983   ISBN 90-277-1525-4
164. D. Gabbay and F. Guenthner (eds.), *Handbook of Philosophical Logic.* Vol. I: Elements of Classical Logic. 1983   ISBN 90-277-1542-4
165. D. Gabbay and F. Guenthner (eds.), *Handbook of Philosophical Logic.* Vol. II: Extensions of Classical Logic. 1984   ISBN 90-277-1604-8
166. D. Gabbay and F. Guenthner (eds.), *Handbook of Philosophical Logic.* Vol. III: Alternative to Classical Logic. 1986   ISBN 90-277-1605-6
167. D. Gabbay and F. Guenthner (eds.), *Handbook of Philosophical Logic.* Vol. IV: Topics in the Philosophy of Language. 1989   ISBN 90-277-1606-4
168. A. J. I. Jones, *Communication and Meaning.* An Essay in Applied Modal Logic. 1983
                                                                                          ISBN 90-277-1543-2
169. M. Fitting, *Proof Methods for Modal and Intuitionistic Logics.* 1983   ISBN 90-277-1573-4
170. J. Margolis, *Culture and Cultural Entities.* Toward a New Unity of Science. 1984
                                                                                          ISBN 90-277-1574-2
171. R. Tuomela, *A Theory of Social Action.* 1984   ISBN 90-277-1703-6
172. J. J. E. Gracia, E. Rabossi, E. Villanueva and M. Dascal (eds.), *Philosophical Analysis in Latin America.* 1984   ISBN 90-277-1749-4
173. P. Ziff, *Epistemic Analysis.* A Coherence Theory of Knowledge. 1984
                                                                                          ISBN 90-277-1751-7
174. P. Ziff, *Antiaesthetics.* An Appreciation of the Cow with the Subtile Nose. 1984
                                                                                          ISBN 90-277-1773-7
175. W. Balzer, D. A. Pearce, and H.-J. Schmidt (eds.), *Reduction in Science.* Structure, Examples, Philosophical Problems. 1984   ISBN 90-277-1811-3
176. A. Peczenik, L. Lindahl and B. van Roermund (eds.), *Theory of Legal Science.* Proceedings of the Conference on Legal Theory and Philosophy of Science (Lund, Sweden, December 1983). 1984   ISBN 90-277-1834-2
177. I. Niiniluoto, *Is Science Progressive?* 1984   ISBN 90-277-1835-0
178. B. K. Matilal and J. L. Shaw (eds.), *Analytical Philosophy in Comparative Perspective.* Exploratory Essays in Current Theories and Classical Indian Theories of Meaning and Reference. 1985   ISBN 90-277-1870-9
179. P. Kroes, *Time: Its Structure and Role in Physical Theories.* 1985   ISBN 90-277-1894-6
180. J. H. Fetzer, *Sociobiology and Epistemology.* 1985   ISBN 90-277-2005-3; Pb 90-277-2006-1
181. L. Haaparanta and J. Hintikka (eds.), *Frege Synthesized.* Essays on the Philosophical and Foundational Work of Gottlob Frege. 1986   ISBN 90-277-2126-2
182. M. Detlefsen, *Hilbert's Program.* An Essay on Mathematical Instrumentalism. 1986
                                                                                          ISBN 90-277-2151-3
183. J. L. Golden and J. J. Pilotta (eds.), *Practical Reasoning in Human Affairs.* Studies in Honor of Chaim Perelman. 1986   ISBN 90-277-2255-2
184. H. Zandvoort, *Models of Scientific Development and the Case of Nuclear Magnetic Resonance.* 1986   ISBN 90-277-2351-6
185. I. Niiniluoto, *Truthlikeness.* 1987   ISBN 90-277-2354-0
186. W. Balzer, C. U. Moulines and J. D. Sneed, *An Architectonic for Science.* The Structuralist Program. 1987   ISBN 90-277-2403-2
187. D. Pearce, *Roads to Commensurability.* 1987   ISBN 90-277-2414-8
188. L. M. Vaina (ed.), *Matters of Intelligence.* Conceptual Structures in Cognitive Neuroscience. 1987   ISBN 90-277-2460-1

SYNTHESE LIBRARY

189. H. Siegel, *Relativism Refuted*. A Critique of Contemporary Epistemological Relativism. 1987
ISBN 90-277-2469-5
190. W. Callebaut and R. Pinxten, *Evolutionary Epistemology*. A Multiparadigm Program, with a Complete Evolutionary Epistemology Bibliograph. 1987　　ISBN 90-277-2582-9
191. J. Kmita, *Problems in Historical Epistemology*. 1988　　ISBN 90-277-2199-8
192. J. H. Fetzer (ed.), *Probability and Causality*. Essays in Honor of Wesley C. Salmon, with an Annotated Bibliography. 1988　　ISBN 90-277-2607-8; Pb 1-5560-8052-2
193. A. Donovan, L. Laudan and R. Laudan (eds.), *Scrutinizing Science*. Empirical Studies of Scientific Change. 1988　　ISBN 90-277-2608-6
194. H.R. Otto and J.A. Tuedio (eds.), *Perspectives on Mind*. 1988　　ISBN 90-277-2640-X
195. D. Batens and J.P. van Bendegem (eds.), *Theory and Experiment*. Recent Insights and New Perspectives on Their Relation. 1988　　ISBN 90-277-2645-0
196. J. Österberg, *Self and Others*. A Study of Ethical Egoism. 1988　　ISBN 90-277-2648-5
197. D.H. Helman (ed.), *Analogical Reasoning*. Perspectives of Artificial Intelligence, Cognitive Science, and Philosophy. 1988　　ISBN 90-277-2711-2
198. J. Woleński, *Logic and Philosophy in the Lvov-Warsaw School*. 1989　ISBN 90-277-2749-X
199. R. Wójcicki, *Theory of Logical Calculi*. Basic Theory of Consequence Operations. 1988
ISBN 90-277-2785-6
200. J. Hintikka and M.B. Hintikka, *The Logic of Epistemology and the Epistemology of Logic*. Selected Essays. 1989　　ISBN 0-7923-0040-8; Pb 0-7923-0041-6
201. E. Agazzi (ed.), *Probability in the Sciences*. 1988　　ISBN 90-277-2808-9
202. M. Meyer (ed.), *From Metaphysics to Rhetoric*. 1989　　ISBN 90-277-2814-3
203. R.L. Tieszen, *Mathematical Intuition*. Phenomenology and Mathematical Knowledge. 1989
ISBN 0-7923-0131-5
204. A. Melnick, *Space, Time, and Thought in Kant*. 1989　　ISBN 0-7923-0135-8
205. D.W. Smith, *The Circle of Acquaintance*. Perception, Consciousness, and Empathy. 1989
ISBN 0-7923-0252-4
206. M.H. Salmon (ed.), *The Philosophy of Logical Mechanism*. Essays in Honor of Arthur W. Burks. With his Responses, and with a Bibliography of Burk's Work. 1990
ISBN 0-7923-0325-3
207. M. Kusch, *Language as Calculus vs. Language as Universal Medium*. A Study in Husserl, Heidegger, and Gadamer. 1989　　ISBN 0-7923-0333-4
208. T.C. Meyering, *Historical Roots of Cognitive Science*. The Rise of a Cognitive Theory of Perception from Antiquity to the Nineteenth Century. 1989　　ISBN 0-7923-0349-0
209. P. Kosso, *Observability and Observation in Physical Science*. 1989　ISBN 0-7923-0389-X
210. J. Kmita, *Essays on the Theory of Scientific Cognition*. 1990　　ISBN 0-7923-0441-1
211. W. Sieg (ed.), *Acting and Reflecting*. The Interdisciplinary Turn in Philosophy. 1990
ISBN 0-7923-0512-4
212. J. Karpiński, *Causality in Sociological Research*. 1990　　ISBN 0-7923-0546-9
213. H.A. Lewis (ed.), *Peter Geach: Philosophical Encounters*. 1991　　ISBN 0-7923-0823-9
214. M. Ter Hark, *Beyond the Inner and the Outer*. Wittgenstein's Philosophy of Psychology. 1990
ISBN 0-7923-0850-6
215. M. Gosselin, *Nominalism and Contemporary Nominalism*. Ontological and Epistemological Implications of the Work of W.V.O. Quine and of N. Goodman. 1990　ISBN 0-7923-0904-9
216. J.H. Fetzer, D. Shatz and G. Schlesinger (eds.), *Definitions and Definability*. Philosophical Perspectives. 1991　　ISBN 0-7923-1046-2
217. E. Agazzi and A. Cordero (eds.), *Philosophy and the Origin and Evolution of the Universe*. 1991　　ISBN 0-7923-1322-4

218. M. Kusch, *Foucault's Strata and Fields*. An Investigation into Archaeological and Genealogical Science Studies. 1991   ISBN 0-7923-1462-X
219. C.J. Posy, *Kant's Philosophy of Mathematics*. Modern Essays. 1992   ISBN 0-7923-1495-6
220. G. Van de Vijver, *New Perspectives on Cybernetics*. Self-Organization, Autonomy and Connectionism. 1992   ISBN 0-7923-1519-7
221. J.C. Nyíri, *Tradition and Individuality*. Essays. 1992   ISBN 0-7923-1566-9
222. R. Howell, *Kant's Transcendental Deduction*. An Analysis of Main Themes in His Critical Philosophy. 1992   ISBN 0-7923-1571-5
223. A. García de la Sienra, *The Logical Foundations of the Marxian Theory of Value*. 1992
                                        ISBN 0-7923-1778-5
224. D.S. Shwayder, *Statement and Referent*. An Inquiry into the Foundations of Our Conceptual Order. 1992   ISBN 0-7923-1803-X
225. M. Rosen, *Problems of the Hegelian Dialectic*. Dialectic Reconstructed as a Logic of Human Reality. 1993   ISBN 0-7923-2047-6
226. P. Suppes, *Models and Methods in the Philosophy of Science: Selected Essays*. 1993
                                        ISBN 0-7923-2211-8
227. R. M. Dancy (ed.), *Kant and Critique: New Essays in Honor of W. H. Werkmeister*. 1993
                                        ISBN 0-7923-2244-4
228. J. Woleński (ed.), *Philosophical Logic in Poland*. 1993   ISBN 0-7923-2293-2
229. M. De Rijke (ed.), *Diamonds and Defaults*. Studies in Pure and Applied Intensional Logic. 1993   ISBN 0-7923-2342-4
230. B.K. Matilal and A. Chakrabarti (eds.), *Knowing from Words*. Western and Indian Philosophical Analysis of Understanding and Testimony. 1994   ISBN 0-7923-2345-9
231. S.A. Kleiner, *The Logic of Discovery*. A Theory of the Rationality of Scientific Research. 1993
                                        ISBN 0-7923-2371-8
232. R. Festa, *Optimum Inductive Methods*. A Study in Inductive Probability, Bayesian Statistics, and Verisimilitude. 1993   ISBN 0-7923-2460-9
233. P. Humphreys (ed.), *Patrick Suppes: Scientific Philosopher*. Vol. 1: Probability and Probabilistic Causality. 1994   ISBN 0-7923-2552-4
234. P. Humphreys (ed.), *Patrick Suppes: Scientific Philosopher*. Vol. 2: Philosophy of Physics, Theory Structure, and Measurement Theory. 1994   ISBN 0-7923-2553-2
235. P. Humphreys (ed.), *Patrick Suppes: Scientific Philosopher*. Vol. 3: Language, Logic, and Psychology. 1994   ISBN 0-7923-2862-0
                        Set ISBN (Vols 233–235) 0-7923-2554-0
236. D. Prawitz and D. Westerståhl (eds.), *Logic and Philosophy of Science in Uppsala*. Papers from the 9th International Congress of Logic, Methodology, and Philosophy of Science. 1994
                                        ISBN 0-7923-2702-0
237. L. Haaparanta (ed.), *Mind, Meaning and Mathematics*. Essays on the Philosophical Views of Husserl and Frege. 1994   ISBN 0-7923-2703-9
238. J. Hintikka (ed.), *Aspects of Metaphor*. 1994   ISBN 0-7923-2786-1
239. B. McGuinness and G. Oliveri (eds.), *The Philosophy of Michael Dummett*. With Replies from Michael Dummett. 1994   ISBN 0-7923-2804-3
240. D. Jamieson (ed.), *Language, Mind, and Art*. Essays in Appreciation and Analysis, In Honor of Paul Ziff. 1994   ISBN 0-7923-2810-8
241. G. Preyer, F. Siebelt and A. Ulfig (eds.), *Language, Mind and Epistemology*. On Donald Davidson's Philosophy. 1994   ISBN 0-7923-2811-6
242. P. Ehrlich (ed.), *Real Numbers, Generalizations of the Reals, and Theories of Continua*. 1994
                                        ISBN 0-7923-2689-X

243. G. Debrock and M. Hulswit (eds.), *Living Doubt*. Essays concerning the epistemology of
Charles Sanders Peirce. 1994   ISBN 0-7923-2898-1
244. J. Srzednicki, *To Know or Not to Know*. Beyond Realism and Anti-Realism. 1994
                                                             ISBN 0-7923-2909-0
245. R. Egidi (ed.), *Wittgenstein: Mind and Language*. 1995   ISBN 0-7923-3171-0
246. A. Hyslop, *Other Minds*. 1995   ISBN 0-7923-3245-8
247. L. Pólos and M. Masuch (eds.), *Applied Logic: How, What and Why*. Logical Approaches to
Natural Language. 1995   ISBN 0-7923-3432-9
248. M. Krynicki, M. Mostowski and L.M. Szczerba (eds.), *Quantifiers: Logics, Models and Computation*. Volume One: Surveys. 1995   ISBN 0-7923-3448-5
249. M. Krynicki, M. Mostowski and L.M. Szczerba (eds.), *Quantifiers: Logics, Models and Computation*. Volume Two: Contributions. 1995   ISBN 0-7923-3449-3
                                        Set ISBN (Vols 248 + 249) 0-7923-3450-7
250. R.A. Watson, *Representational Ideas from Plato to Patricia Churchland*. 1995
                                                             ISBN 0-7923-3453-1
251. J. Hintikka (ed.), *From Dedekind to Gödel*. Essays on the Development of the Foundations of
Mathematics. 1995   ISBN 0-7923-3484-1
252. A. Wiśniewski, *The Posing of Questions*. Logical Foundations of Erotetic Inferences. 1995
                                                             ISBN 0-7923-3637-2
253. J. Peregrin, *Doing Worlds with Words*. Formal Semantics without Formal Metaphysics. 1995
                                                             ISBN 0-7923-3742-5
254. I.A. Kieseppä, *Truthlikeness for Multidimensional, Quantitative Cognitive Problems*. 1996
                                                             ISBN 0-7923-4005-1
255. P. Hugly and C. Sayward: *Intensionality and Truth*. An Essay on the Philosophy of A.N. Prior.
1996   ISBN 0-7923-4119-8
256. L. Hankinson Nelson and J. Nelson (eds.): *Feminism, Science, and the Philosophy of Science*.
1997   ISBN 0-7923-4162-7
257. P.I. Bystrov and V.N. Sadovsky (eds.): *Philosophical Logic and Logical Philosophy*. Essays in
Honour of Vladimir A. Smirnov. 1996   ISBN 0-7923-4270-4
258. Å.E. Andersson and N-E. Sahlin (eds.): *The Complexity of Creativity*. 1996
                                                             ISBN 0-7923-4346-8
259. M.L. Dalla Chiara, K. Doets, D. Mundici and J. van Benthem (eds.): *Logic and Scientific Methods*. Volume One of the Tenth International Congress of Logic, Methodology and Philosophy
of Science, Florence, August 1995. 1997   ISBN 0-7923-4383-2
260. M.L. Dalla Chiara, K. Doets, D. Mundici and J. van Benthem (eds.): *Structures and Norms
in Science*. Volume Two of the Tenth International Congress of Logic, Methodology and
Philosophy of Science, Florence, August 1995. 1997   ISBN 0-7923-4384-0
                                        Set ISBN (Vols 259 + 260) 0-7923-4385-9
261. A. Chakrabarti: *Denying Existence*. The Logic, Epistemology and Pragmatics of Negative
Existentials and Fictional Discourse. 1997   ISBN 0-7923-4388-3
262. A. Biletzki: *Talking Wolves*. Thomas Hobbes on the Language of Politics and the Politics of
Language. 1997   ISBN 0-7923-4425-1
263. D. Nute (ed.): *Defeasible Deontic Logic*. 1997   ISBN 0-7923-4630-0
264. U. Meixner: *Axiomatic Formal Ontology*. 1997   ISBN 0-7923-4747-X
265. I. Brinck: *The Indexical 'I'*. The First Person in Thought and Language. 1997
                                                             ISBN 0-7923-4741-2
266. G. Hölmström-Hintikka and R. Tuomela (eds.): *Contemporary Action Theory*. Volume 1:
Individual Action. 1997   ISBN 0-7923-4753-6; Set: 0-7923-4754-4

267. G. Hölmström-Hintikka and R. Tuomela (eds.): *Contemporary Action Theory.* Volume 2: Social Action. 1997   ISBN 0-7923-4752-8; Set: 0-7923-4754-4
268. B.-C. Park: *Phenomenological Aspects of Wittgenstein's Philosophy.* 1998
ISBN 0-7923-4813-3
269. J. Paśniczek: *The Logic of Intentional Objects.* A Meinongian Version of Classical Logic. 1998
Hb ISBN 0-7923-4880-X; Pb ISBN 0-7923-5578-4
270. P.W. Humphreys and J.H. Fetzer (eds.): *The New Theory of Reference.* Kripke, Marcus, and Its Origins. 1998   ISBN 0-7923-4898-2
271. K. Szaniawski, A. Chmielewski and J. Woleński (eds.): *On Science, Inference, Information and Decision Making.* Selected Essays in the Philosophy of Science. 1998
ISBN 0-7923-4922-9
272. G.H. von Wright: *In the Shadow of Descartes.* Essays in the Philosophy of Mind. 1998
ISBN 0-7923-4992-X
273. K. Kijania-Placek and J. Woleński (eds.): *The Lvov–Warsaw School and Contemporary Philosophy.* 1998   ISBN 0-7923-5105-3
274. D. Dedrick: *Naming the Rainbow.* Colour Language, Colour Science, and Culture. 1998
ISBN 0 7923 5239-4
275. L. Albertazzi (ed.): *Shapes of Forms.* From Gestalt Psychology and Phenomenology to Ontology and Mathematics. 1999   ISBN 0-7923-5246-7
276. P. Fletcher: *Truth, Proof and Infinity.* A Theory of Constructions and Constructive Reasoning. 1998   ISBN 0-7923-5262-9
277. M. Fitting and R.L. Mendelsohn (eds.): *First-Order Modal Logic.* 1998
Hb ISBN 0-7923-5334-X; Pb ISBN 0-7923-5335-8
278. J.N. Mohanty: *Logic, Truth and the Modalities from a Phenomenological Perspective.* 1999
ISBN 0-7923-5550-4
279. T. Placek: *Mathematical Intiutionism and Intersubjectivity.* A Critical Exposition of Arguments for Intuitionism. 1999   ISBN 0-7923-5630-6
280. A. Cantini, E. Casari and P. Minari (eds.): *Logic and Foundations of Mathematics.* 1999
ISBN 0-7923-5659-4   set ISBN 0-7923-5867-8
281. M.L. Dalla Chiara, R. Giuntini and F. Laudisa (eds.): *Language, Quantum, Music.* 1999
ISBN 0-7923-5727-2; set ISBN 0-7923-5867-8
282. R. Egidi (ed.): *In Search of a New Humanism.* The Philosophy of Georg Hendrik von Wright. 1999   ISBN 0-7923-5810-4
283. F. Vollmer: *Agent Causality.* 1999   ISBN 0-7923-5848-1
284. J. Peregrin (ed.): *Truth and Its Nature (if Any).* 1999   ISBN 0-7923-5865-1
285. M. De Caro (ed.): *Interpretations and Causes.* New Perspectives on Donald Davidson's Philosophy. 1999   ISBN 0-7923-5869-4
286. R. Murawski: *Recursive Functions and Metamathematics.* Problems of Completeness and Decidability, Gödel's Theorems. 1999   ISBN 0-7923-5904-6
287. T.A.F. Kuipers: *From Instrumentalism to Constructive Realism.* On Some Relations between Confirmation, Empirical Progress, and Truth Approximation. 2000   ISBN 0-7923-6086-9
288. G. Holmström-Hintikka (ed.): *Medieval Philosophy and Modern Times.* 2000
ISBN 0-7923-6102-4
289. E. Grosholz and H. Breger (eds.): *The Growth of Mathematical Knowledge.* 2000
ISBN 0-7923-6151-2

SYNTHESE LIBRARY

290. G. Sommaruga: *History and Philosophy of Constructive Type Theory*. 2000
ISBN 0-7923-6180-6
291. J. Gasser (ed.): *A Boole Anthology*. Recent and Classical Studies in the Logic of George Boole.
2000   ISBN 0-7923-6380-9
292. V.F. Hendricks, S.A. Pedersen and K.F. Jørgensen (eds.): *Proof Theory*. History and Philosophical Significance. 2000   ISBN 0-7923-6544-5
293. W.L. Craig: *The Tensed Theory of Time*. A Critical Examination. 2000 ISBN 0-7923-6634-4
294. W.L. Craig: *The Tenseless Theory of Time*. A Critical Examination. 2000
ISBN 0-7923-6635-2
295. L. Albertazzi (ed.): *The Dawn of Cognitive Science*. Early European Contributors. 2001
ISBN 0-7923-6799-5
296. G. Forrai: *Reference, Truth and Conceptual Schemes*. A Defense of Internal Realism. 2001
ISBN 0-7923-6885-1
297. V.F. Hendricks, S.A. Pedersen and K.F. Jørgensen (eds.): *Probability Theory*. Philosophy,
Recent History and Relations to Science. 2001   ISBN 0-7923-6952-1
298. M. Esfeld: *Holism in Philosophy of Mind and Philosophy of Physics*. 2001
ISBN 0-7923-7003-1
299. E.C. Steinhart: *The Logic of Metaphor*. Analogous Parts of Possible Worlds. 2001
ISBN 0-7923-7004-X
300. To be published.
301. T.A.F. Kuipers: *Structures in Science Heuristic Patterns Based on Cognitive Structures*. An
Advanced Textbook in Neo-Classical Philosophy of Science. 2001   ISBN 0-7923-7117-8
302. G. Hon and S.S. Rakover (eds.): *Explanation*. Theoretical Approaches and Applications. 2001
ISBN 1-4020-0017-0
303. G. Holmström-Hintikka, S. Lindström and R. Sliwinski (eds.): *Collected Papers of Stig Kanger
with Essays on his Life and Work*. Vol. I. 2001
ISBN 1-4020-0021-9; Pb ISBN 1-4020-0022-7
304. G. Holmström-Hintikka, S. Lindström and R. Sliwinski (eds.): *Collected Papers of Stig Kanger
with Essays on his Life and Work*. Vol. II. 2001
ISBN 1-4020-0111-8; Pb ISBN 1-4020-0112-6
305. C.A. Anderson and M. Zelëny (eds.): *Logic, Meaning and Computation*. Essays in Memory
of Alonzo Church. 2001   ISBN 1-4020-0141-X
306. P. Schuster, U. Berger and H. Osswald (eds.): *Reuniting the Antipodes – Constructive and
Nonstandard Views of the Continuum*. 2001   ISBN 1-4020-0152-5
307. S.D. Zwart: *Refined Verisimilitude*. 2001   ISBN 1-4020-0268-8
308. A.-S. Maurin: *IF Tropes*. 2002   ISBN 1-4020-0656-X
309. H. Eilstein (ed.): *A Collection of Polish Works on Philosophical Problems of Time and Space-
time*. 2002   ISBN 1-4020-0670-5
310. Y. Gauthier: *Internal Logic*. Foundations of Mathematics from Kronecker to Hilbert. 2002
ISBN 1-4020-0689-6
311. E. Ruttkamp: *A Model-Theoretic Realist Interpretation of Science*. 2002
ISBN 1-4020-0729-9
312. V. Rantala: *Explanatory Translation*. Beyond the Kuhnian Model of Conceptual Change. 2002
ISBN 1-4020-0827-9

Previous volumes are still available.

KLUWER ACADEMIC PUBLISHERS – DORDRECHT / BOSTON / LONDON